Electronic Troubleshooting

Second Edition

Jerome E. Oleksy

Dean of Technology
ETI Technical College
Cleveland, Ohio

Glencoe/McGraw-Hill

A Macmillan/McGraw-Hill Company

Mission Hills, California ■ New York, New York

Sponsoring Editor: **John J. Beck**

Editing Supervisor: **Carole Burke**

Design and Art Supervisor: **Joseph Piliero**

Production Supervisor: **Kathleen Donnelly**

Text Designer: **Mike Benoit**

Cover Designer: **Lee Goodman Design**

Cover Photo: **Dick Luria, Art Montes de Oca, and T. Tracy, all F.P.G. International Corp.**

Library of Congress Cataloging-in-Publication Data

Oleksy, Jerome E.
    Electronic troubleshooting/Jerome E. Oleksy.—2nd ed.
      p. cm.
    ISBN 0-02-676390-7
    1. Electronic apparatus and appliances—Maintenance and repair.
    I. Title.
    TK7870.2.038   1990
   621.381′028′8—dc20

90-2898
CIP

Send all inquiries to:

Glencoe/McGraw-Hill
29th floor
1221 Avenue of the Americas
New York, NY 10020

ISBN 0-02-676390-7

1 2 3 4 5 6 7 8 9 0   HALHAL   9 8 7 6 5 4 3 2 1 0

Table of Contents

4 Field Effect Transistors 71

5 Multistage Amplifiers 93

6 Power Amplifiers 135

Preface

Almost all electronics technologists test and troubleshoot electronic systems at one time or another. Whether their jobs involve field service, production testing, quality assurance, or engineering development, they must be able to analyze and troubleshoot circuits and systems. This book was written for those people who must work with electronic systems and who want to gain a better understanding of what they are doing.

Many students who do well in a mathematically rigorous course in transistor circuit theory are frustrated when confronted with a large system. They feel that the straightforward, single-stage circuits, with which they are familiar, are buried among many other unknown parts. They don't know where to begin.

In this text we start with simple circuits and analyze them using only the math necessary for understanding their operation. Much of the theory that is needed only by design engineers has been purposely left out.

After discussing the operation of a circuit, the book shows a larger system that contains the simple circuit. To make the discussion realistic, circuit diagrams of commercially available systems—such as tape recorders, guitar amps, and radios—are shown and discussed. The simple circuit is highlighted, and its relationship to other circuits is emphasized. Then, typical malfunctions are discussed, and pointers are given showing how to localize faults in the system.

Numerous problems appear at the end of each chapter. They should help crystallize the concepts covered in the chapter. Each chapter also has at least two experiments. Here students can observe circuit operation or malfunction and try out their troubleshooting skills.

The first edition of this text was well received, and it prompted comments to give more examples of tracing through an electronic system to pinpoint a fault. This has been done in this edition.

Two new chapters, 12 and 13, have been added to this edition. These new chapters discuss the use of sensors and transducers (Chapter 12) and power control in a-c circuits (Chapter 13). They should make the text even more useful to schools offering industrially related courses, as well as to in-house training courses in industry.

The text was designed to be used in a troubleshooting course following a first course in solid-state electronics. However, since essential solid-state theory is reviewed, the text can also be used as a stand-alone course to train students who have no previous solid-state background to analyze and troubleshoot electronic systems. Before beginning this text, the student does need to know the basics of a-c and d-c circuit theory and how to use simple test equipment, such as multimeters, oscilloscopes, and signal generators.

The author thanks the lab instructors at ETI Technical College for their assistance and helpful criticism, the students for their patience and enthusiasm, and especially former lab director Roger Gipson for his many valuable troubleshooting tips.

The author wishes also to acknowledge the constructive recommendations provided by Robert D. Bloompott, P.E., Associate Professor, Electronics Technology, Illinois Central College, East Peoria, Illinois; and John Smart, Electronics Instructor, Eastland Career Center, Groveport, Ohio.

Jerome E. Oleksy

Introduction 1

Everyone who works with electronic hardware must be able to troubleshoot electronic equipment. The service technician, the production tester, the customer engineer, and the engineering designer, all do some troubleshooting as part of their jobs. The material in this book will help to make troubleshooting less frustrating and more rewarding.

First, you will analyze a small circuit, such as an amplifier stage, to get familiar with its normal operation, biasing, etc. Then we will discuss typical faults, such as open or shorted components, and how to zero in on these faults with a minimum of measurements, unsoldering, and part swapping. After that, you will look at actual systems, such as tape recorders, audio power amps, and digital frequency counters, in which smaller circuits are used. We will discuss the relationship of small circuits to other circuits and how to determine quickly if the stage is working normally.

To test larger systems, you will use signal injection and signal tracing to find the defective stage. For example, when testing an amplifier system of a tape recorder, you will inject a small signal into the input near the tape head, using a signal generator. Then you will use a scope and look at the final output of the system to see if normal signal is present. If it isn't, you will scope a point approximately half way between input and output to see if the signal there is normal. You will continue scoping check points, splitting the system into smaller and smaller sections, until you find a stage that has a normal input signal, but an abnormal output signal. Having isolated the defective stage, you will use a voltmeter to look for symptoms, such as biasing changes, which will narrow down the possible problems. Finally, if necessary, you will make resistance checks or check the suspect transistor or chip. We'll use flowcharts and tables to outline typical test procedures.

Unless otherwise stated, all voltages on diagrams are measured with respect to the power supply common (ground) point. Whenever you read voltages on a commercial diagram, such as a tape recorder, the voltages shown are usually those measured by a test technician on a particular piece of equipment. If your voltages differ slightly, your circuit still may be OK. For example, suppose a schematic shows voltages in an NPN transistor biasing circuit that read + 1.72 V (volts) to ground at the base and + 1.01 V to ground at the emitter. You measure + 1.84 V to ground at the base and + 1.15 V to ground at the emitter. Is your circuit OK? Probably. The *difference* between the base and emitter in both cases is about 0.7 V, with the base positive with respect to the emitter. That's the important thing. The actual voltage to ground differs by only a tenth of a volt or so, which is certainly satisfactory. Also, for power supply voltages or collector-to-ground voltages, a few percent variation is usually acceptable.

When the direction of current flow is shown on a diagram, the arrow points in the direction of *conventional* current flow, that is, from positive to negative. This convention is consistent with the direction of arrows on solid-state symbols. The terms *current source* and *current sink*, used frequently with integrated circuits, also assume conventional current flow. If you prefer to use electron flow, simply point the current direction arrow in the opposite direction. The end result will be the same no matter which direction you use, as long as you are consistent.

You will need several special instruments to do the experiments. The signal generator is one of the more widely used pieces of equipment. Any good generator capable of being adjusted from about 20 Hz (hertz) to 500 KHz (kilohertz) or so and having a moderate output impedance of, say, 600 Ω (ohms) or less is adequate. Fig. 1-1 shows a waveform generator capable of producing sine waves, square waves, or triangular waves. It also can be used as a sweep generator. It has a maximum output frequency of 200 KHz, which is adequate for testing any audio equipment.

1-1   Dynascan Model 3015 waveform generator. Courtesy of B&K Precision.

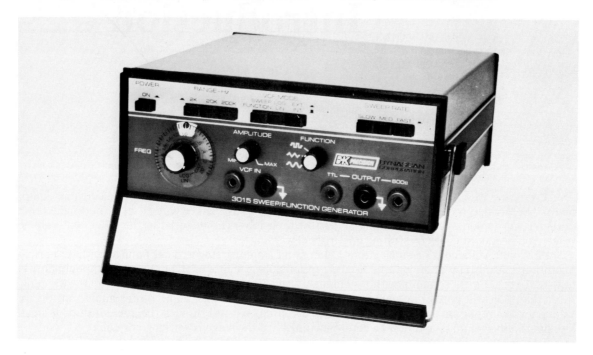

You will also need a good oscilloscope, preferably a triggered time base scope with d-c input. Fig. 1-2 shows a dual-trace oscilloscope with adequate frequency response for practically any testing.

1-2   Dynascan Model 1530 30-MHz, dual-trace, delayed-sweep oscilloscope. Courtesy of B&K Precision.

A multimeter is essential to measure voltage, current, and resistance. You can use either the conventional D'Arsonval movement type of meter or a more modern digital multimeter, such as the Keithley Model 169 shown in Fig. 1-3. The Model 169 is intended for bench work, but many service

1-3 Keithley Model 169 digital multimeter. Courtesy of Keithley Instruments, Inc.

personnel prefer a smaller hand-held version, such as Keithley's model 128 shown in Fig. 1-4. The Model 128 also has a special "beeper" feature. An internal beeper will sound whenever the voltage being measured exceeds a preset threshold value. The threshold is adjustable by means of a pot. For example, you could use the beeper to test for logic HIGH levels in a digital system. The beeper can be set to sound whenever the measured voltage exceeds, say, 2.5 V. By using the beeper feature, you do not have to keep looking back at the instrument each time you move the probe to a new test point.

Other special instruments that make troubleshooting more efficient are introduced later in the text. These include the transistor tester, logic probe, logic monitor, digital pulser, and signature analyzer.

1-4 Keithley Model 128 digital mutltimeter. Courtesy of Keithley Instruments, Inc.

Power Supplies

Nearly every electronic system, such as a tape recorder, a radio, or a computer, needs a source of direct-current (d-c) power. Of course batteries can be and are used in portable equipment, but in larger systems, where considerable power is needed, as well as in most stationary systems, batteries are an inconvenience and expensive. Therefore, some means is needed to convert readily available alternating-current (a-c) line voltage into dc at some desired voltage. This is the job of the converter or, simply, the d-c power supply.

SEMICONDUCTOR DIODES

Alternating current, by definition, periodically alternates or reverses direction. In order to produce dc from ac, you must use a device that permits current flow in only one direction. Such a device is the *diode*. Converting ac into dc is called *rectification*.

Fig. 2-1A is a simplified picture of how P-type and N-type crystals are joined together to form a P–N *junction* diode. Fig. 2-1B is the schematic symbol for the diode. The arrow indicates the direction of conventional current flow.

2-1   Solid-state diode.

P-TYPE CRYSTAL   N-TYPE CRYSTAL

(A) Diode formed by join-ing P-type and N-type material.

ANODE   CATHODE

(B) Symbol.

When the P material, called the *anode*, is made positive with respect to the N material, called the *cathode*, the diode conducts and is said to be *forward biased*. This is shown in Fig. 2-2A. Similarly, when the polarity of the power source is reversed, as in Fig. 2-2B, the diode does not conduct and is said to be *reverse biased* or *back biased*. For simplicity, a forward-biased diode acts like a closed switch, and a back-biased diode acts like an open switch.

Typical packages for silicon diodes are shown in Fig. 2-3. Diode part numbers usually begin with 1N (such as, 1N4002). A more accurate representation of how the diode acts in a circuit is shown in Fig. 2-4A. The *characteristic curve*, as it is called, shows how the diode current increases as the voltage across it is increased. Note that as the voltage E in Fig. 2-4B is increased, the forward voltage V_F across the diode increases to about 0.7 V before any forward current I_F begins to flow. However, once V_F exceeds 0.7 V, I_F increases rapidly for slight increases in forward voltage. For this reason,

2-2 Diode operation.

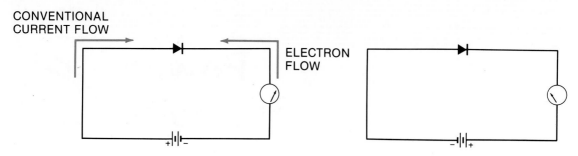

(A) Forward-biased diode conducts.

(B) Backward-biased diode does not conduct.

2-3 Diode packages.

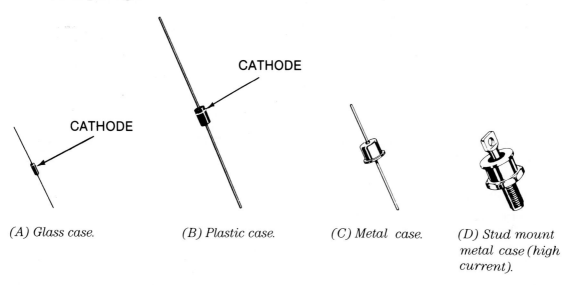

(A) Glass case.

(B) Plastic case.

(C) Metal case.

(D) Stud mount metal case (high current).

we say that the diode is nonconducting until V_F reaches 0.7 V. Beyond 0.7 V, the diode is conductive. The resistor in series with the diode limits the forward current through the diode.

When a small reverse voltage is applied, as in Fig. 2-4C, practically no current flows. Increasing reverse voltage V_R is plotted to the left of center, and reverse current I_R is plotted downward to show that they are in the opposite direction of V_F and I_F.

If the reverse voltage is increased to an excessive value, the diode will *break down*. That means it will suddenly start conducting. However, as long as the current is limited by sufficient external resistance, it will not be damaged after breakdown. The breakdown voltage V_B is the maximum safe voltage that can be applied before breakdown occurs. Diodes are available with breakdown voltages from about 25 V up to several thousand volts. Normally, diodes are chosen so that the maximum reverse voltage that will ever appear across the diode is less than V_B.

HALF-WAVE RECTIFIER

Now that we have a device that conducts in only one direction, let's see how it is used to build a power supply. Fig. 2-5 shows a diode used as a *half-wave rectifier*. Note that the input signal, measured at point X with respect to ground, is a sine wave with a peak value of 12 V. But since the diode only conducts when it is forward biased, the waveform at point Y only contains the positive half cycles, as in Fig. 2-5C.

If the diode were *ideal*, voltage would not drop across it when it was forward biased. The peak voltage across the load would then be equal to the peak of the input voltage. However, as explained previously, real forward-biased diodes do have a small voltage drop of about 0.7 V or so. So the actual

2-4   Measuring and plotting diode characteristics.

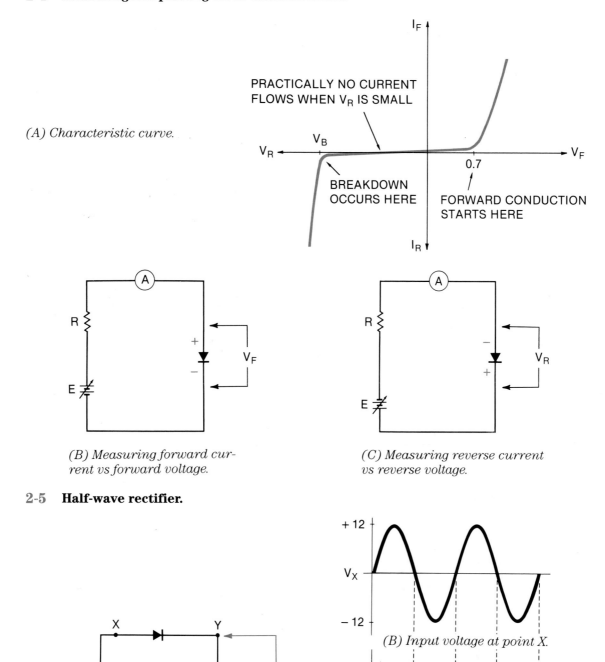

(A) Characteristic curve.

PRACTICALLY NO CURRENT
FLOWS WHEN V_R IS SMALL

BREAKDOWN
OCCURS HERE

FORWARD CONDUCTION
STARTS HERE

(B) Measuring forward current vs forward voltage.

(C) Measuring reverse current vs reverse voltage.

2-5   Half-wave rectifier.

(A) Basic circuit.

(B) Input voltage at point X.

(C) Output voltage at point Y.

voltage across the load in Fig. 2-5C measures a peak value of about 11.3 V instead of 12 V. Usually we ignore the small forward voltage drop across the diode when the input source voltage is relatively large, say more than ten times the forward drop. But when we work with low input voltages, on the order of just a few volts, we must take the diode drop into account.

FILTER CAPACITORS

Although the rectifier of Fig. 2-5 does not allow alternating current to flow through the load, the load voltage still pulsates. We need to smooth out the pulsations to produce steady, d-c flow. Fig. 2-6A shows how a *filter* capacitor is used to smooth out the pulsations. As shown from time t_0 to t_1, in Fig. 2-6B, when point X is driven positive, diode D conducts, causing capacitor C to charge to approximately the peak of the input voltage. Then, as the voltage at point X begins to drop below the capacitor voltage, D becomes back biased and acts like an open switch.

2-6   Filtering rectified a-c.

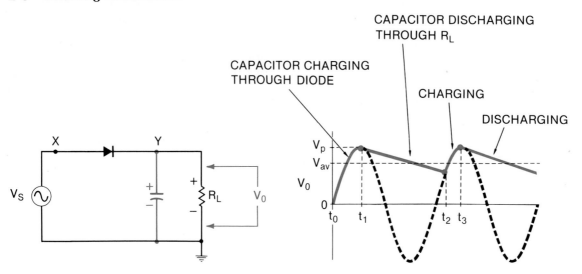

(A) Filter capacitor added to rectifier circuit.   *(B) Output voltage V_0 at point Y.*

From time t_1 until t_2, the diode is nonconducting, and capacitor C maintains a nearly constant voltage across load R_L. (R_L simulates the normal load circuits connected to the supply.) Of course, since charge flows from C through R_L, the voltage across C does not remain entirely constant. Instead it decays somewhat. However, as soon as the input voltage once again begins to exceed the voltage across the capacitor (which occurs at t_2), the diode conducts again, recharging C to the peak of the input voltage. Recharging takes place from t_2 to t_3. This process repeats each cycle.

If the load resistance is made smaller, it will draw more current, causing the voltage across C to droop even more between recharging cycles. The dashed line ($- \cdot -$) in Fig. 2-7 shows this effect. Using a smaller filter cap will have the same effect. On the other hand, either increasing the value of C or decreasing the load current will cause less variation in the load voltage, as shown by the dashed line ($- - -$) in the figure. You will see this effect in the experiments.

2-7   Effects of different values of load current and filter capacitance on V_0.

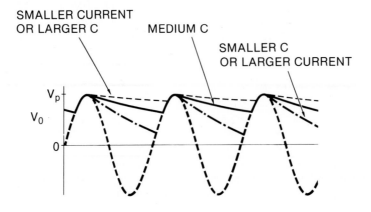

The average value of d-c voltage, as read on a voltmeter, is approximately half way between the peak voltage and the minimum capacitor voltage reached before recharging begins.

The variaton in d-c output voltage is called *ripple*. The peak-to-peak ripple voltage can be approximately determined by the formula

$$V_{rip} = \frac{IT}{C} \qquad\qquad \text{2-1}$$

where

V_{rip} = peak-to-peak ripple voltage in volts
I = load current in amps
T = time between recharging pulses, and
C = filter capacitance in farads

The time T in the formula is actually the period of the input wave. Normally power supplies operate from an a-c input voltage at a frequency of 60 Hz, so T = 1/60 s (seconds), or about 16.7 ms (milliseconds).

EXAMPLE 2-1   In the circuit of Fig. 2-6, assume that C = 200 μF (microfarads), I = 20 mA (milliamperes), T = 16.7 ms, and V_P = 15 V. Also assume that the diode is ideal. Find the value of V_{rip} and V_{ave}.

SOLUTION

$$V_{rip} = \frac{20 \times 10^{-3} \times 16.7 \times 10^{-3}}{200 \times 10^{-6}} = 1.67 \text{ V}$$

$$V_{ave} = V_P - \tfrac{1}{2} \times V_{rip} = 15 - \tfrac{1}{2} \times 1.67 = 14.2 \text{ V}$$

TRANSFORMERS IN D-C POWER SUPPLIES

As you have seen, the d-c output voltage from the rectifier-filter circuit is approximately equal to the peak of the a-c input voltage. Therefore, transformers are usually needed to step the line voltage up or down to a peak value needed as an input to the diode.

Fig. 2-8A is a simplified drawing of a transformer, showing a primary winding and a secondary winding wrapped on a common iron core. The actual construction uses a laminated core to minimize eddy current losses, and the insulated windings are usually wrapped one on top of the other around a center post of the core. But the figure shows the general idea. Fig. 2-8B shows the schematic symbol for an iron core transformer.

2-8   Power supply transformer.

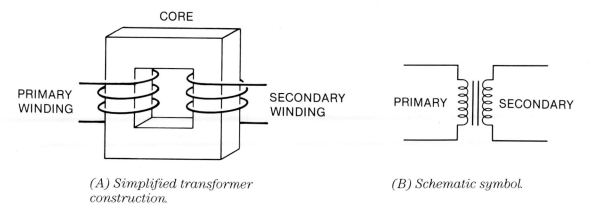

(A) Simplified transformer construction.

(B) Schematic symbol.

Depending on the ratio of the number of turns of wire on the primary to the number of turns on the secondary, the secondary voltage can be either higher or lower than the primary voltage. The formula for finding secondary voltage is

$$V_{sec} = V_{pri} \frac{N_{sec}}{N_{pri}}$$

2-2

where

V_{sec} = transformer secondary (output) voltage
V_{pri} = transformer primary (line) voltage
N_{sec} = number of secondary turns
N_{pri} = number of primary turns

Manufacturers of power supply transformers usually specify primary and secondary effective (rms) *voltages*, rather than turns ratios. So the designers simply choose a transformer with a secondary voltage suitable for their needs. However, since the voltage is specified as an rms value, the peak secondary voltage, and hence the peak d-c output voltage from the rectifier, is 1.41 times the secondary rms value. Fig. 2-9 shows a complete half-wave rectifier supply including the transformer, rectifier, and filter. Also included are miscellaneous components, such as the power switch S, line cord, indicator lamp L, fuse F, and plug P.

2-9   Complete half-wave power supply.

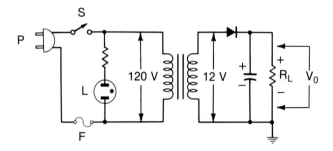

If the diode connections are reversed, the output voltage will be negative with respect to ground. Of course, since the capacitor is electrolytic, the connections to the filter capacitor must also be reversed to observe proper polarity.

> **EXAMPLE 2-2**   For the circuit of Fig. 2-9, determine the transformer turns ratio and the peak value of d-c output voltage.
>
> **SOLUTION**   The figure shows that the primary voltage is 120 V and the secondary voltage is 12 V. So from equation 2-2, the turns ratio is
>
> $$\frac{N_{pri}}{N_{sec}} = \frac{V_{pri}}{V_{sec}} = \frac{120}{12} = 10$$
>
> and
>
> $$V_0 = V_P = 1.41 \times V_{sec} = 1.41 \times 12 = 16.92 \text{ V}$$

Getting back to the diode for a moment, you know that the diode is chosen so that the reverse voltage appearing across it will be less than the breakdown voltage. How much voltage appears across the diode in Fig. 2-9 when it is back biased? When the top of the transformer is negative with respect to the bottom, D is back biased, so it acts like an open switch. As shown in Fig. 2-10, voltage V_R appearing across D is the sum of the voltage across C plus the voltage across the transformer secondary. This voltage, called the *peak reverse voltage* (PRV), has a maximum value of *twice* the d-c output voltage. Diodes used in half-wave rectifier circuits must have a breakdown voltage V_B greater than twice the d-c voltage.

2-10 **Reverse voltage across diode is equal to the sum of the capacitor and transformer voltages.**

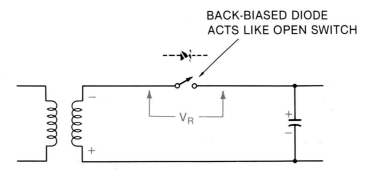

BACK-BIASED DIODE
ACTS LIKE OPEN SWITCH

FULL-WAVE RECTIFIER

As in the half-wave rectifier circuit, current flows through the diode for only a short time each cycle, during which the filter capacitor is recharged. For the remainder of the cycle, the capacitor tries to maintain a constant voltage across the load. If the power supply must deliver much current to the load, very large capacitors are needed. Otherwise the ripple voltage would be excessive.

Using a *full-wave* rectifier, which recharges the capacitor *twice* each cycle, is a better solution. In this way, the capacitor needs to be half as large as would be necessary in the half-wave circuit.

Fig. 2-11 shows how the full-wave rectifier works. On half of the input cycle, the top of the transformer (point A) is driven positive with respect to the bottom (point B). Here, point A is also positive with respect to the center tap, which is grounded. This causes diode D_1 to be forward biased, acting like a closed switch. Simultaneously, D_2 is back biased, acting like an open switch. Current flows through D_1, through R_L, to the center tap, as shown in Fig. 2-11A.

Then on the next alternation, point B is driven positive with respect to point A. At this time D_2 becomes forward biased, and D_1 becomes back biased. Current now flows through D_2, through R_L, to the center tap, as shown in Fig. 2-11B.

Note that on both alternations, current flows in the *same direction* through R_L. Figs. 2-11C and 2-11D show how the current flows through R_L; therefore, the voltage across R_L is related to the transformer secondary voltage.

Once again we need a filter capacitor to smooth out the pulsations and maintain a constant d-c voltage across the load. Fig. 2-12, drawn in a more conventional way than Fig. 2-11, shows how the filter capacitor is connected across R_L and how the load voltage is held reasonably constant by the capacitor.

The amount of ripple voltage can be determined by equation 2-1, as it was for the half-wave circuit. But, since the capacitor is recharged every half cycle, T is equal to about 8.35 ms with a line frequency of 60 Hz, instead of 16.7 ms. So C can be smaller in the full-wave circuit than in the half-wave circuit. The d-c output voltage for the full-wave rectifier is equal to the peak of the transformer secondary voltage from the center tap to one end. Most often, the *total* transformer secondary voltage is listed by the manufacturer. A transformer listed as 12.6 VCT has 6.3 V from center tap to each end.

Like the half-wave circuit, the peak reverse voltage across either diode is twice the d-c output voltage. And like the half-wave rectifier, reversing the connections to the diodes as well as to the filter capacitor produces a negative d-c output voltage.

Diodes intended for full-wave rectifier supplies are available in rectifier *assemblies*, which contain two diodes in one package. Fig. 2-13 shows the internal connections for both positive and negative assemblies. Also shown are typical packages.

Let's take a look at a typical power supply used in consumer electronics equipment. Fig. 2-14 shows the power supply for a Panasonic Model RQ-309DS tape recorder. The highlighted section is that of a full-wave rectifier, which produces a negative d-c output voltage with respect to ground. The diodes are included in the rectifier assembly (part number 2SRO5K), shown in the dotted lines.

2-11 Full-wave rectifier operation.

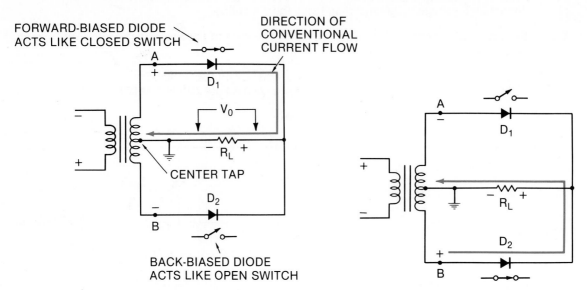

FORWARD-BIASED DIODE
ACTS LIKE CLOSED SWITCH

DIRECTION OF
CONVENTIONAL
CURRENT FLOW

CENTER TAP

BACK-BIASED DIODE
ACTS LIKE OPEN SWITCH

(A) Current flow on positive alternation.

(B) Current flow on negative alternation.

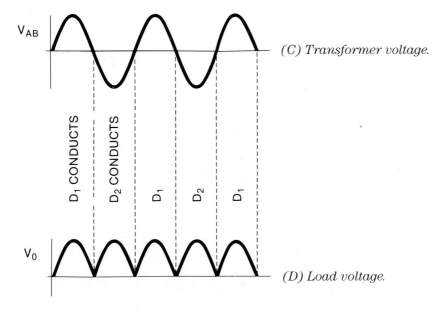

(C) Transformer voltage.

D_1 CONDUCTS

D_2 CONDUCTS

D_1

D_2

D_1

(D) Load voltage.

2-12 Filtering a full-wave rectifier.

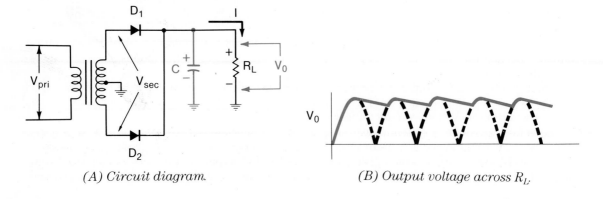

(A) Circuit diagram.

(B) Output voltage across R_L.

2-13 **Full-wave rectifier assemblies.**

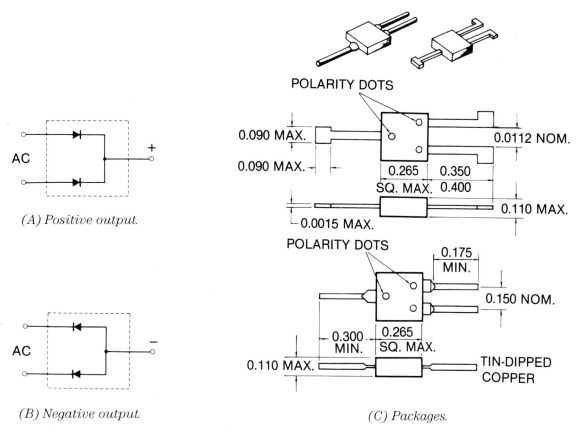

(A) Positive output.

(B) Negative output.

POLARITY DOTS

0.090 MAX.
0.090 MAX.
0.265
SQ. MAX.
0.350
0.400
0.0015 MAX.
0.0112 NOM.
0.110 MAX.

POLARITY DOTS

0.175
MIN.
0.150 NOM.
0.300
MIN.
0.265
SQ. MAX.
0.110 MAX.
TIN-DIPPED
COPPER

(C) Packages.

2-14 **Power supply for Panasonic Model RQ-309DS tape recorder. From _Tape Recorder Series No. 191_, Howard W. Sams & Company, Inc.**

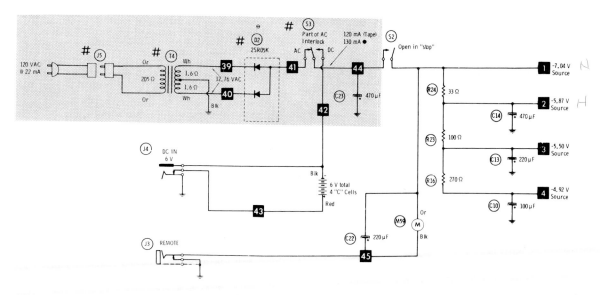

The a-c voltage across the entire transformer secondary is given as 12.76 VAC, which means that there is about 6.4 V either side of the center tap. When the a-c line cord is plugged in, the switch to the right of TP (test point) 41 is in the a-c position. The d-c supply voltage is measured at TP 44, which is the top of filter capacitor C_{23}.

The main components of the full-wave rectifier circuit are highlighted to help you recognize the essential parts of the circuit. Whenever you are analyzing a circuit, look for the essential parts first. Don't get confused by the various additional parts. This is not to say that the additional "hang on" parts are unimportant; they are important. But you should look for some familiar basic parts of the system and work up from there.

As an exercise, find out how the interlock switch S_3 works and determine the purpose of connector jacks J_4 and J_3.

When S_2 is closed, the full d-c output voltage appears at test point TP 1. This is listed as -7.04 V. Resistor R_{24} and capacitor C_{14} form an additional filter, called a *decoupling* filter. Decoupling filters are used primarily to prevent variations in current in one part of the system from affecting the supply voltage in another part. For example, even if the circuitry connected to TP 1 causes slight rapid variations in the voltage at that point, the filter circuit of R_{24}–C_{14} will smooth out the variations, maintaining a clean d-c voltage at TP 2. Similarly, the decoupling filters connected to TP 3 and TP 4 keep those voltages well filtered.

Circuits connected to TP 4 draw current through R_{16}, which results in the voltage at TP 4 being slightly lower than the voltage at TP 3. The same is true for the voltages at the other test point.

FULL-WAVE BRIDGE

It is possible to build a full-wave rectifier supply without using a center-tapped transformer. Fig. 2-15 shows four diodes connected in a *bridge* arrangement. As shown in Fig. 2-15A, when the top of the transformer is positive with respect to the bottom, diodes D_2 and D_3 become forward biased, and current flows with the arrows. In Fig. 2-15B, diodes D_1 and D_4 become forward biased, and current flows with the arrows in the other direction. Note that current flows through R_L in the same direction on both half cycles.

2-15   Full-wave bridge rectifier.

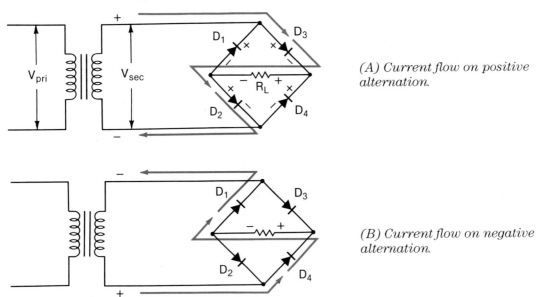

(A) Current flow on positive alternation.

(B) Current flow on negative alternation.

It is interesting to note that the PRV across any nonconducting diode in the bridge is equal to the peak voltage across the load, and not twice the peak, as in the center-tapped version. For example, in Fig. 2-15A, the peak reverse voltage appearing across back-biased diode D_1 is equal to the sum of the peak voltage across R_L, plus the forward voltage drop across D_3. Since the forward drop across D_3 is very small, the PRV is approximately equal to the peak load voltage. The same is true for the other back-biased diodes.

As with the other supplies, a filter capacitor is used to smooth out the pulsations. Fig. 2-16 shows a complete full-wave bridge circuit. Since this is a full-wave circuit, the equation for filter capacitor C is the same as that for the circuit of Fig. 2-12. The d-c output voltage of the bridge rectifier is equal to the peak secondary voltage.

2-16 **Complete full-wave bridge power supply.**

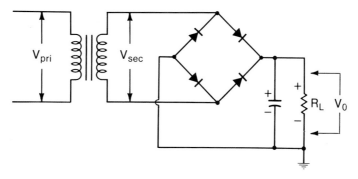

2-17 **Bridge rectifier assemblies.**

(A) Circuit.

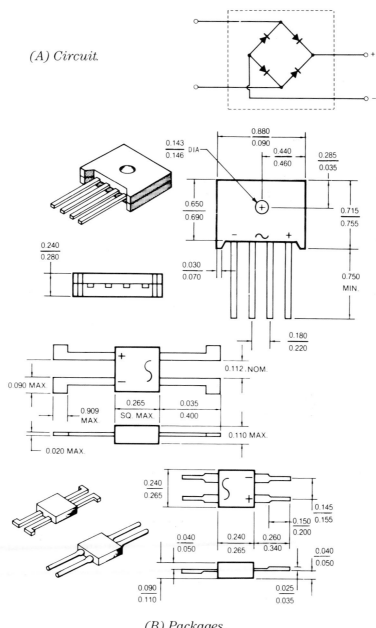

(B) Packages.

EXAMPLE 2-3   In the circuit of Fig. 2-16, assume that $V_{sec} = 20$ Vrms, $I_L = 100$ mA, and $C = 400\ \mu\text{F}$. Find V_0, V_{rip}, and the PRV.

SOLUTION

$$V_0 = V_P = 1.41 \times 20 = 28.2\ \text{V}$$

$$V_{rip} = \frac{IT}{C} = \frac{0.1 \times 8.35 \times 10^{-3}}{400 \times 10^{-6}} \approx 2\ \text{V}$$

and

$$\text{PRV} = V_0 = 28.2\ \text{V}$$

Since there are two diodes conducting in series when C is being charged, the peak capacitor voltage will be less than the peak transformer voltage by two diode drops, or about 1.4 V. In low-voltage circuits, the difference is significant, but it is usually ignored in higher voltage circuits, above 15 V or so. Bridge rectifier assemblies are available with four diodes mounted in a single package, as shown in Fig. 2-17.

VOLTAGE DOUBLERS

The circuit of Fig. 2-18, called a *voltage doubler*, produces a d-c output voltage equal to *twice* the peak amplitude of the transformer secondary voltage. As shown by path 1, when the top of the transformer is positive with respect to the bottom, D_1 conducts, charging C_1 to the peak secondary voltage. On the other alternation of the input cycle as shown by path 2, D_2 conducts, charging C_2 to the peak transformer secondary voltage in the polarity shown. The voltage across R_L is equal to the sum of the voltages across C_1 and C_2, which is equal to twice the peak secondary voltage. This circuit is called a full-wave doubler, because the ripple frequency across R_L is twice the input frequency.

A different type of doubler is shown in Fig. 2-19. When the top of the transformer is negative, as shown by path 1, D_1 conducts, charging C_1 to the peak secondary voltage. Then on the other half

2-18   Full-wave voltage doubler.

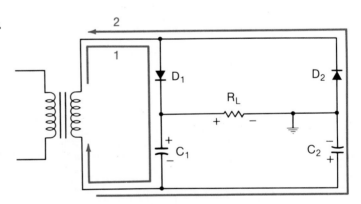

2-19   Half-wave voltage doubler.

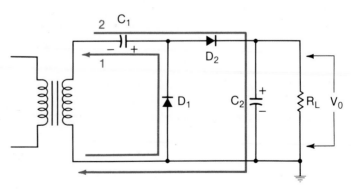

cycle as shown by path 2, D_2 conducts, charging C_2 to a value equal to the peak secondary voltage *plus* the voltage across C_1. So the voltage across C_2 is *twice* the peak secondary voltage. The circuit of Fig. 2-19 only charges C_2 once each cycle, so it is a half-wave doubler. However, this type of circuit can be expanded by adding more sections, making a voltage tripler or quadrupler.

Voltage multipliers are not often used, except in fairly low-power circuits, because they are inefficient.

TYPICAL PROBLEMS WITH POWER SUPPLY COMPONENTS

Now let's discuss what can go wrong with major power supply components and how to test them. Solid-state diodes can either open or short internally. They don't simply get weaker. This was true of the old vacuum tube diodes. The simplest way to test a suspect diode is to measure it with an ohmmeter. As shown in Fig. 2-20A, when the negative lead of the ohmmeter is connected to the cathode and the positive lead, to the anode, the internal battery in the ohmmeter forward biases the diode. The meter should be on the $R \times 1$ scale. Current flows through the diode and causes the meter to deflect, which indicates a low resistance. However, when the leads of the meter are reversed, as in Fig. 2-20B, the diode is back biased, so practically no current flows. The meter then indicates a very high resistance.

2-20 **Testing a diode with an ohmmeter.**

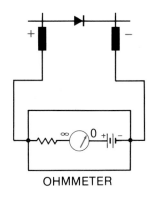
OHMMETER

(A) Forward-biased diode reads low resistance.

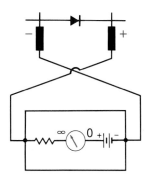

(B) back-biased diode reads high resistance.

The actual value of resistance measured is not important. The important thing is that a good diode measures a high resistance in one direction, and a low resistance in the other direction.

The small amount of current that flows in the reverse direction is called *leakage* current. If the diode were ideal, no reverse current would flow. However, all real diodes have some leakage. But as long as the leakage current is negligible compared to the average forward current, it doesn't present a problem.

You must consider two important characteristics of power supply diodes when replacing a defective diode. These are (1) forward current I_F and (2) breakdown voltage V_B, also listed as PRV or PIV (peak inverse voltage). As long as the replacement diode has values of I_F and PRV, which are equal to or greater than those of the original part, it should work well as a replacement.

Electrolytic filter capacitors are a common source of problems in power supplies. They can either become leaky, or they can open, which sometimes is the result of the dielectric's drying out.

When an electrolytic capacitor becomes leaky, it acts as though there were a low resistance shunting the capacitor. Leakage resistance of electrolytics, which are used as power-supply filters, can be measured with an ohmmeter and should read in the neighborhood of a few hundred or more kiloohms. When you first place an ohmmeter across the capacitor, you should see the needle "kick" toward the low-resistance end of the scale. Then, as the capacitor becomes charged from the battery in the meter, the needle gradually moves up the scale toward the high-resistance end. This is shown in Fig. 2-21. Occasionally, capacitors become so leaky that they begin to act like short circuits. This,

2-21 Checking electrolytic capacitors with an ohmmeter.

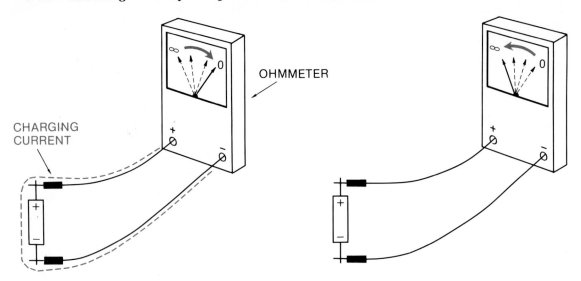

OHMMETER

CHARGING CURRENT

(A) Needle kicks downscale as surge current charges capicitor.

(B) Needle slowly returns upscale to high resistance reading as capacitor becomes charged.

of course, can blow fuses. At worst, it can damage other components, such as the transformer or diodes.

When electrolytic capacitors dry out, they lose capacitance. This loss causes poor filtering of the supply, resulting in excessive ripple. The excessive ripple can cause abnormal hum in the output of an audio amplifier system. It can also interfere with the proper operation of any instrument or computer circuit. You can see excessive ripple by connecting a scope across the filter cap with power on. The best way to test a suspected capacitor for loss of capacitance is to bridge the old capacitor with a known good one with about the same capacitance. If the ripple problem clears up, replace the old capacitor.

Consider both (1) capacitance and (2) working voltage when replacing capacitors. The capacitance of filters ranges anywhere from about 20 μF to many thousands of microfarads, depending on the output current of the supply. Normally, if an exact replacement is not available, you can use a larger capacitance value, up to twice the original size without any problems. Larger values of C filter better. However, it is not wise to use excessively large values, because a large surge current flows into the capacitor when power is first turned on. Excessive surge current may damage other components.

The working voltage of the replacement part should be equal to or greater than that of the original part. The d-c working voltage for the capacitor is usually abbreviated WVDC. The designer chooses a working voltage slightly greater than the d-c output voltage.

Transformers do not often go bad in power supplies. However, one of the windings may open if excessive current is drawn. It is also possible that an internal short, or shorted turn may develop as a result of a break in the insulation.

Open windings are easy to find. Simply measure each winding with an ohmmeter. A likely place for an open to occur is at or near where the external wires are connected to the windings. You may be able to repair the open by carefully cutting away part of the insulating paper or tape near where the external wires enter. Next, scrape away the insulating varnish from the last ¼ inch (7 millimeters) of each broken end of wire. Then twist and solder the ends together. Finally, wrap the connection with insulating tape, and you are back in business.

An ohmmeter measurement may not find an internal short in a transformer. Manufacturers often list typical resistance values for the windings on the schematic. A significant difference between the measured value and the listed value indicates the problem. But sometimes only one or two turns short, possibly as a result of a break in the insulation between adjacent turns. This may not show up as anything significant in an ohmmeter test. It does, however, cause the transformer to draw excessive current and possibly blow a fuse.

Fig. 2-22 shows a good way to test a transformer for a suspected shorted turn. Connect a low-wattage lamp, say 15 to 25 W (watts), in series with the primary winding. Then, with the secondary disconnected from the diodes, apply line voltage to the primary as shown. If the transformer is good, the inductive reactance of the primary will be so high that the lamp will be dim. But if a short in *either* the primary *or* secondary winding exists, the inductive reactance will drop so low that the current flow into the primary will be excessive, causing the lamp to glow brightly.

2-22 Testing a transformer for a shorted turn.

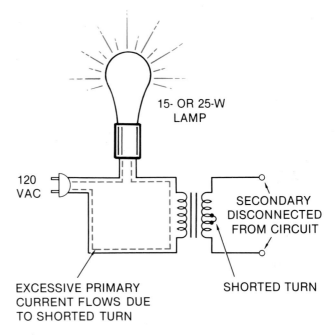

You should replace a transformer with an identical part if possible. However, you can use transformers capable of higher secondary current, as long as the secondary voltage is the same as that of the original. Also make sure that the insulation rating of the windings is at least as high in the replacement transformer as in the original part.

LOCATING PROBLEMS IN POWER SUPPLIES

Power supply problems are some of the most common causes of system failure. Whenever a system malfunctions, the power supply should be one of the first things you check. Now let's develop a systematic procedure for troubleshooting a power supply. We'll use the full-wave rectifier of Fig. 2-23 as a typical supply.

2-23 Full-wave power supply.

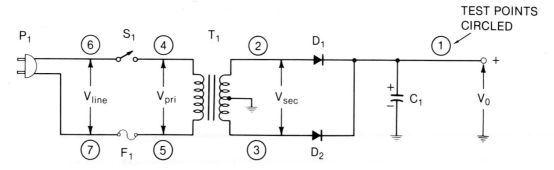

The first quick check for normal operation is to measure the d-c output voltage at TP 1. If the d-c output voltage at TP 1 measures normal and there is no excessive ripple voltage, the power supply is probably OK.

Fig. 2-24 is a flowchart showing a possible procedure to follow. Start at the top, at the oval labeled Abnormal Power Supply Voltage. Then go to the diamond-shaped block, which represents a test to be made. It will lead you to a decision about what to look for next. In flowcharts, diamond-shaped blocks indicate decisions to be made, since they have one input path and two or three possible output paths.

2-24   Power supply troubleshooting flowchart.

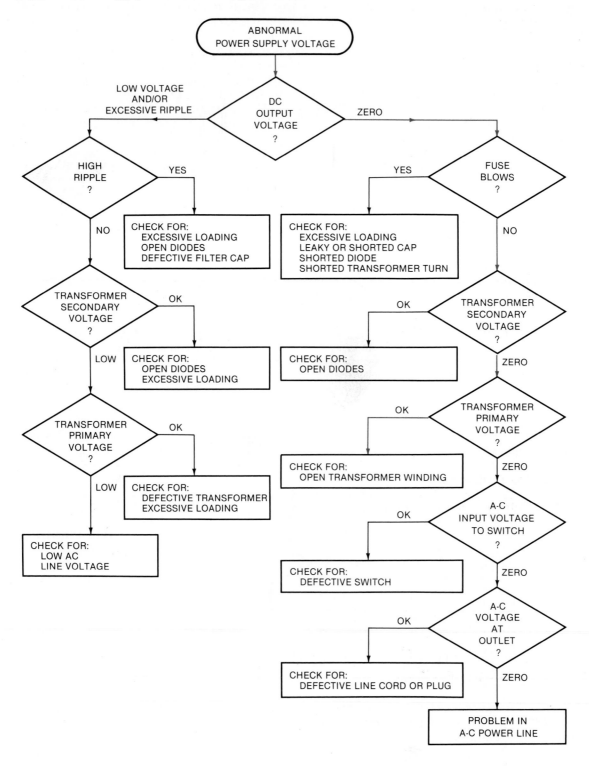

The first measurement indicates whether the output voltage is zero or just lower than normal, possibly with excessive ripple. If the voltage is zero, take the path leading to the right, otherwise go to the left and make further checks.

For practice with using the flowchart, let's assume that the transformer primary winding is open. A measurement at TP 1 reads 0 V, so we go to the right at the first decision block. We then come to the decision box that asks FUSE BLOW? A quick check of the fuse shows that the answer is *no*. So we follow the NO route down to the next decision block. This block asks whether the transformer secondary voltage is OK. We measure the secondary voltage by connecting a voltmeter or an oscilloscope across TP 2 and 3. Here we measure 0 V, so we follow the path labeled ZERO to the next decision box. Here the question is asked, "Is the transformer primary voltage OK?" Putting an oscilloscope or voltmeter across TP 4 and 5 shows that the voltage across the primary is OK. So we follow the arrow to the left of that block and come to a block that tells us to check for an open transformer primary. We make this measurement by pulling the plug and connecting an ohmmeter across the primary winding. An open circuit measurement here confirms the defective component.

Although you will not always have a flowchart when you troubleshoot an electronic system, try to start at the output and work back toward the input, while making voltage or waveform checks to localize the fault. Later we will see how to use signal injection techniques to find faults in circuits other than power supplies. Generally, you should first make waveform and voltage checks to locate a defective area or circuit. Only after that should you make resistance checks to confirm a suspected bad component.

If the symptoms lead to excessive loading, the problem may not be in the power supply at all but rather in the rest of the system. If possible, disconnect the output of the supply from the rest of the system and connect a resistor across the output, which will draw the same current as a normal system would draw. If the output voltage then looks normal, the problem is elsewhere.

PROBLEMS

For the next four problems, refer to Fig. 2-25.

2-25   Circuits for problems 2-1 through 2-4.

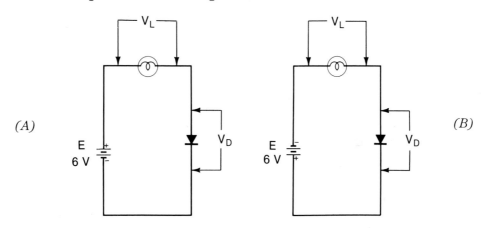

2-1.   If the diode is ideal, what are the values of V_L and V_D in Fig. 2-25A?

$V_L =$

$V_D =$

2-2.   If the diode is ideal, what are the values of V_L and V_D in Fig. 2-25B?

$V_L =$

$V_D =$

2-3. If the diode is a practical silicon rectifier with a breakdown voltage $V_B = 50$ V, what are the approximate values of V_L and V_D in Fig. 2-25A?

$V_L =$

$V_D =$

2-4. If the diode is a practical silicon rectifier with a breakdown voltage $V_B = 50$ V, what are the approximate values of V_L and V_D in Fig. 2-25B?

$V_L =$

$V_D =$

For the next four problems, refer to Fig. 2-26.

2-26 **Circuit for problems 2-5 through 2-8.**

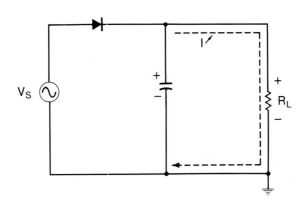

2-5. Assume that $C = 100$ μF, $I = 25$ mA, and V_S is a 60-Hz sine-wave source with a peak value of 30 V. Find the values of V_{ave} and V_{rip}. Assume an ideal diode.

$V_{ave} =$ 28.0

$V_{rip} =$ 4.18 p9
4.2

2-6. Assume that $C = 250$ μF, $I = 30$ mA, and V_S is a 60-Hz source with a peak value of 6 V. Find the values of V_{ave} and V_{rip}. Assume an ideal diode.

$V_{ave} =$ 5

$V_{rip} =$ 2

2-7. Repeat problem 2-5 using a diode whose $V_F = 0.7$ V.

$V_{ave} =$ 27.3

$V_{rip} =$ 4.18

2-8. Repeat problem 2-6 using a diode whose V_F = 0.7 V.

V_{ave} =   4.3

V_{rip} =   2

For the next two problems, refer to Fig. 2-27.

2-27 Circuit for problems 2-9 and 2-10.

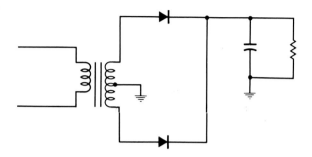

2-9. Assume that the transformer is rated as V_{pri} = 120 VAC 60 Hz and V_{sec} = 20 VCT. If the diodes are ideal, what is the value of V_p, and what will be the amplitude of the peak reverse voltage across each diode?

V_p =

P 16

PRV =

2-10. Referring to problem 2-9, if C = 250 μF and I = 30 mA, what is the peak-to-peak amplitude of V_{rip}, and what is the ripple frequency?

V_{rip} =

f =   120 HZ

For the next two problems, refer to Fig. 2-28.

2-28 Circuit for problems 2-11 and 2-12.

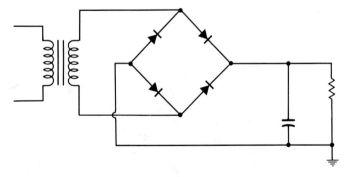

2-11. In the bridge rectifier circuit assume that V_{sec} = 6 Vrms and that V_F for each diode is 0.7 V. What is the peak voltage appearing across C, and what is the peak reverse voltage across each diode?

V_p = 8.5 - 1.4 = 7.08

PRV =

2-12.   Suppose the capacitor is rated at 25 WVDC and the PRV ratings for the diodes is 30 V. What maximum safe value of V_{sec} can be applied?

$V_{sec} =$

TROUBLESHOOTING PROBLEMS

Refer to Fig. 2-23 for the next six problems.

2-13.   The voltage at TP 1 reads zero, and you find that the fuse is blown. When you disconnect C_1 from the circuit and connect an ohmmeter across the capacitor, you get a reading near 20 Ω. Should you replace C_1? (a)Yes. (b)No.

2-14.   Assume that the original rating of C_1 is 250 μF at 50 WVDC. Which of the following replacement capacitors could you use? (a)100 μF at 100 WVDC. (b)300 μF at 75 WVDC. (c)250 μF at 25 WVDC.

2-15.   You are troubleshooting the power supply for excessive ripple. Which of the following would be the best first step? (a)Replace both diodes. (b)Measure V_{pri}. (c)Bridge C_1 with a capacitor of equal value. (d)Test for a shorted turn.

2-16.   Assume that the power supply has excessive ripple and that the output voltage is low. Measuring D_1 with an ohmmeter you read 5 Ω in one direction and near infinite ohms in the other. Measuring D_2 you read near infinite ohms in both directions. You should (a)replace D_1. (b)replace D_2. (c)check C_1. (d)look for a disconnected center tap.

2-17.   When replacing a defective diode with ratings of $I_F = 1$ A (ampere) and PIV = 50 V, which of the following diodes could be used? (a)$I_F = 2$ A, PIV = 25 V. (b)$I_F = 0.5$ A, PIV = 100 V. (c)$I_F = 1$ A, PIV = 25 V. (d)$I_F = 1.5$ A, PRV = 75 V.

2-18.   When checking the supply for zero d-c output voltage, you find that the fuse is OK and that V_{sec} and V_{pri} both measure zero. What would be the best next check? (a)Measure the voltage across TP 6 and 7. (b)Check for a shorted transformer turn. (c)Look for excessive loading. (d)Check the diodes.

Refer to Fig. 2-14 for the next six problems.

2-19.   Assume the voltage at TP 1 measures about normal, the voltage at TP 2 reads higher than normal, and the voltages at TP 3 and 4 both read zero. Which of the following is the most probable cause? (a)Shorted C_{23}. (b)Open R_{23}. (c)Open R_{24}. (d)Shorted C_{10}.

2-20.   Assume that the voltage at TP 44 reads normal but that TP 1 through 4 all read zero. The most probable cause is a (a)defective J_3. (b)defective S_3. (c)defective S_2. (d)defective J_4.

2-21.   Suppose the voltage at TP 1 reads about normal, but the voltages at TP 2 through 4 all read zero. Which of the following could be the cause? (a)Shorted C_{22}. (b)Shorted C_{10}. (c)Shorted C_{13}. (d)Shorted C_{14}.

2-22.   Suppose the voltage at TP 1 is lower than normal and has excessive ripple. The voltages at TP 2 through 4 do not have excessive ripple, but they are lower than normal. Which of the following could be the cause? (a)Loss of capacitance in C_{23}. (b)Excessive resistance in S_2. (c)Shorted R_{24}. (d)Primary voltage lower than normal.

2-23.   The voltage at TP 41 reads zero, but the voltage measures normal across TP 39 and 40. The most likely cause is a (a)leaky or shorted C_{23}. (b)defective S_3. (c)defective rectifier assembly. (d)defective transformer.

2-24.   The voltage at TP 41 is lower than normal and has excessive ripple. The voltage across TP 39 and 40 looks abnormal. Connecting a meter from ground to TP 39, you read 6.4 VAC, and you read zero from ground to TP 40. You could confirm the defective part with an ohmmeter check of (a)C_{23}. (b)the diodes. (c)S_3. (d)the transformer secondary.

EXPERIMENT 2-1   COMPONENT TESTING

In this experiment, you will measure some power-supply components that will help you recognize normal operation.

EQUIPMENT

- (4) silicon rectifier diodes, type 1N4001 or equivalent
- (2) electrolytic capacitors of different capacitance values, possibly 100 and 200 μF, at any voltage rating
- (1) transformer, V_{pri} = 120 V, V_{sec} = 12.6 V
- (1) 25-W, 120-VAC incandescent lamp with socket
- VOM

DIODE CHECKS

Select any one of the rectifier diodes. Using an ohmmeter on the $R \times 1$ scale, connect the negative lead of the meter to the cathode and the positive lead to the anode. Record the measured resistance in Table E2-1, in the column labeled Forward Resistance and the row for diode 1. Next, reverse the meter leads and record the measured resistance in the column labeled Back Resistance. Now divide the high-resistance value by the low-resistance value and record this ratio in the space provided. Repeat this procedure for the remaining three diodes.

Table E-2-1

Diode	Forward Resistance	Back Resistance	High/Low Ratio
1			
2			
3			
4			

CAPACITOR CHECKS

Set the scale of the ohmmeter to $R \times 10$ KΩ or higher. Select one of the electrolytic filter capacitors and connect the positive lead of the ohmmeter to the + terminal of the capacitor and the negative lead to the − terminal. You should see a rapid "kick" of the needle toward the low-resistance end of the scale, then a gradual increase in the resistance reading as the capacitor charges. (A digital ohmmeter will give a similar result. First the reading will go low, but it will gradually increase to a high value.) Record the capacitance and leakage resistance in the space provided. Leakage resistance is measured after several seconds, when the ohmmeter reading stops changing.

$C =$

$R_{\text{leakage}} =$

The actual values of R and C are not important. They will simply serve as a reference for future work. Repeat the above measurements with the other filter capacitor.

$C =$

$R_{\text{leakage}} =$

TRANSFORMER CHECKS

1. With no power applied, measure the d-c resistance of the primary winding.

$R_{pri} =$

Measure the resistance from the center tap to one end of the secondary.

$R_{AC} =$

Measure the resistance from the center tap to the other end of the secondary.

$R_{BC} =$

Measure the resistance across the entire secondary.

$R_{AB} =$

E2-1   Circuits for transformer tests in experiment 2-1.

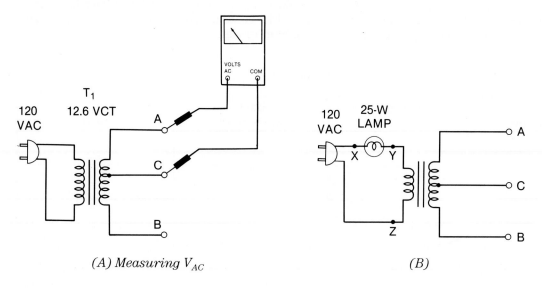

(A) Measuring V_{AC}   *(B)*

2. Next, refer to the circuit of Fig. E2-1. Plug in the line cord and measure the secondary voltage from the center tap to side A.

$V_{AC} =$

Measure the voltage from the center tap to side B.

$V_{BC} =$

Finally, measure the voltage from B to A.

$V_{AB} =$

3. Now unplug the line cord and connect a lamp in series with the primary, as shown in Fig. E2-1B. Plug the line cord in again.

Does the lamp light?

Measure the voltage across the lamp.

$V_{XY} =$

CAUTION You are working with 120-VAC line voltage.

Now measure the voltage across the transformer primary.

$V_{YZ} =$

4. Next, connect a clip lead across half of the secondary winding, from point A to point C. This represents a short across some of the secondary winding.

Does the lamp light?

Measure the voltage across the lamp.

$V_{XY} =$

Measure the voltage across the transformer primary.

$V_{YZ} =$

QUIZ

1. When a diode is forward biased, its resistance is relatively (a)high. (b)low.

2. When a diode is back biased, its resistance is relatively (a)high. (b)low.

3. When an ohmmeter is connected across an uncharged filter capacitor, (a)the meter reads infinite resistance. (b)the meter kicks down to a low reading momentarily.

4. After an ohmmeter has been connected across a good filter capacitor for a long time, (a)the reading goes to a high value, indicating a high leakage resistance. (b)the reading goes to zero.

5. Larger values of C most likely have—(a)higher, (b)lower—values of leakage resistance.

6. In the center-tapped transformer, the resistance of the entire secondary—(a)is, (b)is not—equal to twice the resistance from center tap to one end.

7. In the center-tapped transformer, the voltage across the entire secondary—(a)is, (b)is not—equal to twice the voltage from the center tap to one end.

8. With no jumper across the secondary, the lamp—(a)did, (b)did not—light because (c)there was excessive primary current. (d)there was high primary inductive reactance.

9. When a short was placed across part of the transformer secondary, the lamp—(a)did, (b)did not—light because (c)the excessive secondary current drastically reduced the reactance of the primary. (d)the lamp would only light if the short were across the primary.

10. If the short were place across the entire secondary from point A to point B, (a)the lamp would light with the same brightness. (b)the lamp would burn brighter.

EXPERIMENT 2-2   HALF-WAVE RECTIFIER

In this experiment, you will become familiar with the half-wave rectifier power supply first by measuring a circuit, which is operating normally, and then by simulating malfunctions to do some troubleshooting.

EQUIPMENT
- (2) silicon rectifier diodes, type 1N4001 or equivalent
- (2) capacitors, 200 µF at 25 WVDC
- (1) capacitor, 100 µF at 25 WVDC
- (1) transformer, V_{sec} = 12.6 VCT at 100 mA or greater
- (2) 1-KΩ, ½-W 10% resistor
- (1) line cord
- (1) plug
- (1) power switch
- (1) oscilloscope
- VOM

E2-2   Circuit for experiment 2-2.

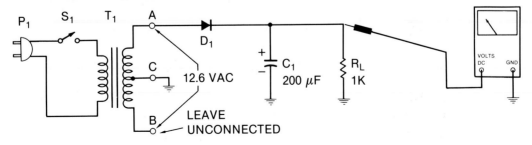

PROCEDURE

1. Build the circuit of Fig. E2-2. Measure the transformer secondary voltage V_{sec} from center tap to point A.

$V_{AC} =$

Then measure the d-c output voltage V_0 across R_L.

$V_0 =$

According to theory, what should be the d-c output voltage? Show the equation and your calculations.

$V_0 =$

Also calculate the load current I_L.

$I_L = V_0/R_L =$

2. Using an oscilloscope, measure the peak-to-peak ripple voltage across R_L.

$V_{rip} =$

Now calculate the ripple voltage according to the equation in the text.

$V_{rip} =$

If the calculated values of V_0 and V_{rip} agree reasonably with the measured values (within 10% or so), go on to the next part. Otherwise recheck your wiring, measurements, and calculations.

Let's try some troubleshooting. In the following steps, you will make changes to various components and note the effect on readings of V_0, V_{sec}, and V_{rip}. Later you will compare the measured voltages to the normal values to see how they differ.

3. Connect another 1-KΩ resistor in parallel with R_L. This simulates excessive current being drawn by the load. It also simulates a leaky filter capacitor. Measure these voltages:

$V_{sec} =$

$V_0 =$

$V_{rip} =$

4. Now remove the extra 1-KΩ resistor, so that R_L = 1 KΩ again. Replace C_1 with a 100-μF capacitor, and measure these voltages:

$V_{sec} =$

$V_0 =$

$V_{rip} =$

The decrease in capacitance simulates loss of capacitance as a result of aging.

5. Remove C_1 entirely from the circuit, and in the space below, sketch the waveform you observe with an oscilloscope across R_L. Removing C_1 simulates an open capacitor.

6. Finally, replace C_1 with the original capacitor, and remove D_1 from the circuit simulating an open diode. Measure these voltages:

$V_{sec} =$

$V_0 =$

$V_{rip} =$

7. The circuit of Fig. E2-2 uses only half of the transformer secondary. By using the entire secondary, you can obtain a d-c output voltage of twice the amplitude as in the circuit of Fig. E2-2. Disconnect the center tap (point C) from ground, and leave it unconnected. Then ground point B of the secondary. Measure the output voltage and ripple.

$V_0 =$

$V_{rip} =$

Why is the ripple voltage higher than that measured in step 2?

8. Sometimes *two* power supplies, one positive and one negative, are needed. By grounding the secondary center tap and using the opposite ends of the secondary (A and B) as separate sources, you can build two power supplies. Draw the diagram of a dual-power supply having both a positive and a negative d-c output. Use two silicon rectifiers and two 200-μF filter capacitors.

Have your lab instructor check your diagram before applying power. Then apply power and measure the two outputs.

$V_{01} =$

$V_{02} =$

QUIZ

To see the effect of component changes, compare your measured values for each change in Table E2-2 with the normal values obtained in steps 1 and 2. In the table write H in the blank if the change causes the measured values of V_{pri}, V_{sec}, V_0, or V_{rip} to be higher than normal, L for lower than normal, N for normal, and 0 for 0 V. Some of the changes were not simulated in your experiment, but you should be able to determine the effects. See line 1 of the table for an example.

	Change	V_{pri}	V_{sec}	V_0	V_{rip}
1	Open primary	N	0	0	0
2	Open secondary				
3	Open D_1				
4	Open C_1				
5	Open R_L				
6	Decrease R_L 50%				
7	Decrease C_1 50%				

EXPERIMENT 2-3   FULL-WAVE RECTIFIER

You will now compare the full-wave rectifier to the half-wave rectifier.

EQUIPMENT

Same as in experiment 2-2.

E2-3   Circuit for experiment 2-3.

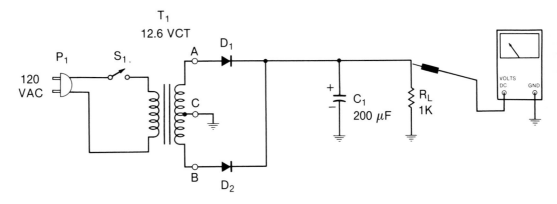

PROCEDURE

1. Build the circuit of Fig. E2-3. Measure the following voltages:

$V_{AB} =$

$V_{AC} =$

$V_0 =$

2. Using an oscilloscope, measure the ripple voltage.

$V_{rip} =$

Calculate the ripple voltage according to equation 2-1.

$V_{rip} =$

Let's try some more troubleshooting, as we did on the half-wave circuit.

3. Simulate excessive loading by connecting another 1-KΩ resistor in parallel with R_L. Measure these voltages:

$V_{sec} =$

$V_0 =$

$V_{rip} =$

4. Remove the extra 1-KΩ resistor, and replace C_1 with a 100-μF capacitor. Measure these voltages:

$V_{sec} =$

$V_0 =$

$V_{rip} =$

5. Remove C_1 entirely from the circuit, and sketch the waveform you observe across R_L.

6. Finally, replace C_1, and remove D_1 from the circuit. Measure these voltages:

$V_{sec} =$

$V_0 =$

$V_{rip} =$

QUIZ

1. The d-c output voltage of the full-wave supply is—(a)higher than, (b)lower than, (c)about equal to—the output voltage of the half-wave supply in step 1 of experiment 2-2.

2. The ripple voltage of the full-wave supply is—(a)higher than, (b)lower than, (c)about the same as—that in the half-wave circuit.

3. Additional loading in step 3 made the ripple voltage—(a)higher, (b)lower—than in step 1. Compared to the half-wave circuit, the ripple is (c)higher. (d)lower. (e)about the same.

4. With a smaller filter capacitor, the full-wave circuit has—(a)more, (b)less—ripple than the half-wave circuit.

5. The waveform sketched in step 5 showed that the current pulses flowed through R_L—(a)once, (b)twice—each cycle.

6. Simulating an open diode by removing D_1 in the full-wave circuit (a)increases the ripple. (b)has no effect on the output. (c)makes the output go to zero.

EXPERIMENT 2-4   FULL-WAVE BRIDGE

The bridge rectifier is one of the most commonly used types of power supplies. This experiment will familiarize you with it.

EQUIPMENT
- (4) silicon rectifiers, or a bridge rectifier assembly
- (2) capacitors, 200 µF at 25 WVDC
- (1) transformer, V_{sec} = 12.6 VCT at 100 mA or greater
- (2) 1-KΩ, ½-W 10% resistor
- oscilloscope
- VOM
- (1) line cord
- (1) plug
- (1) power switch

PROCEDURE
1. Build the circuit of Fig. E2-4A. Apply power and measure the output voltage and ripple.

$V_0 =$

$V_{rip} =$

2. Modify your circuit as shown in Fig. E2-4B, and apply power. Measure V_0.

$V_0 =$

3. As was demonstrated with the half-wave circuit, you can build *two* d-c supplies using a single center-tapped transformer. Build the circuit of Fig. E2-4C. Be sure to observe polarity of the diodes and capacitors. Measure V_0 for each supply.

$V_{01} =$

$V_{02} =$

QUIZ

1. How does the d-c output voltage of the bridge circuit compare with that of the circuit of Fig. E2-3?

2. The bridge circuit has—(a)more, (b)less, (c)the same amount of—ripple as the half-wave circuit.

E2-4 Circuits for experiment 2-4.

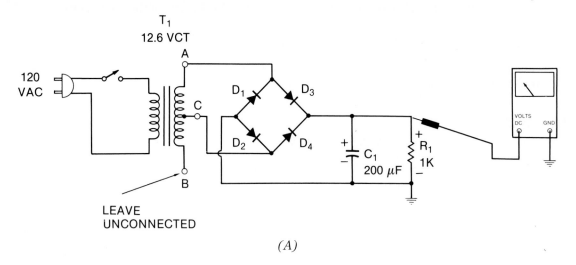

(A)

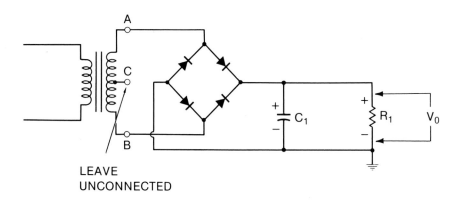

(B)

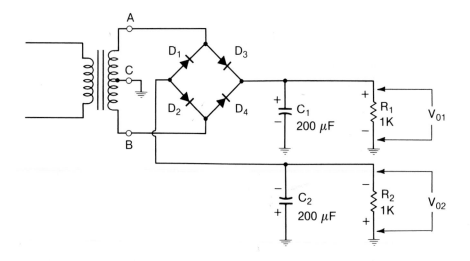

(C) Dual supply.

3. The bridge circuit has—(a)more, (b)less, (c)the same amount of—ripple as the circuit of Fig. E2-3.

4. By observing the output voltage with an oscilloscope, how can you tell that the bridge circuit is a full-wave rectifier?

5. Why is the output voltage higher in step 2 than in step 1?

6. What is the advantage of the circuit of Fig. E2-4C over that of the half-wave version of the dual supply?

7. The dual supply of Fig. E2-4C acts as two full-wave supplies, both operating from a common center-tapped transformer. The diodes are no longer acting as a bridge. Which two diodes are used as rectifiers for the positive supply?

8. Which two diodes are used for the negative supply?

9. If D_1 in Fig. E2-4C opened, would the positive supply be affected? If so, how?

10. Referring to question 9, would the negative supply be affected? If so, how?

EXPERIMENT 2-5   SIMPLE COMPONENT TESTER

You have already learned how to test components, such as diodes, with an ohmmeter, and you already know how to check fuses with your ohmmeter. You will now construct a simple test rig which can be used to test diodes, fuses, and other devices very quickly and efficiently without the use of your ohmmeter.

EQUIPMENT

- (1) 9-V transistor radio battery
- (1) LED (any general-purpose type)
- (2) 1-KΩ resistors
- (2) 4-inch lengths of heavy bare copper wire, #12 gauge or so
- (1) piece of pegboard, or similar material, for mounting

E2-5a   Circuit for experiment 2-5.

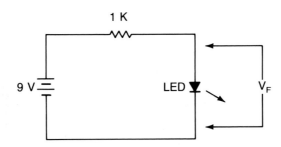

PROCEDURE

1. In order to get familiar with the operation of an LED, build the circuit of Fig. E2-5A. When power is applied, the LED should light up. Measure the forward voltage drop across the counducting LED.

$V_F =$

You should notice that the forward drop is higher than that for a silicon diode.

2. Now bridge the resistor with another 1-KΩ resistor. What happens to the brightness of the LED?

Normally, no more than about 20 mA should pass through an ordinary LED, or it may be damaged. However, the light output of the LED is almost directly proportional to the current flow through it. You can adjust the brightness to suit your application by adjusting the series resistor.

3. Now reverse the connections to the diode. Apply power and describe what you observe.

4. Now that you know how easy it is to use an LED, let's get to our special tester. Examine Fig. E2-5B. This is the complete circuit for our tester. The two pieces of heavy gauge wire should

be mounted, by staples or whatever, in the shape of a V onto some convenient size of insulating board. Notice that the wires do not touch, so the circuit is not complete.

To test a fuse, simply lay the ends of the fuse across the bare wires. If the fuse is good, the LED will light. The reason for the V shape is to accommodate different sizes of fuses.

E2-5B   Component tester.

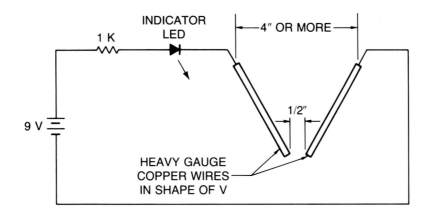

QUIZ

Now figure out how to test diodes. Answer the following questions.

If a good diode is placed across the wires with its anode to the left, will the LED light?

If a good diode is placed across the wires with its anode to the right, will the LED light?

If an open diode is placed across the wires, explain what you will observe.

If a shorted diode is placed across the wires, explain what you will observe.

So you see that fuses or diodes can easily be tested with this rig. The rig can be permanently built, say on a small piece of pegboard, and mounted in any convenient place—under one of the shelves of your workbench, for example, to be used whenever you need it. You don't even need a power switch because the circuit is normally open, so the battery will probably last for a couple of years. Also notice that the diode test also identifies which end of the diode is the anode and which end is the cathode, because the LED only lights when the anode is connected to the positive terminal. You might want to label (+) and (−) signs on the board so that you don't forget which is which.

Bipolar Transistors

In this chapter, we will discuss the bipolar transistor, its characteristics, and its typical uses as small signal a-c amplifiers. Even though the trend in new electronic circuit design is geared toward using integrated circuits (IC), a thorough understanding of transistors is necessary in order to use IC intelligently. In addition, practically all designs use at least a few discrete (individual) transistors to "help out" IC and to customize their operation.

BIPOLAR TRANSISTORS

In chapter 2, you learned that a diode is formed by the junction of P-type and N-type semiconductor material. *Bipolar* or *junction* transistors are made in a similar manner, except that the transistor is composed of *two* junctions, forming a "sandwich" of semiconductor material, as shown in Figs. 3-1A and 3-1C. Note that two versions are possible. One is called an *NPN* transistor; the other is called a *PNP* transistor. Both work essentially the same. However, the NPN-type ordinarily uses a power supply that is positive with respect to ground, while the PNP-type generally uses a negative power supply. Note the schematic symbols for each in Figs. 3-1B and 3-1D. As with the diode symbol, the arrow indicates the direction of conventional current flow from positive to negative.

3-1   Bipolar transistor equivalent construction and symbols.

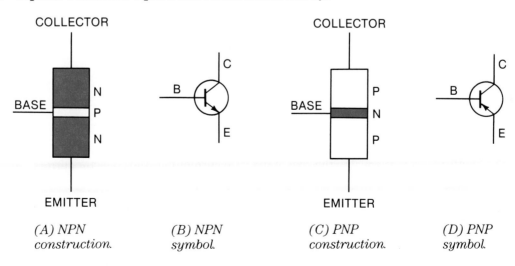

(A) NPN
construction.

(B) NPN
symbol.

(C) PNP
construction.

(D) PNP
symbol.

Early bipolar transistors made extensive use of germanium as the semiconductor material, but silicon has much better temperature characteristics. Silicon is, therefore, the most commonly used material for modern transistors.

The three terminals connected to the transistor tie to the *collector, base*, and *emitter*, as shown in Fig. 3-1. The input to the transistor is the diode formed by the base–emitter junction. When the base of the NPN is made positive with respect to the emitter, the diode becomes forward biased. Electrons start to flow from the N material (emitter) to the P material (base). However, the transistor is so constructed that the base region is very thin. For this reason, most of the electrons emitted into the base actually flow right through the thin base region into the collector region. If the collector is connected to a positive supply voltage, the electrons will flow out of the collector to the collector power supply. This is shown in Fig. 3-2. The base supply is labeled V_{BB}, and the collector supply is labeled V_{CC}.

3-2 Current flow in an NPN transistor.

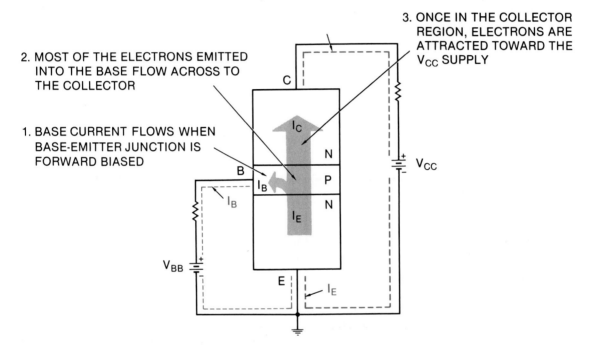

2. MOST OF THE ELECTRONS EMITTED INTO THE BASE FLOW ACROSS TO THE COLLECTOR

1. BASE CURRENT FLOWS WHEN BASE-EMITTER JUNCTION IS FORWARD BIASED

3. ONCE IN THE COLLECTOR REGION, ELECTRONS ARE ATTRACTED TOWARD THE V_{CC} SUPPLY

The key to transistor operation is this: the *base current I_B controls the collector current I_C.* That is, an increase in base current causes a proportional increase in collector current. Likewise, if the base current is decreased or cut off, the collector current will decrease or cut off accordingly. The ratio of I_C to I_B is called beta (β) or h_{FE}. That is,

$$\beta = h_{FE} = \frac{I_C}{I_B} \qquad\qquad 3\text{-}1$$

or

$$I_C = \beta\, I_B \qquad\qquad 3\text{-}1A$$

Typical β values range from about 20 to over 200.

The emitter current is equal to the sum of the base current plus the collector current, or

$$I_E = I_B + I_C \qquad\qquad 3\text{-}2$$

and if β is large

$$I_E \cong I_C \qquad\qquad 3\text{-}3$$

EXAMPLE 3-1   If $I_B = 10\ \mu A$ and $\beta = 99$, find I_C and I_E.

SOLUTION

$$I_C = \beta\,I_B = 99 \times 10\ \mu A = 990\ \mu A$$

$$I_E = I_B + I_C = 10 + 990 = 1000\ \mu A = 1\ mA$$

The input characteristic curve of a silicon transistor is shown in Fig. 3-3. Notice that the curve is that of a silicon diode, where the base–emitter voltage V_{BE} determines I_B. As with the silicon diode, the device is essentially nonconducting until the forward-biasing voltage V_{BE} reaches about 0.7 V. Thereafter, I_B increases rapidly for slight increases in V_{BE}.

3-3   Input characteristic of silicon bipolar transistor.

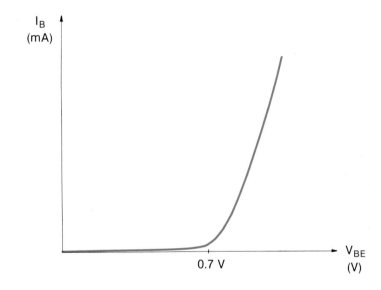

Remember that with the silicon diode there is a maximum reverse voltage that can safely be applied before breakdown occurs. Similarly, there is a maximum collector-to-base voltage that can safely be applied before breakdown occurs. (Note that the collector–base diode is normally reverse biased.) The maximum collector-to-base voltage is labeled V_{CBO}, meaning voltage, collector-to-base, with emitter open. Likewise, there is a maximum safe collector-to-emitter voltage that can be applied before breakdown occurs. This is called V_{CEO}. Transistors are available with ratings of V_{CBO} and V_{CEO} from about 20 to several hundred volts.

The maximum safe collector current $I_{C(max)}$ that the transistor can handle depends on the physical size of the transistor elements. Depending on the heat dissipating characteristics of the transistor, there is a maximum safe collector power dissipation $P_{C(max)}$. If either $I_{C(max)}$ or $P_{C(max)}$ is exceeded, the transistor may be permanently damaged. When replacing a transistor with one of a different part number, always be sure that the replacement part has values of $P_{C(max)}$, $I_{C(max)}$, V_{CBO}, and V_{CEO} are at least equal to or greater than those of the device being replaced.

Transistors are available for a wide variety of applications, such as audio or general purpose, radio frequency (rf), power, and switching. They have device numbers that usually start with 2N, such as 2N2222 and 2N1304.

Depending on the purpose and power dissipation characteristics of the device, transistors come in a variety of packages, a few of which are shown in Fig. 3-4.

THE TRANSISTOR AS A VARIABLE RESISTANCE

A transistor is like a variable resistance. Consider the circuit of Fig. 3-5A. Suppose that resistor R is adjusted so that its resistance is equal to that of lamp L. Also suppose that the current flow

3-4   Common transistor packages.

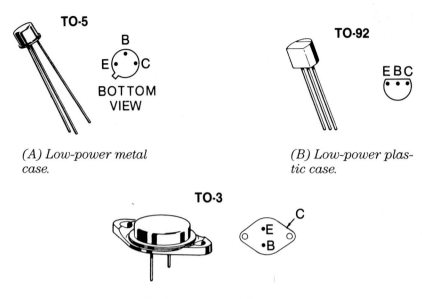

(A) Low-power metal case.

(B) Low-power plastic case.

(C) Power transistor.

through the lamp causes it to light, but not at full brilliance. If we decrease the resistance of R, the lamp will glow brighter. And if we increase the resistance of R, the lamp will go dimmer.

A transistor is used in much the same manner. In Fig. 3-5B, a transistor is used in place of R. As you learned earlier, an increase in base current I_B causes an increase in collector current I_C, and a decrease in I_B causes a decrease in I_C. Since I_C flows through L, we can vary the brightness of the lamp by simply varying I_B. Since I_C is many times larger than I_B, we control a relatively large current through the lamp with a small current fed into the base.

In Fig. 3-5B, I_B flows as a result of supply V_{BB}. If we decrease I_B to zero, say by making $V_{BB} = 0$, no collector current flows. The transistor is said to be *cut off*. A cut-off transistor acts like an open switch, as shown in Fig. 3-6A. If we measure the d-c voltage at the collector when the transistor is cut off, a voltage V_C is equal to the supply voltage V_{CC}.

Actually, a small amount of *leakage* current flows out of the collector, even when $I_B = 0$. This current is usually negligible in a good silicon transistor, but it does increase as the temperature increases. Leakage current can cause problems at high temperatures, unless precautions are taken in designing the circuit. For now, we will assume that the leakage current is negligible.

3-5   A transistor acts like a variable resistor.

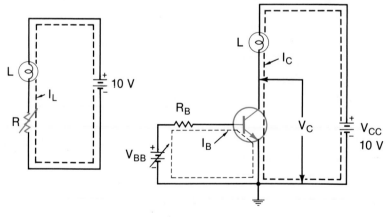

(A) Current through L controlled by R.

(B) Current through L controlled by transistor.

3-6   **Equivalent circuit of transistor used as a switch.**

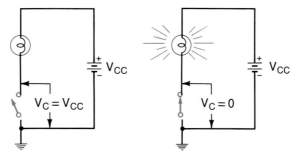

(A) Cutoff transistor acts like an open switch.
   (B) saturated transistor acts like a closed switch.

Getting back to Fig. 3-5B, suppose we make the bias current increasingly larger. As we continue increasing I_B, I_C increases, causing increasingly more voltage to be dropped across L. Eventually, if we turn the transistor on hard enough, making its resistance negligible compared to the resistance of the lamp, practically all of V_{CC} is dropped across L, making $V_C \approx 0$. When practically all of V_{CC} is dropped across the collector load (the lamp in this case), the transistor is said to be *saturated*. A saturated transistor acts like a closed switch, as shown in Fig. 3-6B. So we see that the switch is a common application of the transistor. It is used in this manner in many computers and controls.

BIASING THE TRANSISTOR

When a transistor is used as an amplifier, it is neither cut off nor saturated. Rather, it conducts a moderate amount of collector current. That is, the transistor must be properly biased. I_C is then made to vary by means of an input signal voltage. Biasing a transistor is similar to setting the idle speed on the engine of a car. It must be turned on and running properly before we can expect it to perform.

Suppose we replace the lamp with a collector load resistor R_C, as shown in Fig. 3-7. Next, we will connect resistor R_B from the base to V_{CC}. Since R_B is connected to the positive supply voltage, the base–emitter diode is forward biased, causing I_B, and likewise I_C, to flow. Now let's adjust R_B so that I_C = 1 mA. This will cause a 5-V drop across R_C because I_C flows through it. The voltage from collector to ground will then be equal to the difference between V_{CC} and the drop across R_C, or

$$V_{CE} = V_{CC} - I_C R_C \qquad\qquad 3\text{-}4$$

In this case, V_{CE} = 12 − 1 mA × 5 KΩ = 7 V. Usually, as in this case, the transistor is biased so that V_C is somewhere near half the supply voltage.

3-7   **Base-biased circuit.**

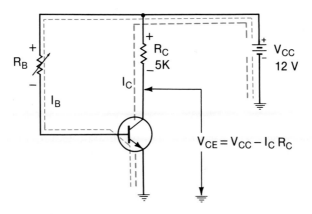

$$V_{CE} = V_{CC} - I_C R_C$$

We are now ready to feed a small signal to the input of the transistor. In Fig. 3-8A, a small a-c signal v_s is applied to the base–emitter junction through coupling capacitor C_C.

We can analyze what takes place in the transistor with the aid of Fig. 3-8B. Because the d-c current I_B flows through R_B, the base–emitter junction is forward biased at about 0.7 V. The sine-wave input signal is superimposed on the d-c base–emitter voltage through C_C. The input signal simply adds to and subtracts from the quiescent base–emitter voltage of 0.7 V. Notice the small sine wave plotted on the projections below the horizontal axis. When the input signal goes positive from point 1 to point 2, the base current increases from point 1' to point 2'. Then when the input signal decreases from point 2 to point 3, the base current decreases from point 2' to point 3'. Likewise for the other points in the figure.

3-8   Applying an a-c signal to the base of an amplifier.

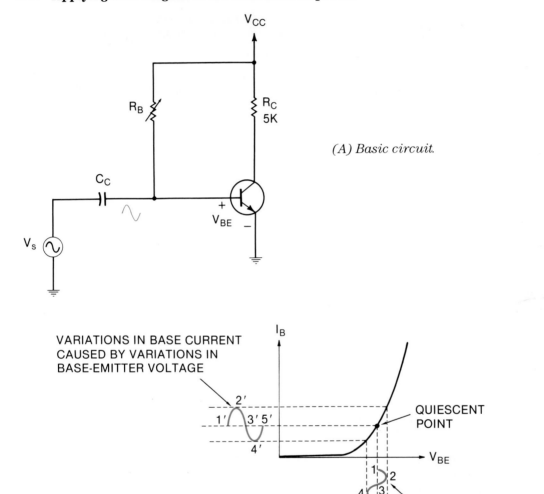

(A) Basic circuit.

(B) Superimposing a signal on the quiescent point.

Now since I_B varies with the input signal , I_C does also. But since I_C is β times larger than I_B, the variations are also β times larger. The variations in I_C flowing through R_C develop an output voltage that is larger than the input voltage. In other words, the a-c output voltage is an amplified version of the a-c input voltage. You can see why we have to bias the transistor somewhere between cutoff and saturation. We do it so that the collector voltage can swing positive and negative around an *operating point* (quiescent point) without clipping off part of the sine-wave signal.

The circuit in Fig. 3-7 uses a single resistor from the base to V_{CC} to bias the transistor, so it is called a *base-biased* circuit. Since the base–emitter diode has very little voltage across it, practically all of the V_{CC} supply is dropped across R_B. As is shown in Fig. 3-9, if $V_{CC} = 12$ V and $V_{BE} = 0.7$ V, the voltage across R_B is $12 - 0.7 = 11.3$ V. I_B through R_B can be easily determined by

$$I_B = \frac{V_{CC} - V_{BE}}{R_B} \approx \frac{V_{CC}}{R_B}$$

3-5

EXAMPLE 3-2   In the circuit of Fig. 3-9, assume $V_{CC} = 12$ V, $R_B = 1$ MΩ, $R_C = 3$ KΩ, and $\beta = 80$. Find I_B, I_C, and V_{CE}.

3-9   Determining the voltage across R_B to find I_B.

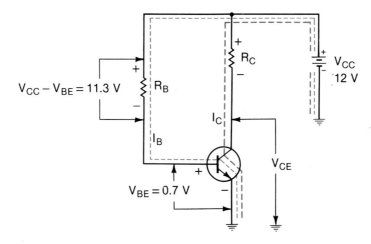

SOLUTION

$$I_B = \frac{V_{CC}}{R_B} = \frac{12 \text{ V}}{1 \text{ M}\Omega} = 12 \text{ μA}$$

$$I_C = \beta I_B = 80 \times 12 \text{ μA} = 0.96 \text{ mA}$$

and

$$V_{CE} = V_{CC} - I_C R_C = 12 - 0.96 \text{ mA} \times 3 \text{ K} = 9.12 \text{ V}$$

The equations help us determine what happens to voltage and current values in the circuit if some component value changes. This will aid you in troubleshooting. For example, if R_B decreases, I_B increases. This causes a corresponding increase in I_C. And when I_C increases, the drop across R_C increases, resulting in a decrease in voltage from collector to emitter.

What difference does it make what the d-c voltage is from collector to emitter? Consider what happens when the amplifier of Fig. 3-10A is driven with a fairly large signal. Assuming $V_{CE} = 9$ V, we see in Fig. 3-10B that the output voltage is a sine wave riding on a d-c level of 9 V. The amplitude of the sine wave is determined by the amplitude of the signal source v_s, multiplied by the *gain* or *amplification* of the amplifier.

In Fig. 3-10B, the sine wave is not distorted. But now suppose v_s increases, so that the output signal amplitude tries to increase to, say, a 6-V peak. On the negative half of the input cycle, I_C decreases, causing V_{CE} to go more positive. (Note that v_0 and v_s are 180° out of phase.) But as shown in Fig. 3-10C, when the transistor reaches cutoff, $V_{CE} = V_{CC}$. The collector voltage *cannot* go positive with respect to V_{CC}. Therefore, part of the positive portion of the output cycle (shown by dotted lines) is clipped off. This, of course, is distortion, since the output signal no longer looks like the input sine-wave signal.

3-10 Distortion can appear on large signals if proper operating point is not used.

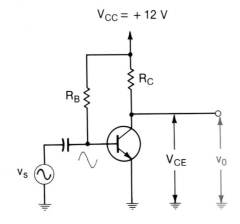

(A) Basic circuit.

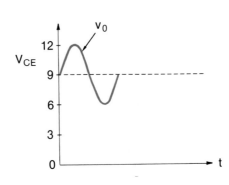

(B) No distortion.

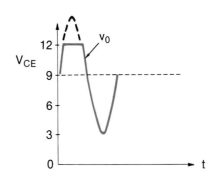

(C) Part of positive half cycle clipped.

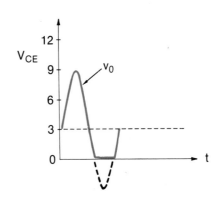

(D) Part of negative half cycle clipped.

Note that if the quiescent value of V_{CE} were 6 V instead of 9 V, no clipping would occur. So R_B is too large in this case. On the other hand, if R_B decreases so that V_{CE} becomes less than 6 V, say perhaps 3 V, as shown in Fig. 3-10D, clipping occurs on the negative peaks of the output cycle. That is, too small a value of R_B also causes clipping.

Generally, when an amplifier clips off the positive peaks of the output signal, the transistor is not conducting enough collector current. On the other hand, clipping the negative peaks indicates that the transistor is turned on too hard.

The base-biased circuit is simple to understand and use. However, it is rather unstable with variations in temperature, which is a disadvantage. When the temperature goes up, the collector current increases. And, as we just saw, if the collector current gets too high, clipping occurs. For this reason, other methods of biasing were developed. We will not look at all possible biasing variations. Instead we will examine the operation of some stabilizing components.

The circuit of Fig. 3-11 uses a voltage divider at the base and an emitter resistor R_E to achieve stabilization. The ratio R_1 to R_2 is chosen to make the base-to-ground voltage V_B equal a few volts, say 4 V in this case. Since the base-emitter junction is forward biased, the drop across it is 0.7 V. This means that the voltage across $R_E = V_B - V_{BE}$ or about 3.3 V. The emitter current through R_E, and likewise the collector current, is then equal to the voltage V_E across R_E divided by R_E. That is

$$I_E = I_C = \frac{V_B - V_{BE}}{R_E}$$

3-6

and if V_{BE} is negligible,

$$I_C \cong \frac{V_B}{R_E}$$

3-6A

Notice in Fig. 3-11 that the voltage V_E is developed when I_E flows through R_E. If I_E increases, due to an increase in temperature, V_E also increases. But if V_E increases while V_B remains constant, the base–emitter voltage V_{BE} decreases. Of course, a decrease in V_{BE} means less forward bias on the transistor, so I_B decreases, causing a proportional decrease in I_C. The net effect is that I_C does not increase as much with an increase in temperature as it would if I_B remained constant. Also, if I_B is much smaller than the current through the voltage divider R_1–R_2, the base-to-ground voltage is determined simply by the V_{CC} supply voltage and the voltage divider. That is,

$$V_B = V_{CC} \frac{R_2}{R_1 + R_2}$$

3-7

So then from equation 3-6A, we get

$$I_C = \frac{V_{CC} R_2}{(R_1 + R_2) R_E}$$

3-8

Notice that we do not need to know the exact value of β to determine I_C. So if β changes, say as a result of a temperature increase, or even replacement of the transistor, I_C will remain the same.

3-11 Voltage-divider bias.

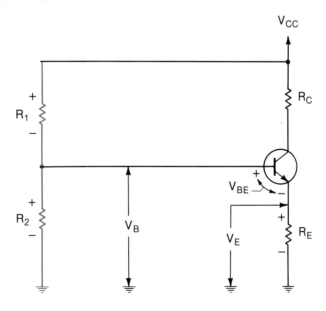

EXAMPLE 3-3 In the circuit of Fig. 3-11, assume $V_{CC} = 20$ V, $R_1 = 30$ KΩ, $R_2 = 10$ KΩ, $R_E = 2$ KΩ, and $R_C = 3$ KΩ. Find I_C and the voltage from collector to ground.

SOLUTION

$$I_C = \frac{V_{CC} R_2}{(R_1 + R_2) R_E} = \frac{20 \text{ V} \times 10 \text{ KΩ}}{(30 \text{ KΩ} + 10 \text{ KΩ}) 2 \text{ KΩ}} = 2.5 \text{ mA}$$

From which $V_C = V_{CC} - I_C R_C = 20 - 2.5 \times 3 = 12.5$ V.

A-C SIGNAL AMPLIFIER

Once the transistor is biased on, it is ready to amplify an input signal. The input resistance seen by the driving source depends on the operating point chosen. For example, if quiescent point 1 is chosen in Fig. 3-12, an input signal v_{in} causes a change ΔV_{BE} in biasing voltage. The variation in bias voltage causes a corresponding change ΔI_B in bias current. However, notice that when higher quiescent values of V_{BE} and I_B are chosen, such as point 2, the same variation ΔV_{BE} causes a *larger* change in ΔI_B. Since we get a higher a-c current at point 2 than at point 1 for the same amplitude of input signal, it is apparent that the a-c input resistance r_{in} seen by the driving source is lower at point 2 than at point 1. The a-c input resistance, in fact, depends on the slope of the line drawn through the operating point and tangent to the curve. We can show that the value of the input resistance for an amplifier whose emitter is grounded to the a-c signal is

$$r_{in} = \beta\, r_e \qquad\qquad 3\text{-}9$$

where

$$r_e = \frac{25\ \text{mV}}{I_E} \qquad\qquad 3\text{-}10$$

3-12 Input resistance is reduced when higher operating current is used.

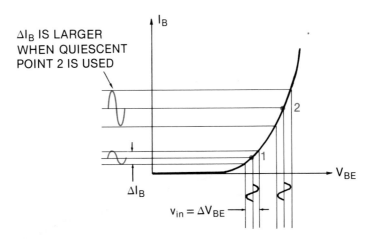

EXAMPLE 3-4   What is the input resistance of an amplifier if $I_E = 1$ mA and $\beta = 100$?

SOLUTION

$$r_e = \frac{25\ \text{mV}}{1\ \text{mA}} = 25\ \Omega$$

so

$$r_{in} = \beta\, r_e = 100 \times 25 = 2500\ \Omega$$

Analyzing the circuit of Fig. 3-13 shows that $I_B = V_{CC}/R_B = 10$ μA. Therefore $I_C = \beta I_B = 1$ mA. Then $V_C = V_{CC} - I_C R_C = 7$ V, and $r_{in} = 2500\ \Omega$ as before. We can use the calculated value of r_{in} to determine the a-c base current i_b. For example, with the 5-mV peak a-c input signal to the base shown, the peak a-c base current is $i_b = 5$ mV/2500 $\Omega = 2$ μA. And since $\beta = 100$, the peak a-c collector current i_c is 200 μA.

As shown in Fig. 3-14A, the total collector current consists of a 200 μA sine wave riding on a 1-mA d-c value. This varying collector current flows through R_C, causing the voltage from collector

3-13 Single-stage amplifier.

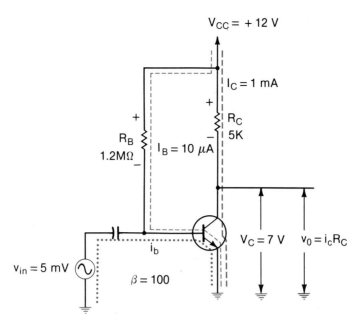

3-14 Collector waveforms for circuit of Fig. 3-13.

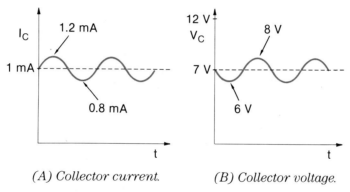

(A) Collector current.   *(B) Collector voltage.*

to ground to be as shown in Fig. 3-14B. The peak a-c voltage developed across R_C is $v_0 = i_c R_C$ = 200 μA × 5 KΩ = 1 V.

We can determine the amplification, or gain, of the amplifier as follows:

$$A_v = \frac{v_0}{v_{in}} = \frac{i_c R_C}{i_b r_{in}} = 200$$

This tells us that the output voltage will be 200 times the input voltage. Of course, this is only true up to the point where clipping of the output signal begins to occur.

Now consider what happens when a load is connected to the output of the amplifier, as shown in Fig. 3-15. Resistor R_L represents the load, which might be a resistor, the input resistance to the following stage, or whatever. Notice that the ac splits up after leaving the transistor, part of it flowing through R_C, the other part through R_L. The equivalent a-c resistance seen by the collector is R_C in parallel with R_L. We will call that equivalent a-c resistance r_L. That is $r_L = R_C \| R_L$. The two resistors are in parallel to the ac, because the V_{CC} supply acts like a short circuit to a-c signals since it is well bypassed.

Since the collector current in Fig. 3-15 flows through a lower value of resistance than in Fig. 3-13, the a-c voltage measured at the collector is lower than before. The gain of the circuit is now

$$A_v = \frac{V_0}{V_{in}} = \frac{i_c r_L}{i_b \beta \, r_e}$$

and since $i_b \times \beta = i_c$, we can say that

$$A_v = \frac{r_L}{r_e} \qquad\qquad\qquad 3\text{-}11$$

Note the two significant points indicated by equation 3-11. First, increasing r_L increases the voltage gain, and vice versa. Second, decreasing r_e increases the voltage gain, and vice versa. *Any circuit change that affects either r_L or r_e will affect the amplitude of the output voltage.*

The input and output coupling capacitors do not affect the voltage gain, since they look like short circuits to ac. They are used simply to block d-c flow, so that the bias of the stage is not upset. The reactances of these capacitors should be negligible at the lowest frequency that the amplifier is to handle.

3-15   Connecting a load to the collector.

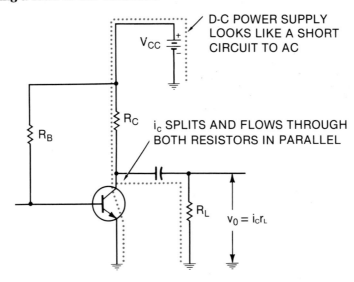

When an emitter resistor is used for stabilization, it is usually bypassed with a capacitor so that the emitter is at a-c ground. See Fig. 3-16. When this is done, the stage has the same gain as if the emitter were directly returned to ground. However, if the emitter bypass capacitor is removed, the gain of the amplifier reduces to approximately

$$A_v = \frac{r_L}{R_E} \qquad\qquad\qquad 3\text{-}12$$

where R_E is the unbypassed emitter resistor.

TYPICAL PROBLEMS WITH TRANSISTOR AMPLIFIERS

Like solid-state diodes, transistors do not simply get weaker with age. They can, however, open or short internally. Sometimes this is a result of an external overload. At other times, it is simply caused by internal fatigue due to repeated heating and cooling. The latter failure is similar to the opening of a filament in an incandescent lamp after many hours of use.

When an open develops between any two terminals in a transistor, it becomes useless and must be replaced. As indicated by Fig. 3-17, an open in the collector of a transistor cuts off all collector current. Therefore, there is no voltage dropped across R_C. *A d-c voltage measurement at the collector of an open transistor will show that $V_{CE} = V_{CC}$.* On the other hand, *voltage across a shorted transistor reads zero.*

3-16 Bypassing the emitter resistor with a capacitor.

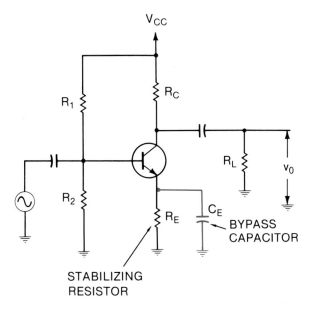

3-17 Open transistor.

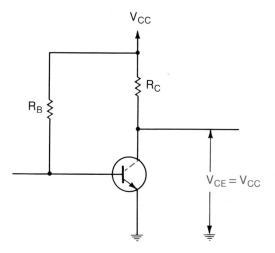

In addition to the transistor's going bad, some of the other components in the circuit can become defective. For example, if any biasing resistor changes value, the quiescent operating point of the transistor changes. If the defective component causes the transistor to turn on harder, V_{CE} decreases. And if the malfunction causes the transistor to conduct less, V_{CE} increases. A significant change in bias can cause the transistor to go into saturation or to cut off. Therefore, you cannot tell by simply measuring V_{CE} whether the transistor is bad or some other component is at fault. *After getting an abnormal reading for V_{CE}, test the transistor for a short or open. If it checks out OK, then look for some problem in the biasing circuit.*

Coupling capacitors can short, or become leaky, or open, causing the stage to malfunction.

Fig. 3-18 represents two stages of transistor amplifiers connected in series (also called *cascaded* stages) for higher gain. An open coupling capacitor would prevent a-c signals at the collector of Q_1 from reaching to the base of Q_2. It would not, however, upset any d-c readings. But if C_c shorted, Q_2 would be turned on harder than normal, because the collector resistor of Q_1 will pull the base of Q_2 more positive than normal.

Fig. 3-19 shows another problem often encountered when testing newly assembled printed circuit boards. The bottom view, or foil side, of the circuit board is shown, with the dashed lines showing the position of components on the other side of the board.

3-18 Cascaded stages coupled with a capacitor.

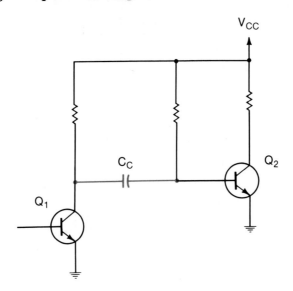

3-19 Solder bridge on printed circuit foil.

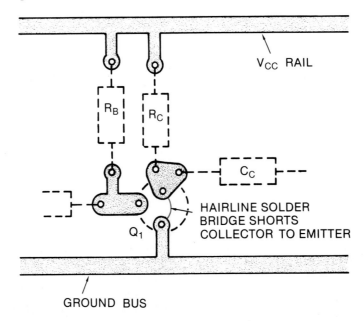

Here is the typical problem. You get no output signal at the collector of the stage. A d-c measurement shows 0 V from collector to ground. So you remove the transistor from the circuit and test it. It tests good. When you measure the voltage from the collector solder pad to ground, it still reads zero, even with the transistor removed. Now you must carefully inspect the foil side of the board. There is often a thin hairline solder "bridge" between two pads or foils, resulting from incomplete etching. You may need a magnifying glass to see these tiny shorts. If you cut away the bridge, the problem clears up. Of course, a break, or open, in a printed circuit foil can cause circuit failure as well.

IDENTIFYING THE BASIC AMPLIFIER

Now that you know how the single-stage transistor amplifier works, let's take a look at a few commercial circuits and try to identify the basic components.

Fig. 3-20 shows part of a SONY Model CF-320 tape recorder. We will concentrate on the driver stage Q_4, which is a single-stage audio amplifier. The highlighted components are the basic parts. Note the voltage divider circuit at the base consisting of R_{35}, R_{36}, and R_{37}. The two resistors R_{35} and R_{36} in series, which connect to the V_{cc} supply, correspond to R_1 in Fig. 3-17. The capacitor from the junction of R_{35} and R_{36} forms a decoupling filter, which keeps power supply variations from affecting the base of Q_4. The small capacitor C_{28} is most likely used to roll off high frequencies or to kill any tendency for the circuit to oscillate.

3-20   Part of SONY Model CF-320. From *Tape Recorder Series No. 161*, Howard W. Sams & Company, Inc.

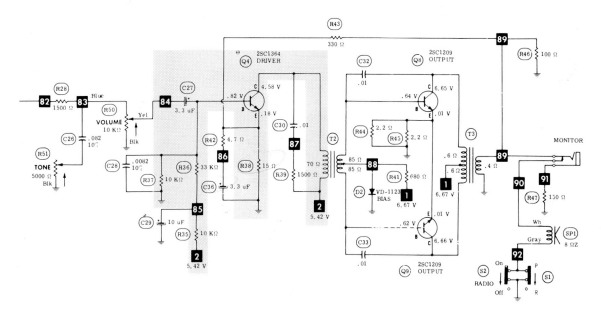

Look at the emitter and find R_{38}, a 15-Ω resistor used like R_E in Fig. 3-16. Notice that R_{43}, a 330-Ω resistor, is connected from the secondary of the output transformer to the emitter of Q_4. This resistor, which is very large compared to R_{38}, does not affect the d-c bias noticeably. Instead, it feeds back some of the output signal to the emitter to help improve the frequency response of the entire circuit. (Feedback is discussed in a later chapter.)

In this circuit, the collector does not work into a resistor but rather into the primary of coupling transformer T_2. Variations in collector current caused by the input signal are coupled to the push–pull power amplifiers Q_8 and Q_9 through this transformer. C_{30} and R_{39} across the transformer are used to improve the frequency response by rolling off some highs. They do not affect the biasing of the circuit.

Fig. 3-21 shows another example of a single-stage amplifier. This is part of a Panasonic Model RF-7400 amplifier. Transistor TR_{12} is the input stage. It gets an input signal through C_{57} from a microphone plugged into the jack. This transistor uses an unbypassed emitter resistor R_{76} and works into collector resistor R_{78}. Notice that the bias resistor R_{74} is tied to the collector, rather than to V_{CC}. Since the collector is positive with respect to the base, electrons flow from the base, through R_{74}, through R_{78} to V_{CC}. This method of biasing is an alternative to the simple base-biased circuit of Fig. 3-13.

By connecting the biasing resistor to the collector rather than to V_{CC}, the circuit becomes more temperature stable. Recall that as the temperature increases, the transistor normally tries to conduct harder, as a result of an increase in leakage current. Then, if the temperature change is large enough and if the base-bias current I_B stays constant, the operating point can shift far enough to cause clipping of the output signal.

But by connecting the base-bias resistor to the collector, rather than to V_{CC}, as the leakage current increases, the drop across the collector resistor increases. This results in a decrease in V_C. And when V_C decreases, the drop across the base-bias resistor decreases. So less bias current flows. The net effect is that as leakage current goes up, I_B goes down, tending to keep I_C more constant.

3-21 **Part of Panasonic Model RF-7400. From *Tape Recorder Series No. 179*, Howard W. Sams & Company, Inc.**

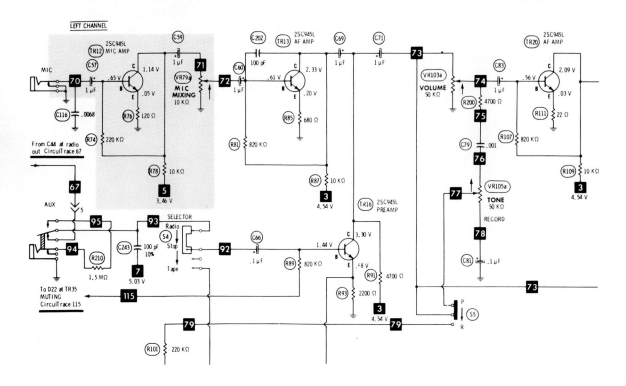

LOCATING PROBLEMS IN SINGLE-STAGE AMPLIFIERS

We have studied how a transistor stage is normally biased for use as an amplifier. We can use the equations listed to design transistor stages and to help us troubleshoot a defective stage. For example, we'll consider the voltage-divider biased circuit of Fig. 3-22. The biasing equations for the circuit are repeated here for your convenience.

$$V_{CE} = V_{CC} - I_C R_C \qquad\qquad 3\text{-}4$$

$$I_C \cong \frac{V_B}{R_E} \qquad\qquad 3\text{-}6\text{A}$$

$$V_B = V_{CC} \frac{R_2}{R_1 + R_2} \qquad\qquad 3\text{-}7$$

3-22 Voltage divider biased amplifier stage.

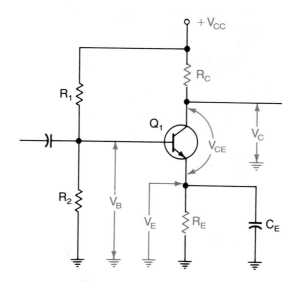

After checking for normal power supply voltage, you should then measure the d-c collector voltage V_C. If V_C is normal, the transistor and all resistors are normal! Think about that. The only way for the d-c collector voltage to be normal is for the collector current through the transistor to be normal. The only way that can be true is for the transistor to be good and for all biasing resistors to be normal. Yes, it can be argued that it is possible to make one resistor larger and another smaller simultaneously and thereby make the collector current normal. But what is the probability of that happening in a circuit that was built correctly and was once operating normally? Pretty slim, right? So, ordinarily, if you measure a normal voltage at the collector of a transistor, you can assume that the stage is operating normally. Then, if you are not getting a normal a-c signal out of it, look for problems in the a-c signal path, including coupling and bypass capacitors.

But let's see what d-c problems might occur. Suppose you measure a collector voltage which is lower than normal. Looking at equations 3-4A, we see that either R_C has gone up or I_C has gone up. Before removing R_C from the circuit to measure it, measure the base voltage V_B. Equation 3-6A shows us that if V_B goes up, the collector current also goes up. This, of course, will cause V_C to go down, which is what we measured. It's also possible that R_E has decreased, which would make I_C larger. Normally, if R_E goes down, V_E will also go down, and probably V_B will be dragged down also. But if V_B goes up, V_E will also go up. The point is that while you are measuring the collector voltage with a voltmeter or scope, it's a good idea to make a couple of other measurements, such as V_B and V_E, to give you a few more clues before removing components to measure resistance.

Let's suppose in the above case that V_B reads higher than normal. According to equation 3-7, this could be the result of R_1 being lower than normal or of R_2 being higher than normal. So now you would shut off the power and unsolder one end of R_1 or R_2 and measure its resistance. It is not a good idea to try to measure the resistance with both ends of the component still soldered in the circuit, because any other components in parallel with the one being measured will alter the reading.

Notice what we did in the above example. By making a few voltage checks, we were able to conclude that either R_1 or R_2 was at fault. We did not have to measure each component in the circuit. That is the key to successful troubleshooting. If you understand how the circuit works, you will understand what each measurement shows you, and you'll find the problem quickly and efficiently.

At the risk of being repetitious, let's look at equations 3-4, 3-6A, and 3-7 again to see what we can conclude. Equation 3-7 tells us that if R_2 goes up, V_B goes up. Equation 3-6A says that if V_B goes up, I_C goes up. Then looking at equation 3-4, we see that if I_C goes up, V_C goes down. The conclusion of all of this is that if R_2 goes up, V_C goes down. Similarly, if R_2 goes down, V_C goes up. Likewise, if R_1 goes up, V_B goes down, which causes I_C to go down and V_C to go up. If R_1 goes down, then V_C goes down.

You can look at the equations and determine the effect on I_C and V_C if either R_E or R_C changed value. Keep in mind that the equations only work as long as the transistor is OK. If the transistor shorts between collector and emitter, V_C will obviously go down. Actually, the voltage between collector and emitter will go to zero, but the voltage from collector to ground will be above zero due to the drop across R_E. Likewise, an open transistor will cause V_C to be equal to V_{CC}.

To summarize the problem possibilities for the circuit of Fig. 3-22, we can use a flowchart like that in Fig. 3-23. Let's go through a test run of the flowchart to see how it works. For this run, let's assume that resistor R_E has increased in value. Equation 3-6A shows us that I_C would be lower than normal. From equation 3-4, we see that this would result in a higher than normal V_C voltage.

Let's suppose that we are troubleshooting the amplifier and we measure a higher than normal V_C. Looking at the flowchart, we proceed past the decision diamond that asks, "Is V_C normal?" Upon arriving at the decision box that asks "Is $V_C >$ normal?" we take the YES route to the right. The next decision box asks, "Does $V_C = V_{CC}$?" Since it does not, we follow the NO route to the next box, which asks, "Is $V_B <$ normal?" At this point we make another voltage measurement, and we find that V_B reads normal, so we follow the NO route. The last box tells us that we should make a resistance check of R_C to see if it is too low in value, or else we should measure R_E to see if it is too high in resistance. The measurement would then confirm that R_E is too high.

Notice that, in most cases, by using a methodical approach to troubleshooting the stage, you will reduce the possibilities to only two or so components. Thus, you will save troubleshooting time and wear and tear on the components and the printed circuit board by not unsoldering many parts.

Here is one last point to think about. We are seeing more and more automatic computerized testing of electronic circuits and systems. The technologists who program the automatic test equipment (ATE) must thoroughly understand what they are testing, how it works, and what faults to look for. This is a fascinating area that many of you will eventually work in. By studying the flowchart of Fig. 3-23, you will begin to understand how a computer can be programmed to find faults in equipment. You may want to try writing a program in BASIC which will prompt the user to enter values for normal readings at various points in a circuit. The computer will then ask the user to enter any abnormal readings. The computer should then come to a decision, based on a program which follows the flowchart, as to the most likely causes for the abnormal readings.

3-23 **Flowchart for troubleshooting the circuit of Fig. 3-22.**

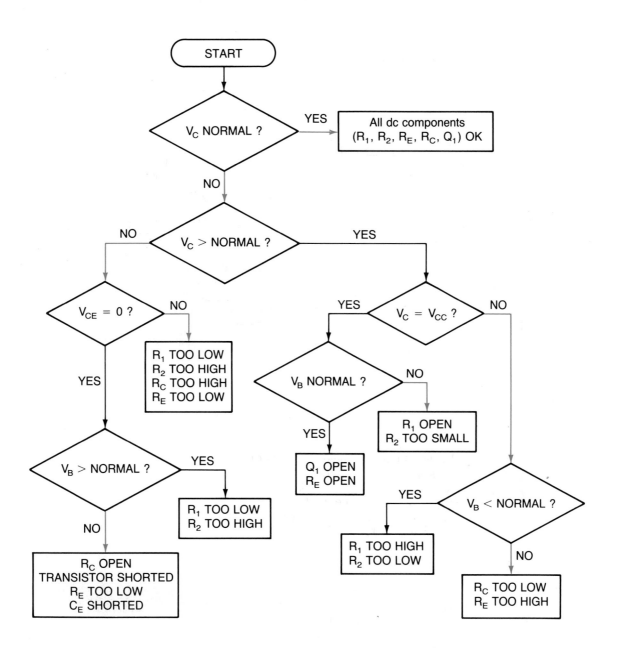

PROBLEMS

3-1. If a transistor has a β of 150 and $I_B = 6$ mA, what is the value of I_C?

$I_C = \beta \cdot I_b = .9A \,/\, 90mA$

3-2. What value of base current will make the collector current 2 mA if β = 80?

$I_B = \dfrac{I_c}{\beta} = 25\mu A$

3-3. A saturated transistor acts like—(a)a closed, (b)an open—switch; so practically all of the V_{CC} supply is dropped across the (c)transistor. (d)load.

3-4. A cutoff transistor acts like—(a)a closed, (b)an open—switch; so practically all of the V_{CC} supply is dropped across the (c)transistor. (d)load.

For the next four problems, refer to Fig. 3-24.

3-5. If $V_{CC} = 15$ V, $R_B = 750$ KΩ, β = 60, and $R_C = 5$ KΩ, find I_B, I_C, and V_{CE}.

$I_B = \dfrac{V_{cc} - V_{ce}}{R_B} \cong \dfrac{V_{cc}}{R_B} \quad 20\mu A$

$I_C = \dfrac{V_B}{R_E} \, \beta \cdot I_B \quad 1.2\,mA = I_b$

$V_{CE} = V_{cc} - V_{Rc} = 6V$

3-6. If $V_{CC} = 10$ V, $R_B = 470$ KΩ, β = 90, and $R_C = 3.3$ KΩ, find I_B, I_C, and V_{CE}.

$I_B = \dfrac{10V}{470k} = 21.3\mu A \qquad V_B = 10V$

$I_C = \beta \cdot I_B = 1.92\,mA$

$V_{CE} = 6.33 - 10 = 3.674V$

3-7. Refer to problem 3-5. What maximum value of R_B will cause the transistor to go into saturation?

$Sat = V_{Rc} = 15V$

$R_B = \quad I_{c(sat)} = V_{cc}/R_c = 3mA$

$I_{B(max)} = I_{c(max)}/\beta = 50\mu A$

$R_B = V_{cc}/I_B \quad 286\,K\Omega$

3-8.   Refer to problem 3-6. What maximum value of R_B will cause the transistor to go into saturation? $Sat = 10 V$

$$R_B = \frac{I_{C(sat)}}{I_{B(max)}} = \frac{V_{cc}/R_c = 3mA}{I_{c(max)}/\beta} = 33.3 \mu A$$

$$R_B = \frac{V_{cc}}{I_B}$$

$$\boxed{R_B = 300 K\Omega}$$

For the next six problems, refer to Fig. 3-25.

3-24   Circuit for problems 3-5 through 3-8.

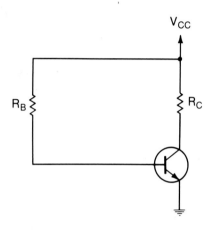

3-25   Circuit for problems 3-9 through 3-14.

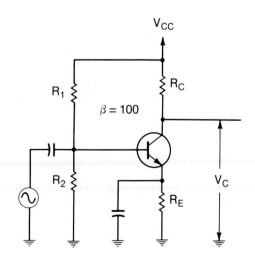

3-9. If R_1 = 70 KΩ, R_2 = 30 KΩ, R_E = 3 KΩ, R_C = 5 KΩ, and V_{CC} = 20 V, find I_C, and V_C.

I_C =

V_C =

3-10. If R_1 = 75 KΩ, R_2 = 25 KΩ, R_E = 5 KΩ, R_C = 3 KΩ, and V_{CC} = 16 V, find I_C, and V_C.

I_C = $8m$

V_C = 134

3-11. Refer to problem 3-9. What value of R_E will make I_C = 1 mA?

R_E =

3-12. Refer to problem 3-10. What value of R_E will make I_C = 2 mA?

R_E = $2K$

3-13. What is the voltage gain of the circuit described in problem 3-9?

A_v =

3-14. Suppose the emitter bypass capacitor is removed from the circuit described in problem 3-9. What is the voltage gain?

$A_v = 1. \, RC/RE = 5K/3K = 1.6$

For the next two problems, refer to Fig. 3-26.

3-26 Circuit for problems 3-15 and 3-16.

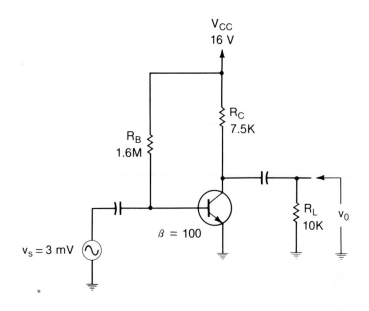

3-15.   Find the voltage gain and the value of v_0.

$A_v =$

$v_0 =$

3-16.   If the transistor is replaced with one where $\beta = 75$, find the voltage gain and the value of v_0.

$A_v = 130$

$v_0 = 390mv$

TROUBLESHOOTING PROBLEMS

For the next six problems, refer to Fig. 3-20.

3-17.   The d-c voltage at the collector of Q_4 reads 5.42 V. Which of the following could be the cause? (a)Open R_{37}. (b)Open R_{36}. (c)Open C_{29}. (d)Shorted C_{36}.

3-18.   The d-c voltage at the collector of Q_4 reads much lower than normal. Which of the following could be the cause? (There may be more than one answer.) (a)Open R_{37}. (b)Open R_{36}. (c)Open C_{29}. (d)Shorted C_{36}.

3-19.   To verify the fault described in problem 3-18, you measure the d-c voltage at the base of Q_4. It reads higher than normal. Which of the following is the most likely cause? (a)Open R_{37}. (b)Open R_{36}. (c)Open C_{29}. (d)Shorted C_{36}.

3-20.   Suppose that the a-c gain of the amplifier stage is lower than normal but that the d-c voltages measure normal. Which of the following could be the cause? (a)Defective Q_4. (b)Shorted C_{29}. (c)Open C_{36}. (d)Open transformer primary.

3-21.   Suppose that the d-c voltages all read normal. There is a normal signal at the slider of the volume control, but there is no a-c signal at the base of Q_4. Which of the following is the most likely cause? (a)Shorted C_{27}. (b)Open C_{27}. (c)Shorted R_{37}. (d)Open R_{37}.

3-22.   Suppose that the voltage at the collector end of T_2 reads zero even with the transistor removed. Which of the following could be the cause? (a)Open R_{38}. (b)Open C_{30}. (c)Open T_2 primary. (d)Open R_{39}.

For the next five problems, refer to Fig. 3-21.

3-23.   Suppose that the d-c collector voltage of TR_{12} is much higher than normal. Which of the following could be the cause? (a)Shorted R_{74}. (b)Shorted R_{76}. (c)Open R_{78} (d)Defective TR_{12}.

3-24.   Suppose that the a-c signal at the collector of TR_{12} is lower than normal but that all the d-c readings are normal. Which of the following could be the cause? (a)Excessive loading below TP 71. (b)Weak transistor. (c)Open C_{59}. (d)Open R_{76}.

3-25. If TR_{12} were replaced with another transistor having a lower β, the d-c collector voltage would probably (a)go up. (b)go down. (c)remain the same.

3-26. Suppose V_B and V_C in the circuit of TR_{12} both read zero. Which of the following could be the cause? (a)Leaky C_{59}. (b)Open R_{78}. (c)Shorted R_{76} (d)Open C_{57}.

3-27. Suppose that the d-c voltages all read normal with a condenser microphone plugged in but that V_B reads lower than normal with a dynamic microphone plugged into the jack. Which of the following could be the cause? (a)R_{74} has changed value. (b)Shorted C_{57}. (c)Gain set too high. (d)Defective microphone.

EXPERIMENT 3-1   TESTING TRANSISTORS

Here you will learn a simple method of checking a bipolar transistor using an ohmmeter. Next you will build a simple GO–NO GO transistor checker, which is useful to have if you do a lot of servicing work. This tester will show you quickly whether or not a transistor is OK, and it will determine for you whether the transistor is an NPN- or a PNP-type.

EQUIPMENT
- ohmmeter
- (6) transistors, preferably some NPN and some
- PNP
- (1) 4- to 6-VDC power supply or (3) 1.5-V C cells
- (1) DPDT (double pole–double throw) switch
- (1) transistor socket
- (2) light-emitting diodes (LED) for indicators, 10- to 15-mA type
- (1) 2.2-KΩ, $\frac{1}{2}$-W $\pm$ 5% resistor
- (1) 220-Ω, $\frac{1}{2}$-W $\pm$ 5% resistor
- (1) momentary pushbutton switch, SPST (single pole–single throw), normally open

PROCEDURE

A bipolar transistor consists essentially of two diodes connected back to back, as shown in Fig. E3-1A. If an ohmmeter is connected between base and emitter of an NPN transistor, with the negative lead on the emitter and the positive lead on the base, the base–emitter diode is *forward biased*. (See Fig. E3-1B.) The ohmmeter in this case reads a *low* resistance. But when the ohmmeter leads are reversed, as shown in Fig. 3-1C, the base–emitter junction is *back biased*. The ohmmeter then reads a *high* resistance. Similarly, if the ohmmeter is connected across the collector–base diode, the meter reads low when the collector is negative with respect to the base, and high in the other direction.

Thus, we can use the ohmmeter to check whether a transistor is good. If we measure a good diode from base to emitter and a good diode from base to collector, the transistor is OK. Incidentally, an ohmmeter connected from collector to emitter should read a fairly high resistance in both directions, because one diode or the other is back biased regardless of the meter polarity.

Since the PNP transistor has both diodes reversed from the NPN, all high/low readings will be reversed from the NPN. The same idea applies here also. You will read diode characteristics between base and emitter, and another diode between collector and base.

Ohmmeter

Get at least six transistors—preferably of different types—some NPN and some PNP. Make the appropriate resistance checks, and enter your readings in Table E3-1. Record the transistor part number (2N _____) and then put a check in the space provided if you think the transistor is OK. Also, determine whether the transistor is an NPN- or a PNP-type.

E3-1 Figures for experiment 3-1.

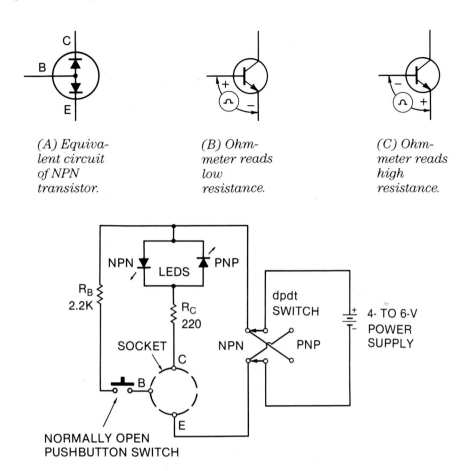

(A) Equiva-
lent circuit
of NPN
transistor.

(B) Ohm-
meter reads
low
resistance.

(C) Ohm-
meter reads
high
resistance.

(D) GO-NO GO transistor checker.

Table E3-1

Unit	Type number	Ohms B pos E neg	Ohms B neg E pos	Ohms B pos C neg	Ohms B neg C pos	Transistor OK?	NPN or PNP?
1							
2							
3							
4							
5							
6							

GO–NO GO **Transistor Checker**

You can also check your transistors using the GO–NO GO circuit of Fig. E3-1D. In this circuit, these two LED are a visual indication of the transistor's condition.

An LED emits light when it is forward biased. By using two LED back to back, you can determine whether the transistor is an NPN or a PNP.

1.   Insert the transistor into the socket. If either LED lights before pressing the button, the transistor is shorted.

2.   Press the button. One LED should light indicating that the transistor is good. The DPDT switch is shown in the NPN position. Simply flip it to the other position to test a PNP-type. The switch can be used to identify whether a transistor is an NPN or a PNP. A good transistor will not be damaged if the switch is in the wrong position—it just won't light the LED.

QUIZ

1.   You measure a low resistance between base and emitter when the negative lead is connected to the base and a high resistance when the leads are reversed. If the transistor is good, you would expect to read a low resistance between collector and base with the—(a)positive, (b)negative—lead connected to the collector.

2.   You read a high resistance in both directions between base and emitter. Is the transistor OK? (a)Yes. (b)No.

3.   You read near-zero resistance in both directions between collector and emitter. Is the transistor OK? (a)Yes. (b)No.

The remaining questions refer to the GO–NO GO checker.

4.   You have the switch in the NPN position and insert a transistor.

Before you press the switch, the base current is _____.

5.   If the NPN LED lights before pressing the switch, the transistor is (a)shorted. (b)open. (c)OK.

6.   When you press the switch, what is the approximate value of base current? (Assume a 6-V power supply.)

$I_B =$ 2.7 mA

7.   If the transistor is OK, the NPN LED should light because (a)base current flowing through the switch causes collector current to flow. (b)LED current flows through the switch.

8.   Suppose you have the switch in the NPN position. When you insert an unknown transistor into the socket, no light comes on. You press the pushbutton, and no light comes on. Is the transistor OK? (a)Yes. (b)No. (c)Uncertain.

9.   Suppose that the switch is in the PNP position and that you insert a transistor, but no light comes on. When you press the button, no light comes on. When you flip to NPN and press the button, no light comes on. Is the transistor OK? (a)Yes. (b)No.

10.   If the forward bias voltage across a conducting LED is 1.5 V and the supply voltage is a 4.5 V, what is the approximate collector current when an LED is lit? (*Hint*: The transistor is driven into saturation.)

$I_C =$ 13.6 m

EXPERIMENT 3-2   BASE-BIASED AMPLIFIER

In this experiment you will learn how to measure β for a transistor. Then you will use β to set up an operating point. Once your circuit is operating, you will test it as an a-c signal amplifier.

EQUIPMENT

- oscilloscope
- VOM
- AF (audio frequency) signal generator
- 12-VDC power supply
- (1) NPN transistor
- (1) 1-MΩ pot
- (2) 10-KΩ, ½-W ± 5% resistors
- (1) 5-KΩ, ½-W ± 5% resistor
- (1) 2-KΩ, ½-W ± 5% resistor
- (1) 1-KΩ, ½-W ± 5% resistor
- (1) 10-Ω, ½-W ± 5% resistor
- (1) 5-μF capacitor at 25 V

E3-2   **Circuits for experiment 3-2.**

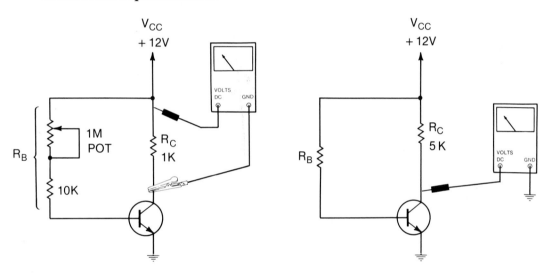

(A) Circuit for determining β.   (B) Circuit for steps 4 and 5.

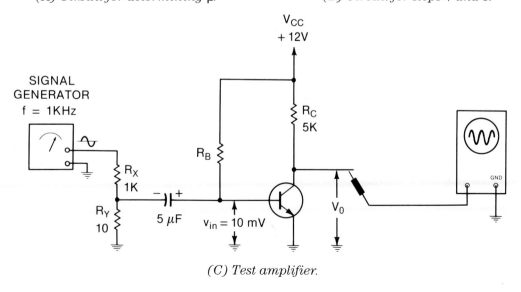

(C) Test amplifier.

PROCEDURE

1. Build the circuit of Fig. E3-2A. With a voltmeter connected across R_C as shown, adjust potentiometer R_B until the voltage across R_C measures exactly 2 V. This makes I_C = 2 mA. (The 10-KΩ resistor in series with the pot is used to protect the transistor from excessive base current in the event that the pot is adjusted to too low a value.)

2. Next, remove the pot from the circuit and measure it with an ohmmeter. Be sure to include the resistance of the series 10 KΩ as part of R_B.

$R_B =$ 240 - 900K   40-150

3. Now calculate I_B, and determine β.

$I_B =$ 50 - 13.3mA

$\beta =$ 40-150

4. Now that you know the value of β for the transistor, determine what value of R_B will make V_{CE} = 7 V in Fig. 3-2B. Show your calculations. Also determine I_C. (Note that R_C = 5 KΩ.)

$I_B = 10m$

$R_B =$ 1.2M

$I_C =$ 1m

5. Now, set the pot so that R_B equals your calculated value in step 4. Build the circuit, and measure V_{CE}.

$V_{CE} =$ 7V

6. If your circuit works OK, apply an input signal, as shown in Fig. E3-2C. Set the generator to 1-V peak to peak (p-p) at a frequency of 1 KHz. The voltage divider R_x–R_y across the signal generator drops the signal level v_{in} to approximately 10 mVp-p.

7. Connect an oscilloscope from collector to ground, and observe the output signal v_0. Draw two cycles of the output signal in the space below. Show the a-c signal riding on a d-c level.

2VPP   7V DC Riding on level

8. Calculate the gain of the amplifier.

$A_v = v_0/v_{in} =$ 200

9. Calculate the value of r_e.

$r_e = 25 \text{ mV}/I_E =$ 25Ω

10. Finally, determine the theoretical gain.

$A_v = r_L/r_e =$ 200

How well does the theoretical gain compare to the measured gain?

11. Change R_C in the circuit to 2 KΩ. Measure v_0 and determine the gain.

$A_v = v_0/v_{in} =$ *80 Down* *s to*

12. Change R_C to 10 KΩ. Observe the output waveform, and draw two of its cycles. Show the ac riding on a d-c level.

400 increase

VCE 9V

nes PP

4V PP

neg volt should just start crimping

QUIZ

1. In step 1, decreasing R_B—(a)increased, (b)decreased—I_B, causing I_C to (c)increase. (d)decrease.

2. In step 3, the voltage across R_B was assumed to be approximately _____ V because the voltage V_{BE} is (a)near zero. (b)near 12 V. (c)unknown.

3. The value of β—(a)changes significantly, (b)remains essentially constant—with different values of I_C.

For the next five questions, refer to Fig. E3-2C.

4. If R_B is decreased, I_B will (a)increase. (b)decrease.

5. This will cause I_C to (a)increase. (b)decrease.

6. The d-c collector voltage will then (a)increase. (b)decrease.

7. This will also cause the voltage gain of the stage to (a)increase. (b)decrease.

8. Generally, whenever a transistor is turned on harder, the a-c output signal will go—(a)up, (b)down—provided that the transistor is not driven into cut off or saturation.

9. If R_C is decreased, the gain will (a)increase. (b)decrease.

10. If R_C is made too large, clipping occurs on the—(a)positive, (b)negative—peaks of the output cycle, because the transistor is being driven into (c)cutoff. (d)saturation.

EXPERIMENT 3-3 AMPLIFIER WITH VOLTAGE-DIVIDER BIAS

In this experiment you will become familiar with a transistor amplifier using voltage-divider bias. You'll see what factors determine its gain and distortion.

EQUIPMENT

- VOM
- oscilloscope
- 12-VDC power supply
- audio signal generator
- NPN transistor
- electrolytic capacitors—15 WVDC or greater
  - (2) 5 µF
  - (1) 50 µF
- resistors ½ W ± 5%
  - (2) 100 KΩ
  - (2) 15 KΩ
  - (3) 5 KΩ
  - (3) 1 KΩ
  - (1) 100 Ω

E3-3 **Circuit for experiment 3-3.**

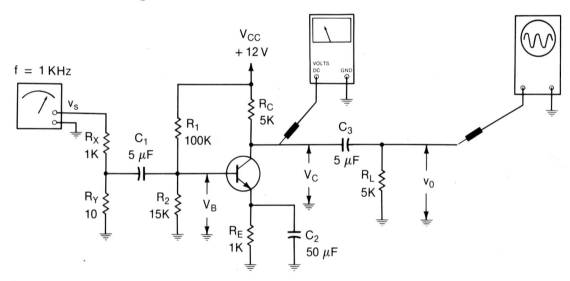

PROCEDURE

1. Build the circuit of Fig. E3-3. Measure the d-c base-to-ground voltage V_B, the collector-to-ground voltage V_C, and the drop across collector resistor R_C. Using the voltage across R_C, determine the value of collector current I_C. Record these values.

$V_B =$ $1.6\ V$

$V_C =$ $7\ v$

$I_C =$ $1m$

2. Next, with an oscilloscope connected across the output, adjust the signal generator until the a-c output amplitude across R_L reads 2 Vp-p.

Consider the measured voltages the *normal* values when the amplifier is operating as it should. Make changes to various circuit components as indicated in Table E3-3, and record whether or not the readings change. Write I if the reading *increases* noticeably (more than 10%), D if the reading decreases, and S if the reading remains essentially the *same* as the normal value. You can determine whether I_C changes by observing the voltage across R_C or R_E. If the voltage increases while the resistance remains the same, the current must have increased. If you think that the transistor is cut off or saturated, place a check in the appropriate blank in the table. Use line 1 of the table as an example.

To *decrease* the value of a resistor by 50%, simply bridge a resistor of equal value in parallel with it. To determine the effect of an *open* component, temporarily remove one end of it from the circuit. Be sure to return the circuit to its original setup before proceeding with the next check.

Table E3-3

Change	V_B	V_C	I_C	v_0	Saturated?	Cut off?
1. open R_1	D	I	D	D	OFF	✔
2. decrease R_1 50%	I	I	I	D	SAT	
3. open R_2	I	D	I	D	SAT	
4. decrease R_2 50%	D	D	D	D		
5. open R_C	S	I	D	D	OFF	
6. decrease R_C 50%	S	I	S	D		
7. open R_E	S	I	D	D	OFF	
8. decrease R_E 50%	S	D	I	I		
9. open R_L	S	S	S	I		
10. decrease R_L 50%	S	S	S	D		
11. open C_1	S	S	S	D		
12. open C_2	S	S	S	D		
13. short C_2	D	D	I	D	SAT	
14. open C_3	S	S	S	D		

QUIZ

Refer to Fig. E3-3. For each symptom, choose one of the following faults as the most probable cause, and write it in the space provided.

Faults
- R_1 open
- R_1 decreased
- R_2 open
- R_2 decreased
- R_C open
- R_C decreased
- R_E open
- R_E decreased
- C_2 open
- C_2 shorted
- open transistor
- shorted transistor

Symptoms

1. The a-c output voltage is lower than normal, but not zero. When you measure the voltage from collector to ground, it reads high, but the voltage across R_E is normal. V_B is also normal.

Probable fault

2. The output voltage is lower than normal. V_C reads high, but less than V_{CC}, and V_B reads lower than normal, but not zero.

Probable fault *Rc Decrer*

3. The a-c output voltage is zero and $V_C = V_{CC}$. V_B measures normal. A resistance check from emitter to ground reads extremely high.

Probable fault *RE Dec*

4. All d-c readings are normal, but the gain is lower than normal.

Probable fault *E2 open*

5. No a-c output voltage. V_C is equal to V_{CC}, and V_B reads zero.

Probable fault *R1 open*

6. V_C measures low. The voltage at the emitter is approximately equal to V_C, and V_B is higher than normal. R_1 measures normal.

Probable fault *R2 open*

7. V_C is lower than normal. Output looks clipped on the negative half cycles when the transistor is driven with a fairly large signal. V_B reads higher than normal.

Probable fault *R1 Decreased*

8. $V_C = V_{CC}$, and the voltage from emitter to ground is zero. V_B measures normal.

Probable fault *Open transistor*

9. Collector-to-emitter voltage reads zero.

Probable fault *Shorted transtor Colector to transmitter*

10. Voltage at collector is equal to V_{CC}. Voltage V_B is lower than normal, and voltage between base and emitter is zero.

Probable fault *Shortc trans Base to Emitter*

Field Effect Transistors

Besides the bipolar transistor, the *field effect* transistor (FET) is also available. The way it operates is very easy to understand, and it offers some features not found in bipolar transistors. The FET is used primarily where an extremely high input impedance is needed.

THE JFET

If a power supply is connected across a bar of lightly doped silicon, as shown in Fig. 4-1A, some current will flow through the bar. The point where electrons enter the bar from the external circuit is called the *source* (S) and where they leave is called the *drain* (D). The bar essentially forms a resistor, and the magnitude of the resistance is determined by the amount of impurities in the crystal, as well as by the length and width of the path.

4-1 Basic JFET construction and operation.

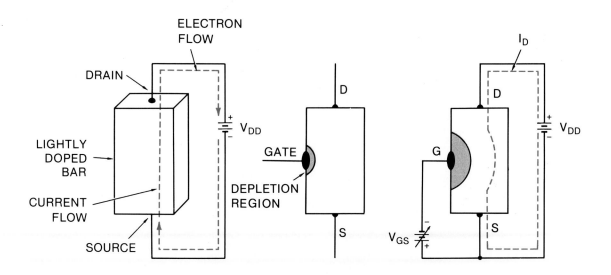

(A) Current flow through bar of silicon.

(B) Gate added to bar.

(C) Decreasing width of channel by applying voltage to gate.

A junction FET (JFET) consists of a P-N junction, which is formed part of the way up the bar. This junction is called the *gate* (G). See Fig. 4-1B. The region near the junction effectively has no free electrons and is therefore called a *depletion region.* This depletion region acts like a perfect insulator. When the junction is reverse biased, as shown in Fig. 4-1C, the depletion region widens. Widening the depletion region effectively reduces the width of the *channel* through which current flows from source to drain. And, of course, if the channel gets narrower, the resistance of the channel goes up. So by varying the reverse-biasing voltage between gate and source, we can vary the current through the FET. This mode of operation is known as the *depletion* mode. Here is the major distinction between a bipolar transistor and an FET. In the bipolar transistor, the *collector current* is controlled by the *base current*, whereas in the FET, the *drain current* is controlled by the *gate voltage.*

In order to compare one FET with another, a parameter called *transconductance* is used. The transconductance, whose symbol is g_m (sometimes Y_{fs}), is the ratio of the change in drain current to the change in gate-to-source voltage. That is,

$$g_m = \frac{\Delta I_D}{\Delta V_{GS}}$$

4-1

Typical values of g_m for a JFET are in the order of a few thousand micromhos (μmho).

EXAMPLE 4-1   If a 2-V change in V_{GS} causes a 5-mA change in I_D, what is g_m?

SOLUTION

$$g_m = \frac{\Delta I_D}{\Delta V_{GS}} = \frac{5 \text{ mA}}{2 \text{ V}} = 2.5 \times 10^{-3} = 2500 \text{ } \mu\text{mho}$$

As shown in Fig. 4-1C, power supply V_{DD} is used to supply drain current I_D, just as V_{CC} supplies collector current I_C for the bipolar transistor. In this figure the *channel* (conductive path) is the bar of N-type crystal, and the transistor is therefore called an *N-channel* JFET. The symbol for the N-channel JFET is shown in Fig. 4-2A.

4-2   Symbols for JFET.

(A) N channel.   (B) P channel.

The channel can also be made of P-type crystal, thereby forming a *P-channel* JFET. The symbol is shown in Fig. 4-2B. The operation of the P-channel device is essentially the same as that of the N-channel, except that all power supply polarities are reversed.

Typical characteristic curves for a JFET are shown in Fig. 4-3. Note that with 0-V V_{GS}, I_D is maximum at about 8 mA for values of V_{DS} above about 4 V. This shows that the current through the JFET is determined essentially by the gate-to-source voltage—not by drain-to-source voltage. When V_{GS} is increased to -1 V, the drain current decreases to 6 mA, etc.

EXAMPLE 4-2   Using the curves of Fig. 4-3, determine g_m at a constant V_{DS} of 10 V.

SOLUTION   If we make some arbitrary change in V_{GS}, say from -2 to -3 V, we find that I_D changes from 4 to 2 mA. Thus,

$$g_m = \frac{\Delta I_D}{\Delta V_{GS}} = \frac{4 \text{ mA} - 2 \text{ mA}}{3 \text{ V} - 2 \text{ V}} = 2000 \text{ } \mu\text{mho}$$

4-3 Typical JFET curves.

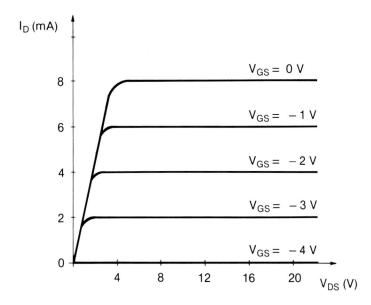

BIASING THE JFET

Just as with bipolar transistors, the FET must be properly biased before it can be used as an amplifier. A separate reverse-bias supply could be used, similar to that in Fig. 4-1C. However, a more economical method of biasing is to use *self bias*, caused by source current flowing through a resistor in series with the source. This is shown in Fig. 4-4.

4-4 Self bias caused by I_S flowing through R_S.

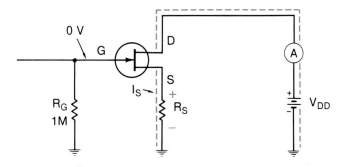

Remember that the gate is essentially a back-biased diode, so no current flows through the gate, or through R_g. Therefore, the voltage from gate to ground is zero.

Current flows from V_{DD} through R_s and through the transistor. As current flows through R_s, it causes a voltage drop across R_s, making the source positive with respect to ground. Since the gate is at ground potential, the gate is therefore negative with respect to the source. In other words, the gate diode is back biased by an amount equal to the voltage drop across R_S.

When power is first applied, current through the JFET begins to increase. But as it increases, the drop across R_S also increases, making the channel narrower. An equilibrium point is eventually reached where I_D levels off, depending on the value of R_S and on the transistor parameters. If the value of V_{GS} required to permit a desired value of I_D is known, the value of R_S can be determined by

$$R_S = \frac{V_{GS}}{I_D}$$

4-2

EXAMPLE 4-3   Determine a suitable value for R_S for the circuit of Fig. 4-4, so that V_{GS} will be -3 V when $I_D = 2$ mA.

SOLUTION

$$R_S = \frac{3 \text{ V}}{2 \text{ mA}} = 1.5 \text{ K}\Omega$$

THE JFET AS AN A-C AMPLIFIER

Now that we know how to bias the JFET, let's use it as a small signal amplifier. Fig. 4-5 shows the complete circuit. Let's assume that the drain current is 2 mA, as before.

4-5   JFET small-signal amplifier.

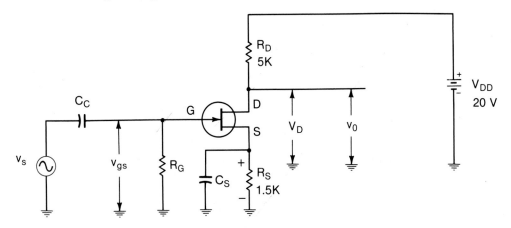

The d-c voltage from drain to ground can be determined by

$$V_D = V_{DD} - I_D R_D \qquad\qquad 4\text{-}3$$

which is similar to the equation for collector voltage of a bipolar type. And as usual, we want to avoid saturation and cutoff. In this case

$$V_D = 20 \text{ V} - 2 \text{ mA} \times 5 \text{ K}\Omega = 10 \text{ V}$$

The incoming signal v_{in} is applied through coupling capacitor C_C to the gate with respect to the source. The signal voltage thus adds to and subtracts from the d-c bias voltage, causing the drain current to increase and decrease accordingly. And, of course, changes in drain current cause changes in drain voltage, resulting in an output signal voltage v_0. The larger the variation in gate voltage, the larger the drain current changes, and thus the larger v_0. Of course, the higher the value of g_m, the greater the variation in I_D for a change in V_{GS}. Calling the a-c drain current i_d and the a-c gate-to-source voltage v_{gs}, we see that

$$g_m = \frac{\Delta I_D}{\Delta V_{GS}} = \frac{i_d}{v_{gs}} \qquad\qquad 4\text{-}4$$

from which

$$i_d = g_m v_{gs} \qquad\qquad 4\text{-}4\text{A}$$

As drain current i_d flows through the load resistance R_D, it causes an a-c output voltage v_0 to occur, where

$$v_0 = i_d R_D \qquad\qquad 4\text{-}5$$

Then substituting equation 4-4A into equation 4-5, we get

$$v_0 = g_m v_{gs} R_D \qquad\qquad 4\text{-}5A$$

From this equation, we see that the voltage gain of the circuit of Fig. 4-5 is

$$A_v = \frac{v_0}{v_{gs}} = g_m R_D \qquad\qquad 4\text{-}5B$$

EXAMPLE 4-4 In the circuit of Fig. 4-5, find the voltage gain if $g_m = 3000$ μmho.

SOLUTION

$$A_v = g_m R_D = 3 \times 10^{-3} \times 5 \times 10^3 = 15$$

The bypass capacitor C_s is used to keep the source at a-c ground.

As with the bipolar amplifier, if an additional load resistor R_L is placed in parallel with R_D, as shown in Fig. 4-6, the total a-c load resistance r_L is equal to R_D in parallel with R_L. So the gain equation changes to

$$A_v = g_m r_L \qquad\qquad 4\text{-}6$$

4-6 JFET amplifier working into load resistor R_L.

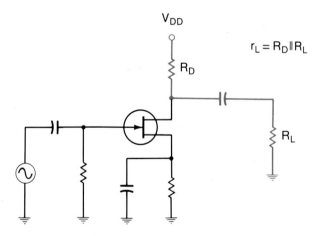

The input resistance seen by the signal generator looking into the amplifier stage of Fig. 4-6 is essentially R_G in parallel with the resistance of a back-biased diode. Since the diode's resistance is extremely high,

$$r_{in} \cong R_G \qquad\qquad 4\text{-}7$$

It is the input resistance that makes the FET valuable as a circuit component. *Compared to bipolar amps, the input resistance of a JFET is much higher, but the gain is generally lower.*

MOSFET

In the JFET, drain current is controlled by the voltage on a reverse-biased diode. There is another type of FET in which drain current is controlled by the voltage on the gate, but no gate diode is used. This device is called the *Metal Oxide Semiconductor Field Effect Transistor*, or MOSFET.

Fig. 4-7 shows how MOSFET are built. Construction starts with formation of a heavily doped substrate materials. The heavy doping is indicated by N+. Next, a layer of very lightly doped (N−) material is deposited onto the substrate. This layer will form the channel. Then, as shown in Fig. 4-7C, two "islands" of N+ material are doped in to form the source and drain. Next, a very thin layer of silicon dioxide is deposited on top; silicon dioxide is an extremely good insulator. Finally, as shown in Fig. 4-7E, metal connections are made to the source and drain, and a metal gate is deposited on top of the insulator. For this reason, this device is also known as an *Insulated Gate* FET, or simply IGFET.

4-7   Fabrication of a MOSFET.

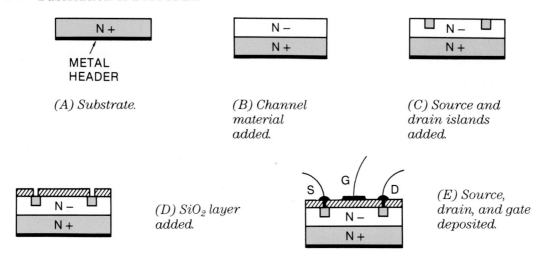

(A) Substrate.

(B) Channel material added.

(C) Source and drain islands added.

(D) SiO₂ layer added.

(E) Source, drain, and gate deposited.

The channel material is so lightly doped that with 0 V between gate and source, practically no current flows from source to drain, as shown in Fig. 4-8A. However, as shown in Fig. 4-8B, if a positive voltage is applied to the gate with respect to the source, electrons from the substrate are drawn up into the channel, making it conductive. So current does flow from source to drain through the channel. The conductivity of the channel is thus *enhanced* by the positive gate voltage, so this is called an *enhancement mode* device. Notice that no current flows in the gate circuit. The gate acts like one plate of a *capacitor*, the channel being the other plate. The silicon dioxide layer forms the dielectric.

4-8   Controlling current flow in a MOSFET.

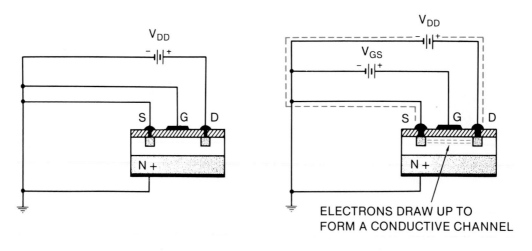

(A) No current flows with 0 V between gate and source.

(B) Current flows when positive voltage is applied to gate.

By increasing and decreasing the gate-to-source voltage, the conductivity of the channel increases and decreases accordingly. Therefore, the drain current in the MOSFET is controlled by the gate-to-source voltage, just as in the JFET. We can see two differences here. First, no reverse-biasing voltage is necessary. Second, the input resistance seen looking into the gate is essentially an open circuit, since it is simply a small capacitor. The value of this capacitance is extremely small, usually in the order of 5–10 pF.

The symbol for the N-channel MOSFET is shown in Fig. 4-9A. If all P-type materials are used instead of N-type, a P-channel device can be built with similar characteristics. See Fig. 4-9B for the symbol.

4-9 N-channel MOSFET.

(A) N channel.

(B) P channel.

Some variations exist in the symbols used by different manufacturers. Sometimes the N-channel symbol is drawn as shown in Fig. 4-10A. And occasionally, instead of being tied internally to the source, the substrate connection is brought out of the package separately. The symbol for a MOSFET with a separate substrate lead is shown in Fig. 4-10B.

4-10 Alternate MOSFET symbols.

(A) N channel.

(B) Separate substrate connection.

Fig. 4-11 shows how a MOSFET can be used as an a-c signal amplifier. Notice that resistors R_1 and R_2 form a voltage divider, thus applying $+5$ VDC between gate and source to bias the device on in the active region. The incoming a-c signal rides on the d-c level, thereby increasing and decreasing the total drain current in accordance with signal variations. The gain of the MOSFET is similar to that of the JFET, and equation 4-6 applies as before.

4-11 MOSFET small-signal amplifier.

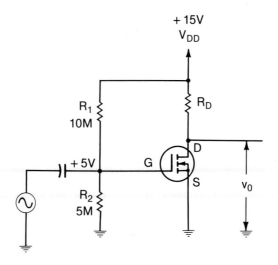

Since practically no drain current flows with 0 V applied to the gate, the MOSFET makes an excellent switching device. Fig. 4-12A shows a MOSFET being driven by a square wave generator. When the input signal, shown in Fig. 4-12B, is at 0 V, the drain voltage, shown in Fig. 4-12C, is equal to V_{DD}, or 15 V. Then when the input is driven sufficiently positive, say to 5 V, the transistor turns on. V_0 then drops to a low value. Most of the V_{DD} supply appears across R_D. How low V_0 drops depends on how large R_D is and on how low a resistance the transistor becomes. The latter is referred to as $R_{DS(ON)}$, meaning resistance, drain-to-source, when on. Both JFET and MOSFET usually have a higher voltage across them when on, as compared to a bipolar transistor. But in low power circuits, MOSFET have an advantage, because no input current is needed to turn them on.

4-12 MOSFET switching circuit.

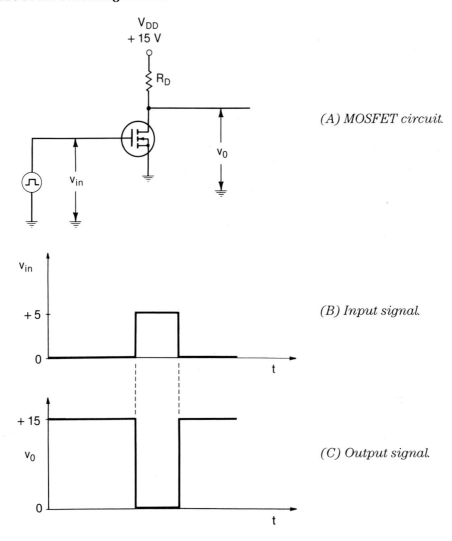

(A) MOSFET circuit.

(B) Input signal.

(C) Output signal.

CAUTION The insulating layer of silicon dioxide between the gate and channel is extremely thin. For this reason, static electricity, even from your fingers, can damage it permanently. Do not handle MOSFET unnecessarily. They usually come from the manufacturer wrapped in metal foil, or stuck into conductive foam. Do not remove them from the packaging until you are ready to install them in a circuit. Also, never insert or remove a MOSFET when the power is on.

In plants where MOSFET parts are assembled, the assembler sometimes wears a metal bracelet which is grounded by means of a small chain. Otherwise, the entire work surface is conductive and grounded.

VMOSFET

In the late 1970s, a new geometry for MOSFET began to be used, called the *Vertical* MOSFET, or simply VMOSFET. The operation of the VMOSFET is basically the same as that of the MOSFET, but it has some other excellent characteristics resulting from its different construction.

Fig. 4-13 shows the internal construction of a typical VMOSFET. Note that a very lightly doped P-region separates the source and drain. Then a V-shaped groove is cut through the P-region, and an insulating layer of silicon dioxide is deposited on the surface. Finally, metal is deposited in the V-groove to form the gate.

4-13 VMOSFET construction.

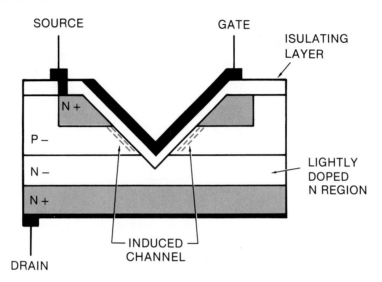

Like the MOSFET, the VMOSFET becomes conductive when a positive potential is applied to the gate with respect to the source pulling electrons into the channel. As the voltage on the gate becomes higher, the channel becomes more conductive and the drain current flows more freely.

Because the surface area in the V-groove is small, the capacitance formed by the gate is extremely small. As a result, the high-frequency performance is excellent. The device can be switched on or off in a few nanoseconds, making the VMOSFET a very efficient switch.

Unlike the bipolar transistor, which tends to conduct more heavily as the temperature increases, the VMOSFET does not. The channel resistance actually tends to increase with an increase in temperature. This increased resistance prevents a condition called *thermal runaway*, which occurs in bipolar transistors and can destroy them.

The characteristics that make the VMOSFET very interesting and valuable compared to the ordinary MOSFET are its high current capability, high transconductance, and low resistance when on. These characteristics make the VMOSFET an excellent choice for such applications as power amplifiers and switching regulators. For example, the IRF-100 N-Channel Power MOSFET can handle up to 16-A I_D, it has a typical transconductance of 3 mhos, and its conducting resistance $R_{DS(ON)}$ is 0.2 Ω. The characteristic curves for the IRF-100 are shown in Fig. 4-14 along with its package.

The symbol for the VMOSFET is the same as that for the MOSFET. Like MOSFET, care must be taken when handling VMOSFET so as not to damage them with static charges. Some types, however, have built in *zener diodes* for gate protection. A zener diode is a special diode that acts like an open circuit when reverse biased, until the reverse voltage across it reaches some specific value. Then the diode breaks down (becomes conductive) and prevents the voltage from getting any higher. The zeners are used to protect VMOSFET break down at 15–18 V, before any damage to the gate can occur. The symbol for the protected gate VMOSFET is shown in Fig. 4-15.

4-14 Package and curves for IRF 100.

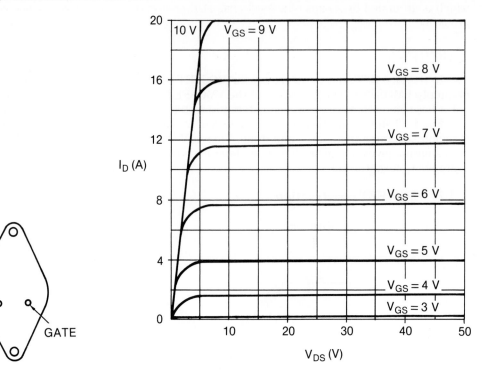

(A) Package (bottom view).   (B) Output characteristics.

When using protected gate devices, take care not to drive the gate negative with respect to the source. If the gate goes negative with respect to the source, the protective diode becomes forward biased and can conduct heavily if driven from a low-impedance source. One protective measure is to put a resistor, say 1 KΩ, in series with the gate to limit the diode current. This resistor also minimizes a tendency for the circuit to oscillate as a result of the inductance and capacitance of connecting leads.

4-15 Protected-gate VMOSFET.

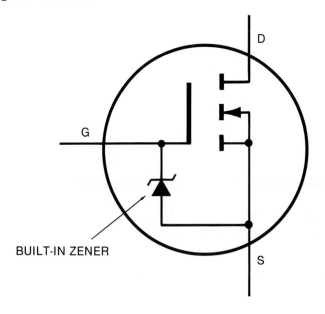

The manufacture of power MOSFETs has been one of the fastest-growing areas in semiconductor technology. The power transistors are often used as switches to control current through high-voltage, high-power loads, such as solenoids and motors. Solenoids are often used to open and close valves in pipelines so as to control the flow of fluids. These types of valves are used in automatic washers and in petroleum and chemical pumping installations.

4-16 MOSFET power inteface.

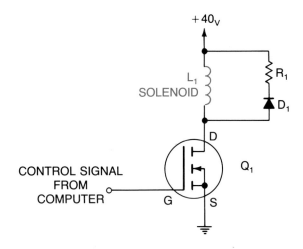

Fig. 4-16 shows a typical circuit. When the control signal at the gate is low (at ground potential), transistor Q_1 is off, so solenoid L_1 is deenergized. At this time, the valve controlled by the solenoid is normally pulled closed by a spring. Then when the control signal at the gate is driven high, say to $+5$ V or so, Q_1 switches on, causing current to flow through the solenoid, which in turn opens the pipeline valve. The control signal fed to the gate, therefore, determines whether the valve is open or closed. But notice that the control signal does not have to supply a lot of power. It is simply a small-amplitude control voltage, which is often generated by a computer or a programmable controller. The power MOSFET used in this manner is called an *interface* device. The interface circuit ties the low-voltage, low-power computer circuitry to the high-voltage, high-power load.

Refer again to Fig. 4-16. L_1 represents the electromagnetic coil of the solenoid. When Q_1 is on, the resistance of L_1 limits the current flow through it, and most of the supply voltage is dropped across L_1. However, when Q_1 switches off, the collapsing magnetic field around L_1 generates a reverse polarity "kickback" voltage, which any inductor does when the current through it is abruptly stopped. This kickback voltage could rise to a value much larger than the power supply voltage and could possibly damage Q_1 or some other components. To prevent the kickback from rising too high, we place diode D_1 and resistor R_1 across the coil. Notice that D_1 is normally reverse biased when Q_1 is conducting, so no current flows through D_1 while the solenoid is on. When Q_1 switches off, the reverse polarity kickback voltage forward biases D_1 as soon as the voltage reaches 0.7 V or so. With D_1 now conducting, a path is provided for current to flow from the coil through R_1 while the magnetic field in the coil is collapsing. This action safely dissipates the energy stored in the coil and prevents the high-voltage spike from occurring.

If you find a bad power MOSFET when troubleshooting a circuit like this one, be sure to check D_1 and R_1 before replacing the transistor and applying power again. Otherwise you may see your new transistor zapped again by the kickback voltage.

TYPICAL FET CIRCUITS

Figure 4-17 shows part of the schematic diagram of a Courier Cruiser citizen's band transceiver. Note that the JFET Q_9 is used as a noise amplifier. Biasing of Q_9 is obtained by current flow through source resistor R_{142}, while the gate voltage is held at 0 V by R_{58}. Coupling into the circuit is through C_{112}, with coil L_{20} providing higher input impedance to higher frequencies. The output signal from Q_9 is fed on to Q_{10}, which can be recognized as a conventional common emitter bipolar amplifier, using voltage divider bias.

4-17 **JFET used in part of Courier Cruiser CB transceiver. From *CB Radio Series No. 79*, Howard W. Sams & Company, Inc.**

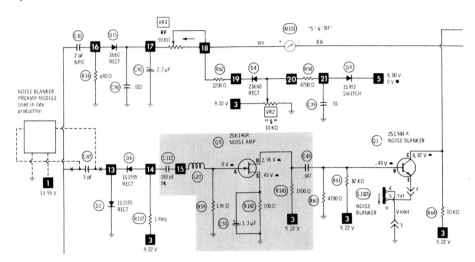

Our purpose here is not to analyze the operation of the Courier Cruiser completely, but rather to identify the individual stages, recognizing how the various components are used.

Fig. 4-18 shows a MOSFET Q_1 in the same transceiver. The 3SK39Q is a *dual-gate* MOSFET, so the total drain current is controlled by the combination of voltages on the two gates.

4-18 **MOSFET used in part of Courier Cruiser CB transceiver. From *Radio Series No. 79*, Howard W. Sams & Company, Inc.**

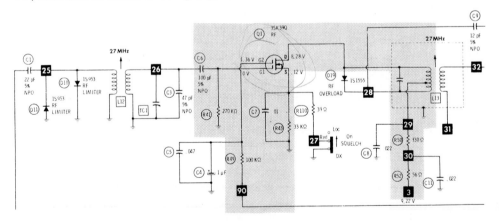

The rf signal to be amplified feeds in through C_6 to gate G_1. The high input impedance of the MOSFET prevents loading of the transformer secondary and, therefore, keeps the bandwidth narrow by keeping the transformer Q high.

The positive voltage on G_2 provides bias. The lower end of R_{49} connects to an AGC (automatic gain control) circuit, which is not shown in Fig. 4-18. As the received signal strength increases, the positive AGC voltage at TP 90 decreases. Similarily, when the received signal amplitude decreases, the voltage at TP 90 goes more positive.

The 3SK39Q transistor has characteristics such that g_m, and hence the amplifier gain, increases or decreases as the magnitude of bias voltage on G_2 increases or decreases, respectively. Hence, if a strong signal is being received, the AGC voltage causes Q_1 to amplify less. And if the received signal gets weaker, the gain of Q_1 is automatically increased. In this manner, the amplitude of the signal out of the amplifier stays nearly constant, regardless of input signal strength variations.

PROBLEMS

4-1.   An N-channel JFET is formed by placing a P-type junction on a bar of N-type silicon. When the gate junction is—(a)forward, (b)reverse—biased, the channel width decreases, causing the resistance from source to drain to (c)increase. (d)decrease.

For the next eight problems, refer to the transistor whose characteristic curves are shown in Fig. 4-19.

4-19   Curves for problems 4-2 through 4-9.

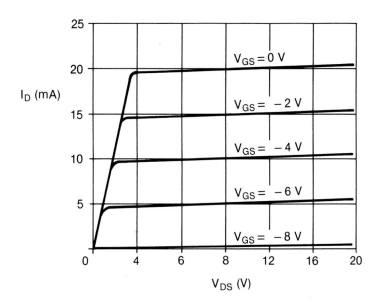

4-2.   When $V_{GS} = -2$ V and $V_{DS} > 4$ V, what is the value of I_D?

$I_D =$

4-3.   What value of gate-to-source voltage would make $I_D = 12.5$ mA?

$V_{GS} =$

4-4.   What value of V_{GS} will cut off the drain current in the transistor?

$V_{GS} =$

4-5.   What is the value of g_m for this transistor?

$g_m =$

For the next seven problems, refer to Fig. 4-20.

4-20 **Circuit for problems 4-6 though 4-12.**

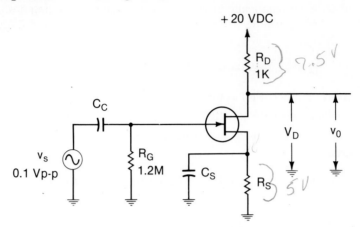

4-6. Assume that the transistor of Fig. 4-20 has characteristics like·those shown in Fig. 4-19. What value of R_S would you use to make I_D = 5 mA?

R_S =

4-7. If the transistor of Fig. 4-20 has those characteristics shown in Fig. 4-19, what value of R_S would you use to make I_D = <u>7.5 mA?</u>

R_S =

4-8. Referring to problem 4-6, what is the value of V_D?

V_D =

4-9. Referring to problem 4-7, what is the value of V_{DS}?

V_{DS} =

4-10. Assume that the transistor of Fig. 4-20 has a g_m of 1500 μmho. What is the voltage gain?

A_v =

4-11. Assume that the transistor of Fig. 4-20 has a g_m of 5000 μmho. What is the voltage gain?

A_v =

4-12. What is the approximate resistance seen by the generator looking into the stage of Fig. 4-20?

r_{in} =

For the next six problems, refer to Fig. 4-21.

4-21 Figures for problems 4-13 through 4-18.

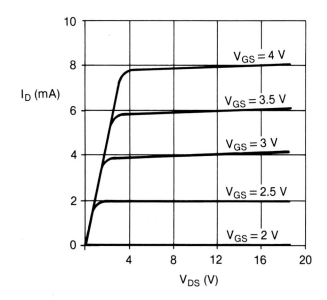

(A) Characteristics.

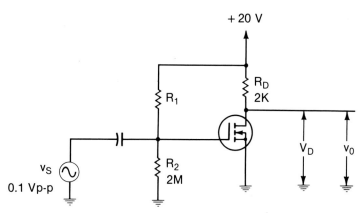

(B) Circuit.

4-13.   Choose a value for R_1 to make I_D = 4 mA.

$R_1 =$

4-14.   Choose a value for R_1 to make V_D = 16 V.

$R_1 =$

4-15.   Referring to problem 4-13, what is the input resistance to the stage?

$r_{in} =$

4-16.   What is g_m for this transistor?

$g_m =$

4-17. What is the voltage gain of the circuit?

$A_v =$

4-18. If R_1 opens what is V_{GS}? What is V_{DS}?

$V_{GS} =$

$V_{DS} =$

4-19. As compared to ordinary MOSFET, VMOSFET have—(a)higher, (b)lower—current carrying capability, —(c)higher, (d)lower—transconductance, and—(e)higher, (f)lower—resistance.

4-20. Zener diodes are sometime built into VMOSFET to (a)regulate the output voltage. (b)protect the input from static overvoltage.

For the next six troubleshooting questions, refer to Fig. 4-17, and choose the most likely cause for each of the following symptoms.

4-21. *Symptom*: The d-c voltage at the drain of Q_9 reads 9.22 V. *Cause*: (a)Shorted Q_9. (b)Open R_{142}. (c)Open R_{143}.

4-22. *Symptom*: V_{DS} reads 1.8 V. *Cause*: (a)R_{58} open. (b)R_{58} shorted. (c)C_{50} open.

4-23. *Symptom*: Normal d-c readings, no a-c output at drain. *Cause*: (a)Open Q_9. (b)Shorted C_{50}. (c)Open L_{20}.

4-24. *Symptom*: Drain of Q_9 reads 9.22 V. *Cause*: (a)Open Q_9. (b)Shorted R_{142}. (c)Open R_{58}.

4-25. *Symptom*: Normal d-c readings, a-c signal at drain of Q_9 very low. *Cause*: (a)Open Q_9. (b)Open R_{143}. (c)Base-emitter short in Q_{10}.

4-26. *Symptom*: V_{DS} lower than normal, $V_S = 0$ V. *Cause*: (a)Shorted R_{58}. (b)Shorted C_{50}. (c)Shorted C_{49}.

For the next six questions, refer to Fig. 4-18.

4-27. The d-c voltage at TP 90 normally should read (a)0 V. (b)1.36 V. (c)<1 V. (d)>2 V.

4-28. The a-c input signal to Q_1 is applied to (a)G_1. (b)G_2. (c)S. (d)D.

4-29. The gain of Q_1 will be higher with the SQUELCH switch in the—(a)up, (b)down—position.

4-30. Which of the following forms a decoupling filter? (a)C_6–R_{41}. (b)C_7–R_{43}. (c)C_{11}–R_{52}.

4-31. Which of the following could cause the d-c voltage at the drain of Q_1 to read 9.22 V? (a)Open R_{50}. (b)Shorted C_4. (c)Shorted C_7. (d)Open D_{19}.

4-32. When the d-c voltage on $G2$ of Q_1 goes more positive, the gain of the stage should (a)increase. (b)decrease. (c)remain the same.

EXPERIMENT 4-1   THE JFET IN D-C CIRCUITS

Here you'll learn how to test a JFET with an ohmmeter. Then you'll see how it can be used as a voltage-variable resistor in applications such as voltmeters.

EQUIPMENT
- VOM
- 9-VDC power supply, or transistor radio battery
- variable low-voltage d-c supply
- 0- to 100-μA meter
- any general purpose N-channel JFET
- (2) 1-MΩ, ½-W ± 5% resistors
- (1) 12-KΩ, ½-W ± 5% resistors
- (2) 10-KΩ, ½-W ± 5% resistors
- (1) 4.7-KΩ, ½-W ± 5% resistor
- (1) 5-KΩ pot
- (1) 2.5-KΩ pot
- other resistors as needed

PROCEDURE
1.   Connect an ohmmeter from source to drain as shown in Fig. E4-1A, and measure the resistance. Reverse the leads, and measure the resistance again. You should read approximately the same resistance in both directions.

$R_{DS} =$

2.   Next, connect the negative lead of the ohmmeter to the source and the positive lead to the gate. Measure the resistance.

$R_{SG} =$

Reverse the leads and measure again.

$R_{GS} =$

Do you see the effect of a diode?

(a)Yes. (b)No.

3.   Build the circuit of Fig. E4-1B. Set the milliameter to read a full-scale value of 50 mA or so. You will probably have to reduce the scale later to get a more accurate reading. Set $V_{GS} = 0$ V. Apply power, and measure I_D.

$I_D =$

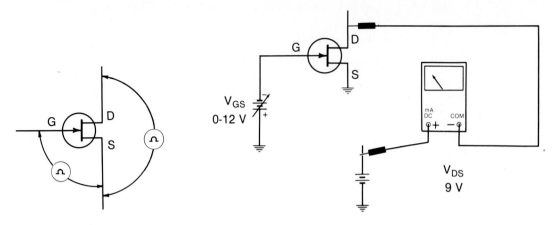

(A) Set-up for steps 1 and 2.   *(B) Circuits for steps 3 through 5.*

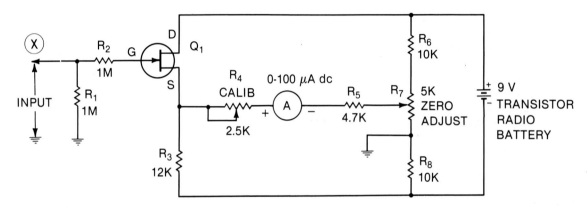

(C) Simple JFET voltmeter.

Assuming practically no drop across the meter so that V_{DS} = 9 V, calculate the drain-to-source resistance.

R_{DS} =

4.   Increase V_{GS} to -2 V, and measure I_D again.

I_D =

Calculate R_{DS} as before.

R_{DS} =

5.   Finally, increase V_{GS} to -4 V, and measure I_D.

I_D =

Calculate R_{DS}.

R_{DS} =

If your experiment worked so far, you should have seen that the resistance of the JFET increases as V_{GS} is made more negative. If it worked as expected, go on to the next step. Otherwise check your circuit and repeat steps 3 through 5.

The JFET is often used in transistor voltmeters to take advantage of its high input impedance. In the circuit of Fig. E4-1C, Q_1 is used as a variable resistor in a bridge arrangement. Its resistance is varied by the voltage being measured. Resistor R_2 is used to protect the transistor from accidental overload.

6. Build the circuit of Fig. E4-1C. Set both pots to midrange. Ground the input terminal (point X), and apply power. Vary the ZERO ADJUST pot until the meter reads zero.

7. Next, calibrate the meter. Apply exactly $+1$ VDC between point X and ground. (The $+1$ VDC must be obtained from some external source or voltage divider.) The meter should deflect upscale. Adjust the CALIB pot until the meter reads full-scale deflection.

With your instrument calibrated for 1-V full-scale deflection, applying ½ V from point X to ground should give half-scale deflection. Try feeding in different known values of voltage, and check the meter accuracy.

QUIZ

1. In step 1, the gate was left open. The ohmmeter reading of R_{DS} was therefore its—(a)minimum, (b)maximum—value.

2. If you did not know whether the JFET was an N-channel or P-channel device, would the measurement of step 1 have shown you which it was? (a)Yes. (b)No.

3. Referring to question 2, would the measurements of step 2 have shown you which it was? (a)Yes. (b)No.
Explain.

4. Steps 3 through 5 demonstrate that the resistance from source to drain—(a)increases, (b)decreases—as V_{GS} is made more negative.

The remaining questions refer to the voltmeter of Fig. E4-1C.

5. With 0 V at point X, adjusting R_7—(a)balanced, (b)unbalanced—the bridge by making the voltage at the pot slider—(c)higher than, (d)lower than, (e)equal to—the voltage at the source of Q_1.

6. In step 7, when $+1$ VDC was applied to the input, Q_1 turned—(a)on, (b)off—more.

7. Referring to question 6, current began flowing through the meter because the left end of R_4 was pulled more (a)positive. (b)negative.

8. The CALIB pot is used in series with R_5 to limit the—(a)maximum, (b)minimum—current through the meter.

EXPERIMENT 4-2   THE JFET AMPLIFIER

Because of its high input impedance, the JFET is very useful as a small signal amplifier. You will now test a simple single-stage amplifier using a JFET. Here you'll see how to bias a JFET, even when you do not have a set of characteristic curves.

EQUIPMENT

- 12-VDC power supply
- oscilloscope
- VOM
- audio signal generator
- (1) N-channel JFET (any general purpose type)
- (1) 10-KΩ pot
- (1) 1-MΩ, 1/2-W $\pm$ 5% resistor
- (1) 2-KΩ, 1/2-W $\pm$ 5% resistor
- (1) 20-μF capacitor
- (1) 5-μF capacitor

E4-2   Circuits for experiment 4-2.

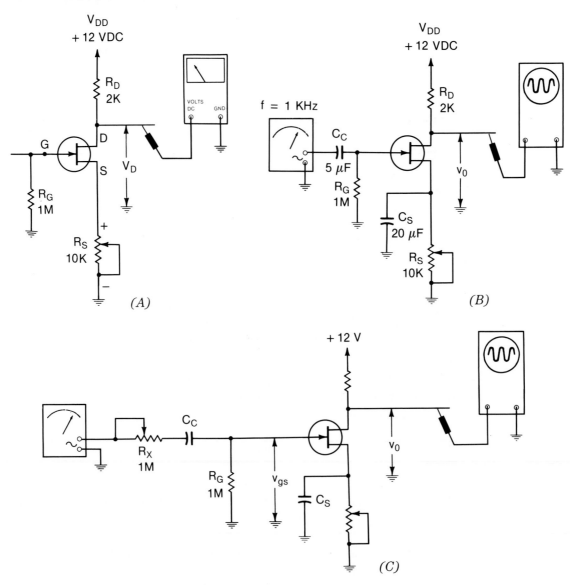

(A)

(B)

(C)

PROCEDURE

1. Build the circuit of Fig. E4-2A. With a voltmeter connected from drain to ground, apply power and adjust R_S until $V_D = 10$ V. This makes $I_D = 1$ mA.

2. Measure the voltage from source to ground and the voltage from gate to ground.

$V_S =$

$V_G =$

If your circuit is working properly, the gate-to-ground voltage should equal zero, so the source-to-ground voltage is the same as the gate-to-source bias voltage.

3. Next readjust R_S until V_D reads 8 V. What is the value of I_D now?

$I_D =$

4. Measure V_S again.

$V_S =$

5. You learned in the text that the transconductance $g_m = \Delta I_D / \Delta V_{GS}$. Using the measured values of steps 1 through 4, determine g_m for your transistor.

$g_m =$

6. Add capacitors C_c and C_s to your circuit, as shown in Fig. E4-2B. Then apply an input signal of 0.5 Vp-p at 1 KHz. Measure the output voltage v_0 with an oscilloscope.

$V_0 =$

Calculate the voltage gain.

$A_v = v_0/v_{gs} =$

7. Now let's measure the input impedance of the circuit. Put a 1-MΩ pot in series with the input coupling capacitor, as shown in Fig. E4-2C. Keep lead lengths short to minimize noise pick up. Keep all of the other circuit components as in step 6, and keep v_s at 0.5 Vp-p. Now while watching v_0 with an oscilloscope, adjust potentiometer R_x until v_0 reads half the value obtained in step 6. Since no amplifier circuit values were changed, the gain of the circuit is the same as before. What did change was the amount of input signal v_{gs}. Since v_0 is half its previous value, v_{gs} must be half of v_s. The rest of v_s must then be dropped across R_x. That means that the resistance of R_x must be equal to the input resistance of the amplifier (assuming that the internal resistance of the generator is very small compared to R_x).

8. Remove R_x from the circuit, and measure it with an ohmmeter.

$R_x = r_{in} =$

QUIZ

1. If you wanted to replace the pot in Fig. E4-2A with a fixed resistor so that $I_D = 1$ mA, what value of R_S would you use? (*Hint*: Use your measurements of steps 1 and 2.)

$R_S =$

2. If you wanted to replace the pot with a fixed resistor in Fig. E4-2A so that I_D = 2 mA, what value of R_S would you use?

$R_S =$

3. Estimate a suitable value for R_S in Fig. E4-2A to make I_D = 1.5 mA.

$R_S =$

4. Calculate the theoretical value of voltage gain using equation 4-5B and your calculated value of g_m from step 4.

$A_v =$

5. Determine the percentage of error between the theoretical value of gain and the measured value, by using

% error = (theoretical value − measured value)/theoretical value × 100 =

If the percentage of error is within the tolerance of your parts and test equipment capabilities, your results are good.

6. In step 7, you measured the a-c input resistance of the amplifier. Why would it not be a good idea to measure the input resistance with a VOM set on the ohms scale?

7. Suppose you decided to measure the input impedance of the amplifier by connecting an oscilloscope from gate to ground, and adjusting R_x until v_{gs} = 0.25 V. If the oscilloscope's input impedance were 1 MΩ, would it affect your measurements as performed in step 7?

(a) Yes. (b) No.

Multistage Amplifiers 5

When more amplification is needed than can be obtained from a single stage, several amplifiers are cascaded (connected in series). In this chapter, we will look at some points to keep in mind when analyzing or testing cascaded stages.

TWO-STAGE CAPACITIVELY COUPLED AMPLIFIER

Fig. 5-1A shows two transistor stages coupled by means of capacitor C_C. The output signal v_{01} of Q_1 is the input signal of Q_2. Capacitor C_C couples the a-c signal from the collector of Q_1 to the base of Q_2, while blocking the d-c collector voltage from upsetting the bias of Q_2.

5-1 Two-stage capacitively coupled amplifier.

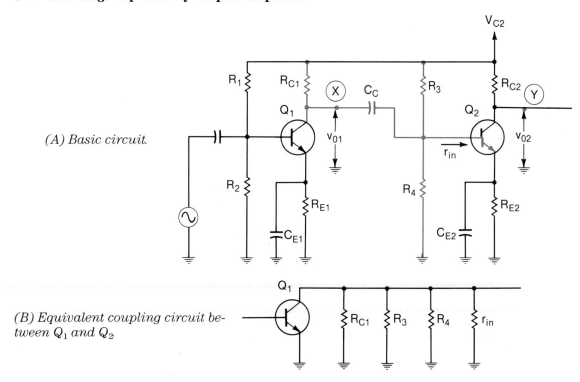

(A) Basic circuit.

(B) Equivalent coupling circuit between Q_1 and Q_2.

Notice that in the a-c equivalent circuit of Fig. 5-1B, the a-c load r_{L1} seen by the collector of Q_1 is the parallel combination of R_{C1}, R_3, R_4, and r_{in} of the second stage. This is an important point to keep in mind, because any changes in R_3, R_4, or r_{in} affect the gain of Q_1. Capacitor C_C is omitted from the equivalent circuit assuming that its reactance is negligible.

The gain of Q_1 is $A_{v1} = r_{L1}/r_{e1}$, and the gain of the second stage is $A_{v2} = r_{L2}/r_{e2}$. The overall gain of the two amplifier stages is the product of the two gains, or

$$A_{v(tot)} = A_{v1} \times A_{v2}$$ 5-1

EXAMPLE 5-1 In Fig. 5-1A, suppose $v_{in} = 2$ mV, $A_{v1} = 40$, and $A_{v2} = 60$. Find the overall gain and the amplitudes of the signals at points X and Y.

SOLUTION The signal at point X is $v_{01} = A_{v1} \times v_{in} = 40 \times 2 = 80$ mV, and the signal v_{02} at point Y is $v_{02} = A_{v2} \times v_{01} = 60 \times 80 = 4800$ mV = 4.8 V.

Alternately, we can find the signal at point Y by multiplying $v_{in} \times A_{v(tot)}$ = 2 mV $\times$ 2400 = 4.8 V.

TESTING AND TROUBLESHOOTING CASCADED STAGES

When testing or troubleshooting a two-stage amplifier like the one in Fig. 5-2, using your senses first is the best way to start. Use your eyes and nose to look for obvious symptoms, such as burned, overheated, or broken parts. If there is nothing obvious, turn on the power supply and test for normal supply voltages. One of the most common sources of problems in electronic systems is power supply failure. It is usually a good idea to scope the supply for excessive ripple.

5-2 **Testing a two-stage amplifier by signal injection.**

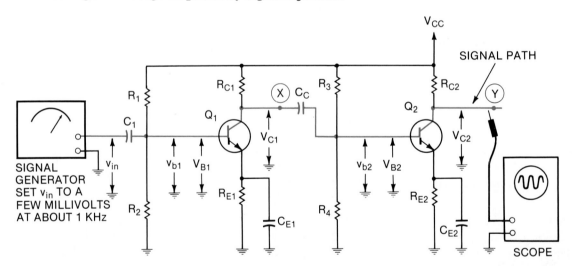

If the power supply checks OK, inject a small signal, say a few millivolts, into the first stage. Have the signal generator set at a frequency within the normal range of the amplifier. Connect an oscilloscope to the output of the last stage, Q_2 in this case, and look for an undistorted, amplified signal.

If no output signal appears, move the oscilloscope to the collector of Q_1. This should isolate the defective stage. If no signal appears at the collector of Q_1, begin making d-c voltage measurements on the first stage. But if the signal at the collector of Q_1 is normal, the trouble is in the second stage. So begin making d-c voltage checks on the second stage.

Lastly, after making voltage measurements, you might have to use a transistor checker or ohmmeter to locate the defective component.

5-3 Flowchart for troubleshooting amplifier of Fig. 5-2.

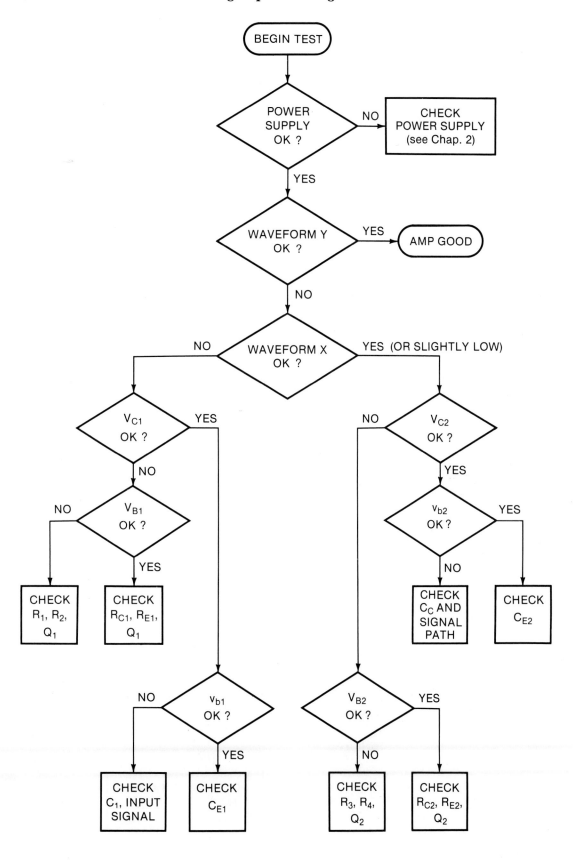

5-4   Flowchart for troubleshooting amplifier of Fig. 5-2.

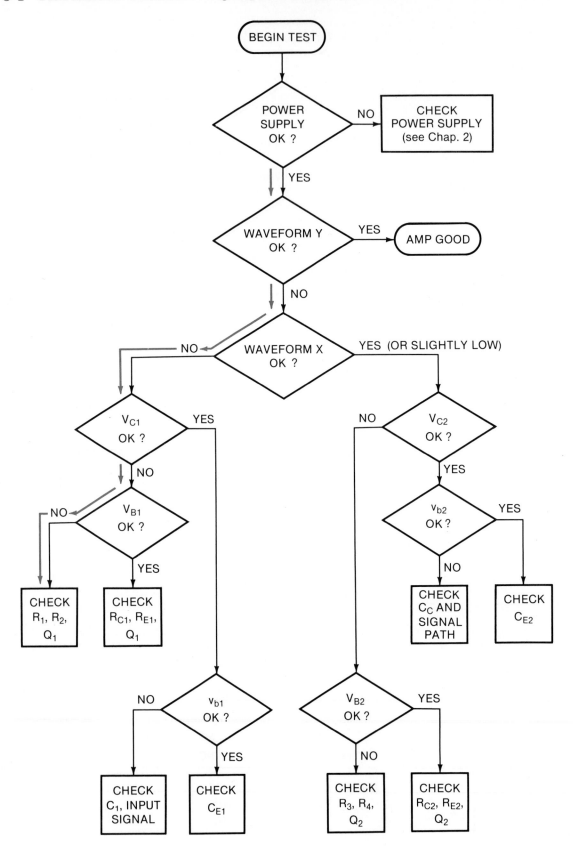

In summary, your troubleshooting procedure should be as follows:
1. Inspect for obvious symptoms.
2. Check for normal power supply voltages.
3. Inject a small signal into the base of Q_1.
4. Make waveform checks, beginning at the output and working back toward the input, to isolate the defective stage.
5. Make d-c voltage checks to find any biasing problems.
6. Use a transistor checker, if available, to test a suspect transistor.
7. Make resistance checks to confirm the defective component.

Fig. 5-3 illustrates how to zero in on a defective component in the circuit of Fig. 5-2. Although you will not always have a flowchart for every circuit, Fig. 5-3 illustrates the thinking and testing process needed in troubleshooting any amplifier. In the flowchart, capital letters, such as V_{B1}, indicate d-c values, such as bias voltages, while lowercase letters, like v_{b1}, indicate a-c or signal voltages.

Let's go through a test run using the flowchart. We will assume for this run that R_2 is open, driving Q_1 into saturation, and that we see no broken or burned parts. The highlighted arrows in Fig. 5-4 show the path that we will follow in this case.

First, we apply power and measure the supply voltage. It reads normal. Next, we inject a small signal into the base of Q_1 and connect an oscilloscope to point Y. In this case, we will not see any output signal. We next look for a waveform at point X. Again we do not see a normal output because Q_1 is saturated. So we take the NO route to the left of the decision box labeled WAVEFORM X OK?

We now take a d-c voltmeter and measure the collector voltage of Q_1. Since the transistor is saturated, the collector voltage will be lower than normal, so we follow the NO route from that decision box.

We measure bias voltage V_{B1}. It will read higher than normal. So we follow the NO route to a block that tells us to check R_1 or R_2 or Q_1. At this point, if a transistor checker is available, we check Q_1 without removing it from the circuit. If no transistor checker is available, unsolder any two of the transistor leads, and check it with an ohmmeter. Next we measure R_1 and R_2 with an ohmmeter. The surest way to check a resistor is to disconnect one end of it from the circuit, then measure it with an ohmmeter. We find that R_2 is open, so we will replace it. By using this logical approach, we have quickly and easily found the problem.

> **CAUTION** Whenever you replace a defective component, try to determine whether that component went bad on its own or was ruined because of some other problem in the system. This is not always easy to determine, but sometimes a bit of investigating will prevent the same component from burning out again when power is applied.

FREQUENCY RESPONSE OF CASCADED STAGES

The coupling capacitor C_C can be omitted from the equivalent circuit of Fig. 5-1B, assuming that its reactance is negligible. But at low frequencies, the reactance of C_C gets too high to ignore.

The coupling circuit between Q_1 and Q_2 is shown on Fig. 5-5A. Resistance $r_{in(stage)}$ represents the parallel equivalent of R_3, R_4, and $r_{in(base)}$ of Q_2. The Thevenin equivalent of Fig. 5-5A is shown in Fig. 5-5B. In the Thevenin equivalent, the transistor is shown as a voltage source in series with resistance R_{C1}.

Because of the reactance of C_C at low frequencies, there is a voltage drop across C_C. Consequently, the voltage v_{b2}, applied to the base of Q_2, is less than the voltage v_{C1} at the collector of Q_1. When the frequency gets so low that v_{b2} drops to 0.707 of its maximum value, the power delivered to the second stage drops to half its maximum value, or -3 dB (decibels). That point will occur when $X_C = R_{C1} + r_{in(stage)}$. And since $X_C = 1/2 \pi fC$, the low-frequency half-power point occurs at

$$f_1 = \frac{1}{2 \pi R_t C} \qquad \text{5-2}$$

5-5 **Interstage coupling circuit at low frequencies.**

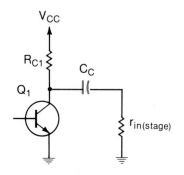

(A) Basic circuit.

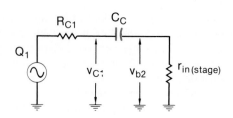

(B) Thevenin equivalent.

where

$$f_1 = -3\text{-dB point}$$
$$R_t = R_{C1} + r_{in(\text{stage})}$$
$$C = \text{capacitance of coupling capacitor } C_C.$$

EXAMPLE 5-2   In the circuit of Fig. 5-1A, suppose $R_{C1} = 5$ KΩ, $R_3 = 40$ KΩ, $R_4 = 10$ KΩ, $r_{in(\text{base})} = 2$ KΩ, and $C_C = 1$ μF. Find the lower half-power frequency f_1.

SOLUTION

$$r_{in(\text{stage})} = 40 \text{ K}\Omega \parallel 10 \text{ K}\Omega \parallel 2 \text{ K}\Omega = 1.6 \text{ K}\Omega$$

so

$$R_t = R_{C1} + r_{in(\text{stage})} = 5 \text{ K}\Omega + 1.6 \text{ K}\Omega = 6.6 \text{ K}\Omega$$

then

$$f_1 = \frac{1}{6.28 \times 6.6 \times 10^3 \times 1 \times 10^{-6}} = 24 \text{ Hz}$$

In order to improve the low-frequency gain of the amplifier, either the size of the coupling capacitor or the resistance or both must be increased.

Besides a lower limit to the circuit's use, there is also an upper limit. At high frequencies, the reactance of coupling capacitor C_C is negligible, as shown in Fig. 5-6, so it can be omitted. But we can now see the shunting effect of several small capacitances, which were ignored at lower frequencies. These capacitances are C_{ce}, the collector-to-emitter output capacitance of Q_1, along with C_w, the distributed wiring capacitance of the circuit, in parallel with C_{in}, the input capacitance of the next stage. We can lump all these capacitances together and call the parallel combination C_s.

As the frequency gets higher, $X_{C(S)}$ gets lower, causing more of the signal to be shunted away from the base of Q_2. Eventually, a frequency is reached where only half as much signal reaches the base of Q_2 as reached it at some midband frequency. At this point, called the upper 3-dB point, $X_{C(S)} = R_{eq}$, where $R_{eq} = R_{C1} \parallel r_{in(\text{stage})}$.

We can find the upper half-power frequency f_2 with

$$f_2 = \frac{1}{2 \pi R_{eq} C_S} \qquad\qquad 5\text{-}3$$

where
$$f_2 = \text{upper 3-dB point}$$
$$R_{eq} = R_{C1} \| r_{in(\text{stage})}$$
$$C_S = C_{ce} \| C_w \| C_{in}$$

5-6   Interstage coupling circuit at high frequencies.

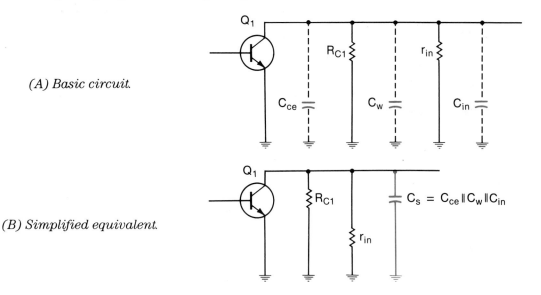

(A) Basic circuit.

(B) Simplified equivalent.

EXAMPLE 5-3   In the circuit of Fig. 5-1, assume $R_{C1} = 5$ KΩ, $R_3 = 40$ KΩ, R_4 = 10 KΩ, $r_{in(\text{base})} = 2$ KΩ, $C_{ce} = 5$ pF, $C_w = 10$ pF, $C_{in} = 500$ pF. (These are typical values for the capacitances.) Find f_2.

SOLUTION

$$R_{eq} = 5 \text{ K}\Omega \| 40 \text{ K}\Omega \| 10 \text{ K}\Omega \| 2 \text{ K}\Omega = 1.2 \text{ K}\Omega$$

$$C_s = 515 \text{ pF}$$

$$f_2 = \frac{1}{6.28 \times 1.2 \times 10^3 \times 515 \times 10^{-12}} = 258 \text{ KHz}$$

Regardless of the values of components used, *all* capacitively coupled circuits have *some* upper and lower cut-off frequencies. The lower cut-off frequency is generally determined by the size of the coupling capacitors, assuming that the emitter-bypass capacitors are large enough. Too small a value of emitter–bypass capacitor can also ruin the low-frequency response.

The high-frequency response of an amplifier is limited by shunt capacitance, as mentioned. In addition, every transistor has an upper frequency limit beyond which the value of β decreases. Transistors specifically designed for very high-frequency use are called radio-frequency (rf) types.

Figure 5-7 shows the frequency response curve of a typical capacitively coupled amplifier. It shows how the output voltage varies as the frequency of the input signal is changed from near zero to some very high frequency, while keeping the input amplitude constant. The span of frequencies between f_1 and f_2, where the output is at least 0.707 of its maximum value, is called the *bandwidth* of the amplifier.

USING NEGATIVE FEEDBACK TO REDUCE DISTORTION

When a transistor is driven with a large signal, say more than 50 mV, the output signal gets distorted. In a high-gain cascaded amplifier like that of Fig. 5-8, a small symmetrical signal applied

5-7 Frequency response curve of typical capacitively coupled amplifier.

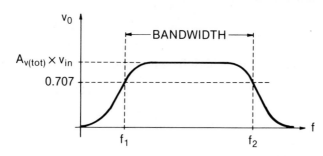

to the base of Q_1 results in a much larger signal being applied to the base of Q_2. The output voltage of Q_2 does not swing symmetrically in the positive and negative directions.

5-8 Distortion of output signal caused by driving output stage with large symmetrical signal.

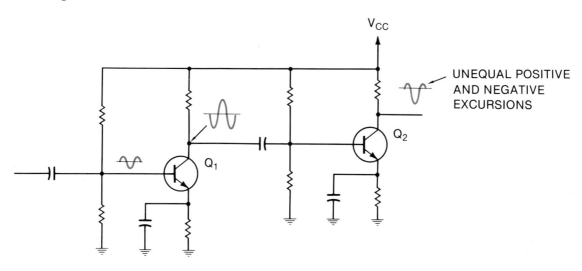

We can see the reason for the nonsymmetry by examining the input characteristics of a transistor as a large signal is applied. See Fig. 5-9. Assume that the transistor is biased at point Q. Apply a large symmetrical signal v_{in}. As v_{in} increases from point 1 to point 2, the base current increases from point 1' to point 2'. The input signal then decreases back to zero amplitude at point 3, and then increases in the negative direction to point 4. From the projections on the characteristic curve, we see that I_B goes through point 3', then to point 4'.

Note that, because of the nonlinearity of the input characteristics, I_B does not *decrease* as much with a negative-going signal as it *increases* with a positive-going signal. Of course, if a-c base current i_b is nonsymmetrical, i_c, and therefore v_c, is also nonsymmetrical. The nonsymmetry introduces qualities in the output that were not present in the input. It can be shown that the distorted output signal contains *harmonics* (multiples) of the input signal frequency, primarily the second harmonic, or twice the input frequency. The amount of harmonics introduced is commonly referred to as the *percent of total harmonic distortion* (% THD). Of course, the smaller the percentage, the better.

Fig. 5-10 shows how *negative feedback* is used to decrease the distortion. Part of the output signal from the collector of Q_2 is fed back to the emitter of Q_1, via voltage divider R_F–R_E. (Note that R_E is not bypassed.) This feedback signal alters the gain of Q_1 differently on positive and negative half cycles.

Here's how it works. As shown in Fig. 5-11A, Q_1 sees input signal v_{in} applied to its base and simultaneously sees feedback signal v_f applied to its emitter. As always, the base current in the transistor is controlled by the *difference* voltage (v_{be}) between its base and emitter. Fig. 5-11B shows v_{in} and v_f superimposed on the same axis. *Note that on the positive half cycle, the difference (v_{be}) between v_{in} and v_f is larger than it is on the negative half cycle.* Thus the a-c base current

5-9 **Projection of large amplitude symmetrical signal on I_B–V_{BE} curve shows cause of distortion.**

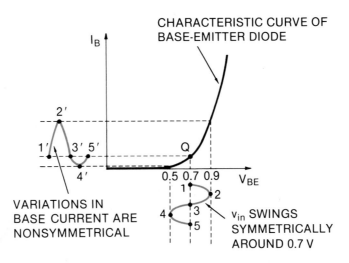

5-10 **Feeding back part of the output signal to reduce distortion.**

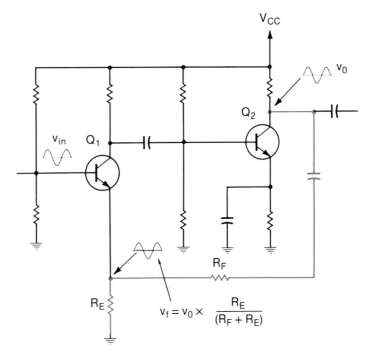

of Q_1 looks like that shown in Fig. 5-11C. And, of course, if the base current of Q_1 is made nonsymmetrical, the a-c collector voltage of Q_1 is also nonsymmetrical.

Thus the signal at the collector of Q_1 has the right shape and phase·relationship to tend to cancel out the nonsymmetry introduced by Q_2 being overdriven in the first place. In other words, using negative feedback tends to cancel out distortion that would have been introduced by the amplifier.

How well it reduces distortion depends on the *amount* of negative feedback used. The larger the percentage of output signal fed back, the less distortion there is.

However, there is one catch. Feeding back part of the output signal, as shown in Fig. 5-10, also reduces the overall gain of the amplifier. The more feedback, the less gain. So the designer must make a trade off between gain and distortion. The overall gain of the amplifier of Fig. 5-10 is equal to the ratio of R_F to R_E, plus 1. That is,

$$A_{v(\text{closed loop})} = \frac{R_F}{R_E} + 1 \qquad\qquad 5\text{-}4$$

where $A_{v(\text{closed loop})}$ is the gain *with* the feedback loop closed. This equation is valid as long as the *closed-loop* gain is small compared to the *open-loop* gain (gain without feedback).

EXAMPLE 5-4   In a circuit like that of Fig. 5-10, $R_F = 20$ KΩ and $R_E = 1$ KΩ. What is the gain?

SOLUTION

$$A_{v(\text{closed loop})} = \frac{20 \text{ KΩ}}{1 \text{ KΩ}} + 1 = 21$$

Negative feedback generally decreases the gain, and also decreases the distortion in an amplifier. Although the gain is reduced, it is very *stable*. In other words, the gain is independent of transistor characteristics or changes in temperature. *The gain is controlled by the ratio of external resistances.*

5-11   How v_f subtracts from v_{in} to reduce overall distortion.

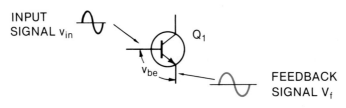

(A) v_{in} and v_f applied to Q_1.

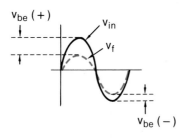

(B) v_{be} is the difference between v_{in} and v_f.

(C) Resultant base current of Q_1.

DIRECT-COUPLED AMPLIFIERS

As you know, the low-frequency response of an amplifier is determined by the size of the coupling and bypass capacitors. Capacitive coupling cannot be used when it is necessary to handle *very* low-frequency signals, or even *d-c* signals such as in power supply regulators or those generated by thermocouples. Instead, the output of one stage is directly coupled to the input of the next stage.

In order to analyze direct-coupled amplifiers, let's take a look at the single stage of Fig. 5-12 first. Note that 5 VDC is applied to the base of the transistor. This positive voltage forward biases the base–emitter diode, dropping about 0.7 V across it. Therefore, the voltage V_E from emitter to ground is $5 - 0.7 = 4.3$ V. Since the voltage across R_E is 4.3 V, the current through the transistor, which is the same as the current through R_E, is

102   Electronic Troubleshooting

$$I_E = \frac{V_E}{R_E} = \frac{4.3 \text{ V}}{4.3 \text{ K}\Omega} = 1 \text{ mA}$$

Keep in mind that *the current through the transistor is controlled by the voltage across R_E*. We can change I_E by changing R_E or V_E or both, as long as the transistor is not saturated.

5-12   The voltage across R_E controls I_E.

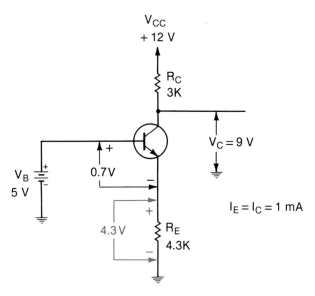

EXAMPLE 5-5   In the circuit of Fig. 5-12, if $V_B = 3$ V, $R_E = 1$ KΩ, and $R_C = 2$ KΩ, find I_C and V_C.

SOLUTION

$$I_C = I_E = \frac{V_B - 0.7}{R_E} = \frac{3 \text{ V} - 0.7}{1 \text{ K}\Omega} = 2.3 \text{ mA}$$

then

$$V_C = V_{CC} - I_C R_C = 12 - 2.3 \text{ mA} \times 2 \text{ K}\Omega = 7.4 \text{ V}$$

Now let's look at Fig. 5-13. Notice that the base of Q_2 is connected directly to the collector of Q_1. Therefore the collector current of Q_2 is controlled by the collector voltage of Q_1, as well as by R_E. The base current of Q_2 flows through the collector resistor of Q_1. It will not, however, affect the collector voltage of Q_1 significantly, as long as I_{B2} is much smaller than I_{C1}. In effect, Q_1 in this circuit acts like resistor R_2 in the voltage–divider bias circuits in chapter 3. R_{C1}, of course, replaces R_1 in the circuit.

EXAMPLE 5-6   Suppose in Fig. 5-13, $R_{C1} = 7.5$ KΩ, $R_{C2} = 2$ KΩ, $R_{E2} = 1.5$ KΩ, and I_{C1} is set at 1 mA by adjustment of R_{B1}. What are the values of I_{C2} and V_{C2}?

SOLUTION

$$V_{B2} = V_{C1} = 12 - 1 \text{ mA} \times 7.5 \text{ K}\Omega = 4.5 \text{ V}$$

So

$$I_{C2} = \frac{4.5\text{ V} - 0.7}{1.5\text{ K}\Omega} = 2.53\text{ mA}$$

Then

$$V_{C2} = 12 - 2.53\text{ mA} \times 2\text{ K}\Omega = 6.94\text{ V}$$

5-13 Simple direct-coupled amplifier.

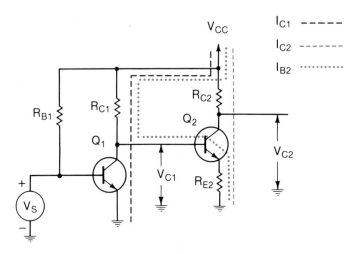

Q₁ AND R$_{C1}$ ACT AS
A VOLTAGE DIVIDER
TO BIAS Q₂

Anything that causes I_{C1} to change, such as a signal at its base, also causes I_{C2}, and therefore V_{C2} to change. So if a signal is applied to the base of Q_1, it is amplified by the product of the gains of the two transistors. The gain of Q_1 is essentially $A_{v1} = R_{C1}/r_{e1}$, but the gain of the second stage is $A_{v2} = R_{C2}/R_{E2}$. Unfortunately, the gain of the second stage is quite small, as a result of the small ratio of R_{C2} to R_{E2}. If we try to increase the gain by increasing R_{C2}, the transistor soon goes into saturation. For this reason, the circuit of Fig. 5-13 is not practical for high gain.

There is another weakness of the circuit in Fig. 5-13. Recall that the collector current in a transistor increases as the temperature increases. Any increase in I_{C1} resulting from a rise in temperature causes a decrease in V_{C1}. The base of Q_2 sees the change in V_{C1} as a signal. Therefore, the change in V_{C1} is amplified by Q_2. Thus any temperature instability of the amplifier is made even worse by direct coupling.

Fig. 5-14 shows a direct-coupled circuit that uses feedback to improve temperature stability. To analyze the circuit, let's see what happens when power is first turned on. Assume that both transistors are initially off.

Note that the forward bias for Q_1 is obtained from the top of R_E through the voltage divider R_1–R_2. If Q_2 is not conducting, Q_1 has no forward bias, so its collector voltage goes positive. This pulls the base of Q_2 positive, turning Q_2 on, and developing a voltage across R_E. As the voltage at the top of R_E goes more positive, forward bias is applied to the base of Q_1, turning it on. The harder Q_1 turns on, the more voltage developed across R_{C1} by I_{C1} flowing through it. Thus, the voltage applied to the base of Q_2 decreases.

Since the emitter voltage of Q_2 is always 0.7 V less positive than its base, the voltage at the top of R_E soon reaches an equilibrium point such that the voltage at the base of Q_1 is held near 0.7 V. If the voltage goes too high, Q_1 turns on too hard, and if it drops too low, Q_1 conducts less. The voltage that eventually is established at the top of R_E depends on the ratio of R_1 to R_2.

Here's how the temperature stabilization works. Suppose the circuit of Fig. 5-14 is operating normally with all voltages as shown. Next, suppose Q_1 heats up, causing I_{C1} to increase. The increase in I_{C1} causes a larger voltage drop across R_{C1}, which in turn decreases the voltage at the base of Q_2. The emitter voltage of Q_2 decreases also because the emitter always stays about 0.7 V less positive

5-14   **Direct-coupled amplifier with d-c feedback for stabilization.**

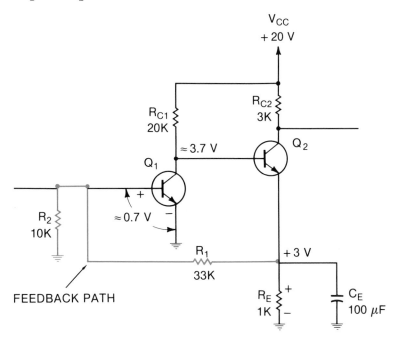

than the base. The decrease in emitter voltage causes a decrease in base voltage of Q_1, due to the voltage divider R_1–R_2. And of course, the decrease in base voltage of Q_1 decreases I_{C1}. The net effect is that the total increase in I_{C1} resulting from a temperature increase is much less than it would have been without feedback.

Capacitor C_E is used as an emitter bypass, as was the case in previously discussed amplifiers. This circuit is not intended to amplify d-c signals, but it does make a good low-frequency audio amp.

Fig. 5-15 is an example of a commercial circuit using direct coupling between two stages. In the highlighted portion, the collector of transistor X_{101} is directly coupled to the base of X_{102}. Notice that the base-bias resistor of X_{101} is R_{104}, which connects to the top of R_{114} in the emitter circuit of X_{102}. In this case, the first transistor of the pair, X_{101}, does not use voltage divider bias, as did Q_1 of Fig. 5-14. Instead, it simply uses a form of base bias that was introduced earlier. Nevertheless, the action is the same as was described for the circuit of Fig. 5-14. Notice also that the indicated base-emitter voltage of X_{101} is only 0.53 V rather than 0.7 V. Keep in mind that 0.7 V is only an *approximate* forward bias voltage for silicon transistors. The actual voltage does vary slightly with different transistors.

Study the highlighted circuit thoroughly to make sure that you see how it operates. Don't let all of the other components in the circuit distract you. Simply concentrate on the basic components of the two-stage direct-coupled amplifier. This is a key point in testing or troubleshooting any complex system. Try to identify the essential parts of the circuit first. Most of the other components usually affect frequency response, gain, tone, as well as some switching of input and output signals.

Incidentally, notice capacitor C_{110}, which is connected from the collector of X_{102}, through some R and C components, to the emitter of X_{101}. What do you think that circuit is used for?

DIFFERENTIAL AMPLIFIERS

When it is necessary to amplify a small d-c signal, such as that developed by a thermocouple, a *differential* or *difference* amplifier is often used because it has good temperature stability.

The circuit shown in Fig. 5-16 uses two identical transistors whose emitters are tied together and fed through a common resistor R_E. Here's how it works. Suppose no input signal is applied to either base. Then $V_1 = V_2 = 0$ V, since both bases are returned to ground through resistors.

Even though there is 0 V on each base, both transistors are forward biased because the lower end of R_E is connected to a *negative* power supply. Remember, to forward bias a transistor, it is only necessary to make the base of an NPN transistor positive with respect to its emitter. It does not

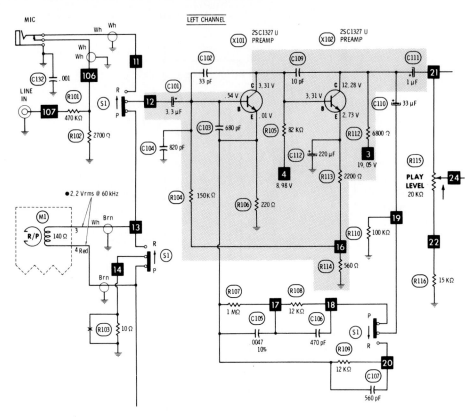

need to be positive with respect to ground. So returning the emitters to a negative supply does the job.

Since the base–emitter diodes are forward biased, the emitters are at *minus* 0.7 V. Therefore, the voltage across R_E is $15 - 0.7 = 14.3$ V. For a good approximation, we can say that the drop across R_E is equal to V_{EE} or about 15 V. Therefore, the current through R_E is approximately equal to $V_{EE}/R_E \cong 15$ V/7.5 K$\Omega \cong 2$ mA.

If the two transistors have identical characteristics, the 2 mA flowing through R_E splits up evenly, with 1 mA flowing through each transistor. Therefore the voltage V_{C1} at the collector of Q_1 is equal to $V_{CC} - I_C R_C = 10$ V. Likewise $V_{C2} = 10$ V. This shows that *with no signal applied to either transistor, the difference voltage V_D between the two collectors is zero.*

Here's where the temperature stability comes in. Even if the temperature goes up, so that both transistors try to conduct harder, the difference between the collectors will remain at 0 V. This will be true as long as the two transistors are identical and $R_{C1} = R_{C2}$.

Suppose we connect a small d-c signal source, like a thermocouple, between the two bases, as shown in Fig. 5-17. When the thermocouple is heated, a voltage is developed at its cold ends, which is proportional to the temperature at the heated junction. Assume that the end connected to the base of Q_1 goes positive with respect to the end connected to the base of Q_2. Since both emitters are tied together, making the base of Q_1 more positive turns Q_1 on harder. Likewise Q_2 conducts less. This causes more voltage to be dropped across R_{C1} than across R_{C2}. Consequently, there is a difference voltage V_D between the two collectors, which is proportional to the differential input voltage, but larger in amplitude.

The gain of the differential amplifier is equal to

$$A_{v(\text{diff})} = \frac{V_D}{V_1 - V_2} = \frac{R_c}{r_e} \qquad \text{5-5}$$

5-16 Basic differential amplifier.

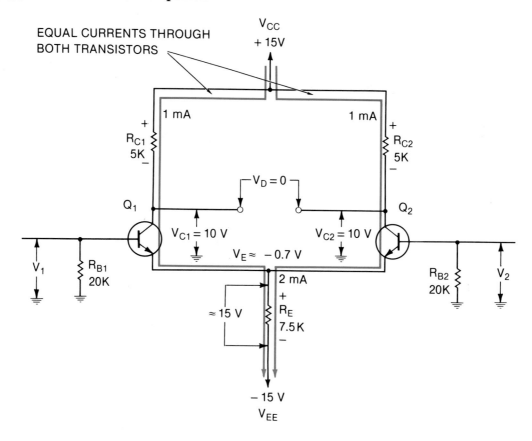

5-17 Using a differential amplifier to amplify the voltage generated by a thermocouple.

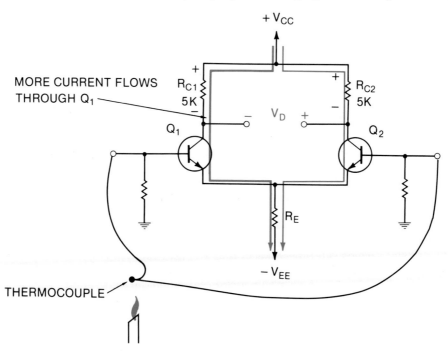

where $A_{v(\text{diff})}$ is the differential voltage gain, V_1-V_2 is the difference between the two base voltages, $R_C = R_{C1} = R_{C2}$ and $r_e = 25 \text{ mV}/I_E$. (I_E is the emitter current of one transistor.)

Notice that equation 5-5 is essentially the same as that for the single-stage a-c amplifier discussed in chapter 3. This circuit can, however, amplify d-c as well as a-c signals.

EXAMPLE 5-7   In the circuit of Fig. 5-17, suppose the signal V_1-V_2 generated by the thermocouple is 1.5 mV. What is the differential output signal V_D?

SOLUTION

$$r_e = \frac{25 \text{ mV}}{1 \text{ mA}} = 25 \text{ }\Omega$$

$$A_{v(\text{diff})} = \frac{5 \text{ K}\Omega}{25 \text{ }\Omega} = 200$$

$$V_D = A_v \times (V_1 - V_2) = 200 \times 1.5 \text{ mV} = 300 \text{ mV}$$

The larger the values of R_E and V_{EE}, the better the temperature stability is. This follows since the current through R_E remains more constant with changes in the transistors. Since it is impractical to use extremely high-voltage supplies, a better solution is to use a *constant current source* driving the emitters. Fig. 5-18A shows a constant current source replacing R_E. This source, once set, always passes the same value of current, regardless of the voltage across it.

One way to build a simple constant current source is shown in Fig. 5-18B. The current through transistor Q_3 is controlled by the voltage across its own emitter resistor R_E. And the voltage across R_E is held constant by the voltage divider R_X–R_Y and diode $D1$. Using the values shown, we see that current flow from ground through R_X–R_Y to the $-V_{EE}$ supply causes 5 V to appear across the series combination of D_1 and R_Y. Therefore, if 0.7 V is dropped across the base–emitter diode of Q_3, the remaining 4.3 V is developed across R_E. With 4.3 V across a resistance of 2.2 KΩ, the current through Q_3 is approximately 2 mA. (R_E can be adjusted slightly if it is necessary to "tweak-up" the current to an exact value.) Diode D_1 further improves the temperature stability of the constant current source, since it has characteristics similar to the base–emitter diode of Q_3. Fig. 5-18C shows the complete differential amplifier with constant current source.

Monolithic (single chip) differential amplifiers are available from various manufacturers. Consequently, you do not need to use matched-pair transistors. In addition, differential amplifiers are used extensively in operational amplifiers (op amps), which will be discussed in more detail in chapter 8.

EMITTER FOLLOWERS

Recall that the emitter voltage of a silicon transistor always stays about 0.7 V less than its base voltage. Consider the circuit of Fig. 5-19. There is no collector resistor. The collector ties directly to V_{CC}, and the output is taken at the emitter. If V_{in} is set at 5 V, we know that the output voltage is about 4.3 V. If V_{in} increases to 6 V, V_0 goes up to 5.3 V. If V_{in} drops to 4 V, V_0 drops to 3.3 V. The point is that the *change* in output voltage is equal to the *change* in input voltage. In other words, the circuit has a voltage gain of 1. Since the output voltage follows (is in phase with) the input voltage, the circuit is called an *emitter follower. An emitter follower has a voltage gain of 1.*

Besides having a unity voltage gain, the emitter follower has a very high-input impedance. As can be seen in Fig. 5-19, the current flowing through the load R_E is equal to I_E, but the signal source only has to supply I_B, which is β times smaller than the load current. The signal source controls the load voltage, but it does not have to deliver the full-load current. *The emitter follower effectively makes the load resistance look larger by a factor of* β, when the signal source sees it. That is

$$r_{in} = \beta \, r_L \qquad \text{5-6}$$

where r_L is the a-c load seen by the emitter.

5-18   Using a constant current source to improve temperature stability.

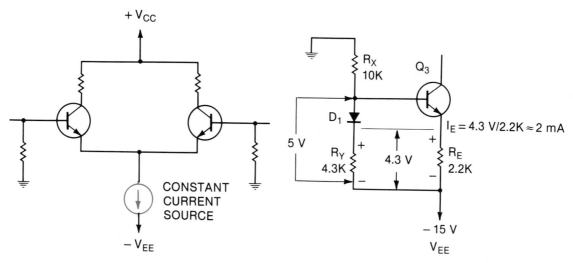

(A) Constant current source replaces R_E.

(B) Simple constant current source.

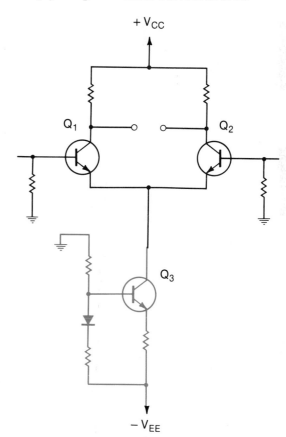

(C) Complete circuit.

EXAMPLE 5-7   In the circuit of Fig. 5-19, what resistance does the source see as it looks into the base of the transistor?

SOLUTION

$$r_{in} = \beta \, R_E = 80 \times 1 \text{ K}\Omega = 80 \text{ K}\Omega$$

5-19   **Basic emitter follower.**

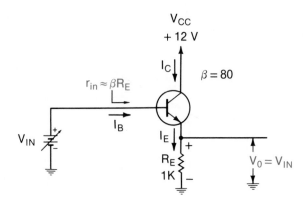

The emitter follower is often directly coupled to the collector of a transistor to minimize loading on the amplifier stage. In Fig. 5-20, Q_2 couples the 300-Ω load R_L to the collector of Q_1, but the a-c load seen by the collector of Q_1 is essentially equal to 5 KΩ in parallel with $\beta \times R_L = 100 \times 300 \ \Omega = 30$ KΩ. This way, the gain of Q_1 remains high, since Q_1 works into a high-resistance load. Yet the full output signal is developed across the relatively small 300-Ω load. Such a circuit can be used in driving a low-impedance transmission line. The emitter follower "transforms" the low-input impedance of the transmission line into a much higher impedance as seen by Q_1. The emitter follower thus takes the place of an impedance-matching transformer.

5-20   **Using an emitter follower to make a small load resistance look much larger.**

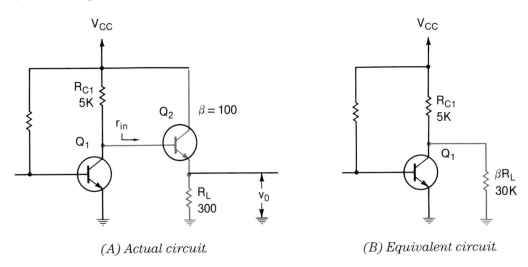

(A) Actual circuit.   *(B) Equivalent circuit.*

If it is necessary to keep dc out of the load, the circuit arrangement of Fig. 5-21 can be used. The input resistance, seen looking into the base of Q_2, is $\beta(R_E \parallel R_L)$. Usually R_E is much larger than R_L, so as not to waste power.

ANALYZING A COMPLETE AMPLIFIER SYSTEM

Fig. 5-22 shows the complete left channel of a tape recorder, starting with the microphone input jack and tape heads at the upper left, and ending with the headphone jack at the upper right. As mentioned earlier, the first thing to do when analyzing a system is to identify the basic building blocks, or circuits. These basic amplifier stages are highlighted for easy identification. Notice the two-stage amplifier at the left, which consists of transistors X_{101} and X_{102}. This is the same stage that is shown in part in Fig. 5-15. A very similar amplifier stage, consisting of transistors X_{103} and X_{104}, is found immediately to the right of the volume control R_{119}.

5-21   Using a coupler capacitor to keep dc out of the load.

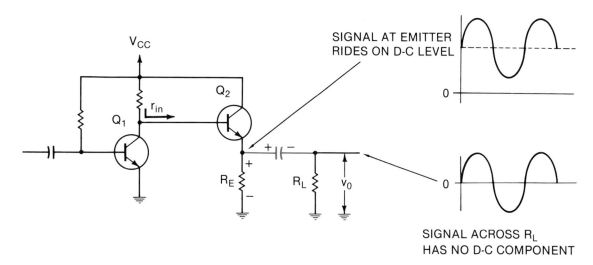

In the upper right corner of the diagram, we see emitter follower X_{105}. The transistor is drawn upside down here, but you should recognize the similarity between this circuit and that of Fig. 5-21. Study it until you do. Capacitor C_{122} couples the a-c signal to the low-impedance headphones, but keeps d-c out of them. Also connected to TP 47 through calibration pot R_{132} is diode D_{101} and capacitor C_{123}. The diode and capacitor form a half-wave rectifier, which rectifies part of the audio output signal. The resulting d-c voltage at TP 49 is fed to a d-c record/play level meter. This meter indicates the amplitude of the audio signal at that point.

Another circuit configuration that you should recognize is the RECORD AMP (X_{106}) from chapter 3. This stage uses voltage-divider bias, obtained from R_{138}–R_{139}, as well as capacitive input and output coupling.

At first glance, a lot of other components seem to be in the system. They should not bother you if you take a few moments to see what they do.

Let's start with the switches. At the upper left of Fig. 5-22, we see part of switch S_1, which is used to select the record R or playback P mode. You can see that the pole of the switch (connected to TP 12) is connected to either the microphone jack or the playback head, depending upon the switch position. S_1 is a multipole switch, having other sets of contacts at the tape head (TP 14), the input to the volume control (TP 28), and the feedback loop of the first amplifier stage (TP 19). All sets of contacts are switched simultaneously by the same slide lever.

Switch S_2 is used to select between various R-C components for tone control. Fig. 5-23 shows the components (redrawn) in the coupling circuit between the PREAMP and AF AMP. S_2 is in the normal position, and S_1 in the playback position. Compare Fig. 5-23 with the actual circuit of Fig. 5-22 until you see that they are the same. It is often helpful to redraw a complex section of a circuit to make it more understandable.

More or less of the signal available at TP 21 is fed on to TP 28 by moving the pot slider of R_{115} up or down. The arrow near the control indicates clockwise rotation of the pot shaft.

Here's how the frequency response (tone control) of the circuit is determined. Fig. 5-24A shows the equivalent circuit between TP 26 and TP 28. At a frequency of approximately 100 Hz, the reactance of C_{114} is about ten times larger than the resistance of R_{117}. So the a-c equivalent circuit can be drawn as shown in Fig. 5-24B. v_0 is some percentage (approximately 0.56%) of v_{in}. But as the frequency of the input signal increases, the reactance of C_{114} gets smaller, coupling a larger percentage of the input signal to the output. Similarly, the reactance of series capacitor C_{113} is approximately equal to 39 KΩ at 100 Hz. However, it gets lower as the frequency goes up. The net effect of the entire circuit in Fig. 5-23 is that more of the input signal is available across R_{119} at higher frequencies, than at lower frequencies. The response curve of the coupling circuit is somewhat like that shown in Fig. 5-24C. This type of circuit is generally called a *high-pass* filter.

But if we switch S_2 down to the chrome position, a different set of components are used in the coupling circuit. Do you think that more or fewer of the low frequencies are passed with S_2 in the chrome position. Why?

5-22 **Complete left channel of Magnavox Model TE3410WA11 with basic circuits high-lighted. From** *Tape Recorder Series No. 173,* **Howard W. Sams & Company, Inc.**

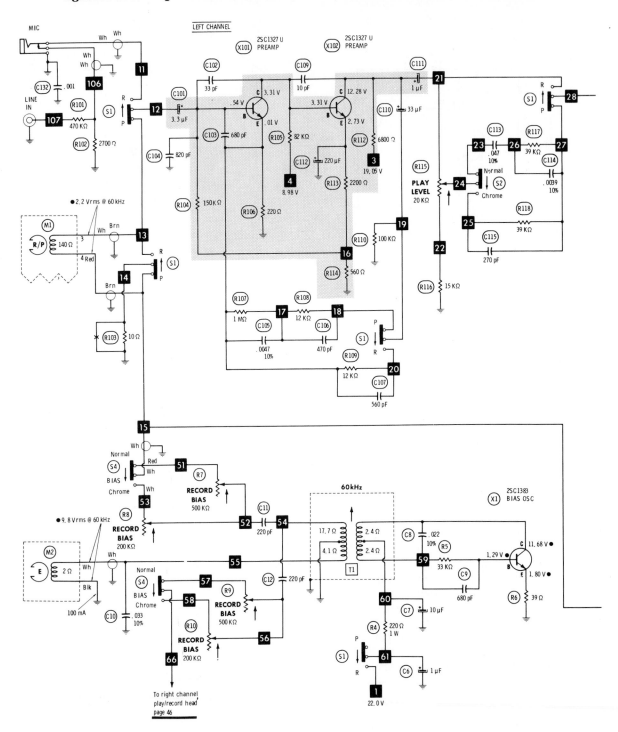

5-22 Continued.

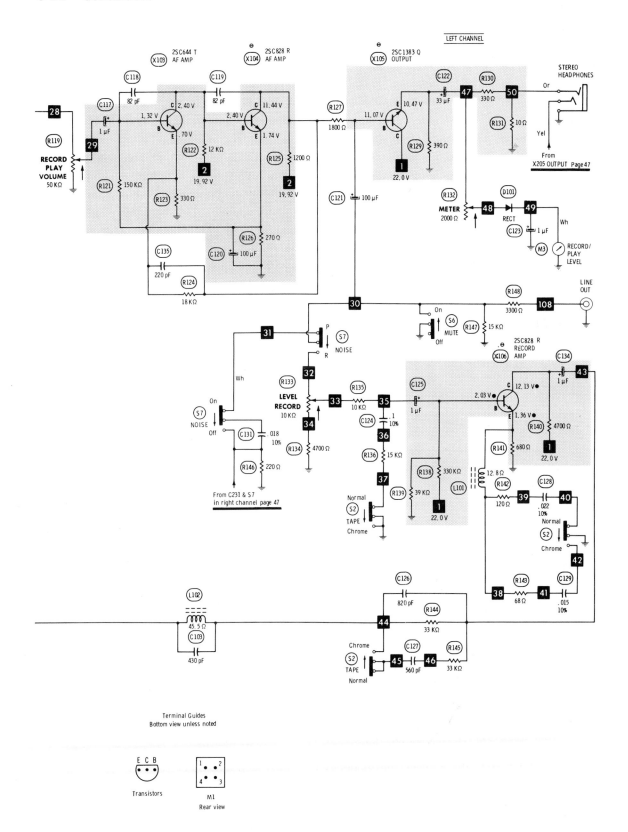

5-23 **Coupling circuit of Fig. 5-22. with S_1 in playback and S_2 in normal positions.**

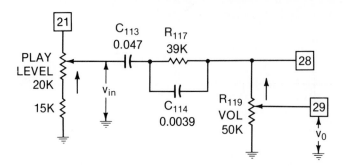

5-24 **Analyzing the tone control.**

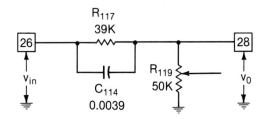

(A) Part of actual circuit.

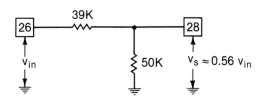

(B) Equivalent circuit at 100 Hz and below.

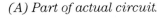
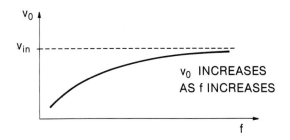

v_0 INCREASES
AS f INCREASES

(C) Response curve.

There are a few other sections of Fig. 5-22 that control frequency response. In the record position, S_7 connects the output of the AF AMP (TP 30) to the input of the RECORD AMP through TP 32. Notice the circuit connecting C_{124} and R_{136} to ground through another section of S_2. In the normal position, some of the higher frequencies are shunted to ground. This does not happen in the chrome position. Also study the components near TP 40 and TP 44. By noting the relative values of the components, you should be able to determine whether the circuits would favor amplification of higher or lower frequencies. Look at the feedback loop from TP 19, through either of two paths, to the emitter of X_{101}. These components affect the gain of the PREAMP, as well as its frequency response.

One final point to notice in the circuit is the presence of low-value capacitors from collector to base on each of the PREAMP and AF AMP transistors. These tiny capacitors roll off the high-frequency gain of these stages. They act as part of the input capacitance to each stage, as discussed in an earlier topic. As a matter of fact, these small capacitances actually "look" much larger than they really are because the stages they bridge are amplifiers. This is known as the *Miller* effect. Here a capacitor from base to collector in any common emitter stage appears larger by a factor equal to the gain of that stage. Thus, if the gain of X_{101} is 100, capacitor C_{102} looks like a 3300-pF capacitance, as seen looking into the stage from TP 12. Rolling off the highs minimizes the amplification of high-frequency noise and also reduces the tendency of a high-gain amplifier system to "peak up" or oscillate at some high frequency.

Fig. 5-25 shows the entire left channel again. This time the *signal path* in playback mode is highlighted. To test this amplifier, you would first inject a signal near the pickup head, say at TP 13. Then with an oscilloscope, you would trace the signal through the entire amplifier chain along the highlighted path. The signal amplitude should get larger as you move from left to right through each stage (depending, of course, on the settings of the play level and volume controls). One exception to the increase in amplitude is that the signal at TP 47 should be the same as that of the base of X_{105}. Why?

Here's an efficient way to begin testing if you suspect trouble in the playback mode.

1. If the power supplies are normal, set S_1 in the playback mode and adjust the play level and volume controls to midrange.

2. Inject a small audio signal into TP 12.

3. Connect an oscilloscope to the output (TP 47 or 50), and look for a signal.

4. If no output is present, move the probe to the top of the volume control, TP 28, and look for a signal.

Step 4 cuts the system in half. If no signal is present at TP 28, you can check TP 21 or the collectors of X_{101}–X_{102}. But if a normal signal is present, you know that the trouble lies somewhere in the AF AMP or OUTPUT stages. Cut the possibilities in half whenever you can, so that you will zero in on the defective stage quickly.

If no signal is present at TP 28, you can quickly move your scope to TP 21. If a signal is present at TP 21, you know that the trouble lies somewhere in the components between TP 21 and TP 28. Before removing a lot of components for measurement, you might save a lot of time by flipping S_1 and/or S_2 back and forth a few times to see what happens. For example, suppose you get no signal at TP 28 with S_2 in the normal position, but you do get a signal when S_2 is thrown to the chrome position. This should tell you that either S_2, C_{113}, R_{117}, or C_{114} is at fault. In other words, always study the schematic to see if you can eliminate some possibilities by flipping a switch or two.

Let's suppose that the complaint is no signal at the output in the playback mode. You connect a scope to TP 50, and you inject a signal of a few millivolts at an audio frequency into TP 13. You adjust the play level and volume controls to midrange, and you switch S_1 to the playback position and S_2 to the normal position. You then see no signal at TP 50. You next move the scope to TP 28 and see a normal signal. That tells you immediately that the trouble lies within the AF AMP or OUTPUT amplifier circuits.

Probably a good test to make would be to measure the collector of X_{104}. This measurement splits the defective section in half. If you have your scope set for d-c input, you can observe the d-c collector voltage as well as the a-c signal. Here is a big clue! If the d-c collector voltage of a transistor is normal, you can assume that the transistor and all of its biasing components are OK! If any biasing component was bad, the collector voltage would be abnormal. Also, if the transistor was open or shorted, the collector voltage would be abnormal. In this circuit, a normal d-c voltage at the collector of X_{104} tells you that X_{104} and X_{103} and resistors R_{121}, R_{122}, R_{123}, R_{125}, and R_{126} are probably all OK, because the two stages are directly coupled. That is, any change in the bias of one stage affects the other stage. Note that this interaction is different from the capacitively coupled circuits, such as in Fig. 5-2. In a capacitively coupled circuit, a change in d-c voltage or current in one stage does not normally affect the d-c readings in any other stage.

If the d-c readings were normal but you still measured very low or no a-c signal at the collector of X_{104}, you should check the a-c path. This would include C_{117}, C_{120}, and possibly the negative feedback components C_{135}–R_{124}.

But let's suppose that the d-c collector voltage of X_{104} was lower than normal. This could be the result of X_{104} conducting more heavily than normal. You should be able to verify this by measuring a higher than normal voltage at the emitter of X_{104}, due to more current flow through R_{126}. Now what might cause X_{104} to turn on hard? Possibly X_{104} is shorted, or perhaps its base voltage is too high, due to a problem in the previous stage. Here's a simple way to test for a shorted transistor: Temporarily short the base of X_{104} to ground, say with a clip lead, and watch its d-c collector voltage with a scope or voltmeter. If the transistor is OK, shorting its base to ground will shut it off and cause its collector voltage to rise toward the supply voltage. On the other hand, if the transistor is shorted, there won't be any change in its collector voltage when you short its base to ground.

Let's suppose in our example that shorting the base of X_{104} to ground caused the collector voltage of X_{104} to rise toward 20 V. We now know that the trouble is not X_{104}, so we measure the collector voltage and base voltage of X_{103}. If the collector voltage of X_{103} is high (which would account for X_{104} turning on too hard) and the base of X_{103} is low or open, then check R_{121}. But if the collector of X_{103} is high and its base voltage is also high (caused by higher than normal current through R_{126}), then you either have an open X_{103} or a defective R_{123}.

5-25 **Magnavox Model TE3410WA11 with playback signal path highlighted. From** *Tape Recorder Series No. 173*, **Howard W. Sams & Company, Inc.**

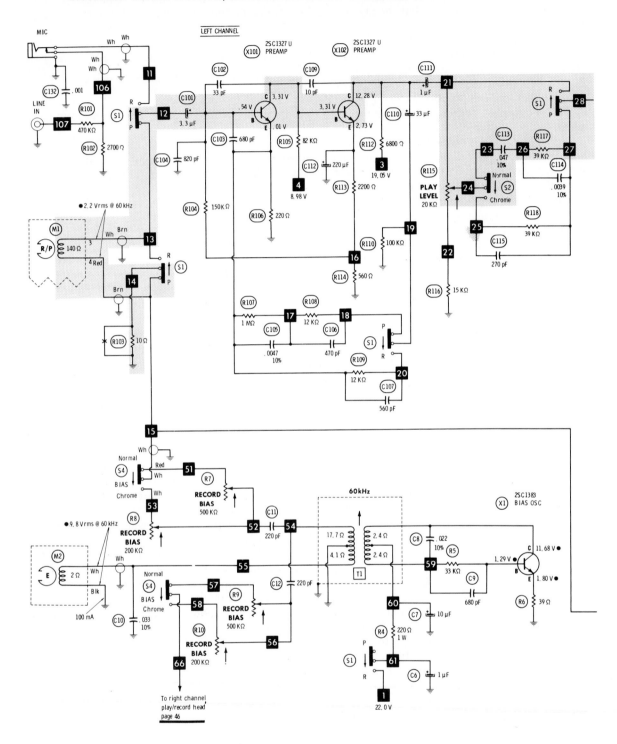

5-25 Continued.

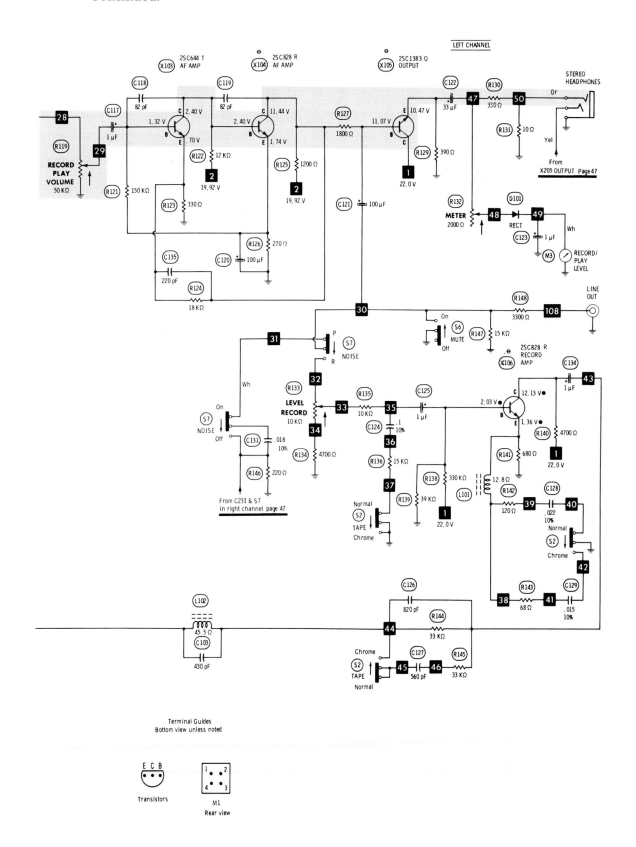

The point of the above discussion is that you can localize the defective component with just a few measurements if you try to understand how the circuit works. First make some scope checks to look for a signal, then make some voltage measurements to determine the most likely components. Then you'll only have to check a few individual components to determine the exact fault.

Another useful practice when troubleshooting is to redraw the circuit if it seems hard to understand. Often you get used to seeing the circuit drawn one particular way when studying theory out of a textbook, but when you look at an actual commercial system the circuits don't look like the ones you studied. Redrawing the circuit can be very helpful.

For example, let's examine the RECORD AMP X_{106}. There seem to be an awful lot of components around it, so it might not look like anything you are familiar with. Remember that capacitors block dc, so any components coupled to the amplifier through capacitors do not affect the bias and are only used to somehow affect the a-c signal components by altering the gain and/or frequency response. Fig. 5-26 shows the RECORD AMP of Fig. 5-25 redrawn like the simple circuits you studied in chapter 3. Notice that X_{106} is a common emitter amplifier using voltage divider bias. The voltage at the base of the transistor is determined by resistors R_{138} and R_{139}. The current through the transistor is controlled by the base voltage and the value of the emitter resistor R_{141}. Components $L_{101}, R_{142},$ and C_{128} replace the single emitter-bypass capacitor seen previously, but they do essentially the same job.

The main point is this: notice how much easier it is to understand how X_{106} works when you redraw it in a more familiar configuration. It will then be much easier to test and troubleshoot, once you see how it works.

5-26   Record amp of Fig. 5-22.

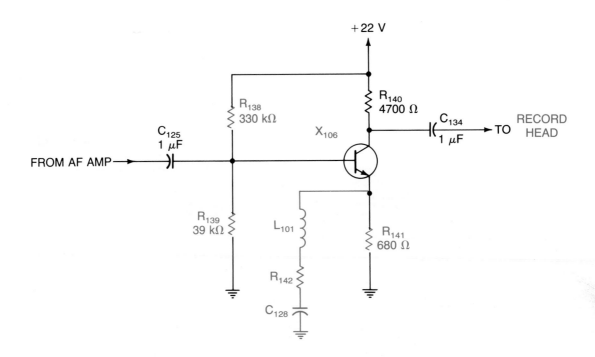

PROBLEMS

For the next two problems, refer to Fig. 5-1A.

5-1. In the circuit, assume that the a-c load r_L seen by the collector of Q_1 is 2 KΩ, r_{e1} = 25 Ω, r_{L2} = 3 KΩ, r_{e2} = 20 Ω, and v_{in} = 1.5 mV. Find the voltage gains of each stage, the overall voltage gain, and the a-c output voltage.

A_{v1} =

A_{v2} =

$A_{v(tot)}$ =

v_0 =

5-2. Repeat problem 5-1, assuming r_{L1} = 1.5 KΩ, r_{e1} = 30 Ω, r_{L2} = 2 KΩ, r_{e2} = 25 Ω, v_{in} = 2 mV.

A_{v1} =

A_{v2} =

$A_{v(tot)}$ =

For the next four problems, refer to Fig. 5-2.

5-3. While testing the amplifier, the power supply voltage reads normal. There is no output at point Y when an input signal is injected. However, the waveform at point X looks OK. What should be your next measurement?

5-4. In testing a circuit like that of Fig. 5-2, the waveform at X is clipped on positive half cycles, V_{C1} reads high, and V_{B1} reads low. Which of the following could be the cause? (a)R_1 decreased in value. (b)R_2 decreased in value. (c)R_{C1} decreased in value. (d)Q_1 shorted collector to emitter.

5-5. Referring to problem 5-3, if V_{C2} and V_{B2} both read normal, which of the following could be the cause? (a)R_3 increased in value. (b)R_{C2} decreased in value. (c)Capacitor C_c open. (d)Capacitor C_{E2} shorted.

5-6. Referring to problem 5-4, if V_{B1} reads normal, but all other measurements are as indicated, which of the following could be the cause? (Choose two.) (a)R_{C1} increased in value. (b)R_{C1} decreased in value. (c)R_{E1} increased in value. (d)R_{E1} decreased in value.

5-7. If the size of the coupling capacitor between two transistor amplifier stages is increased, the low-frequency response will be (a)better. (b)worse.

5-8. If the input resistance to a capacitively coupled transistor amplifier stage is made higher, the low-frequency response will be (a)better. (b)worse.

For the next two problems, refer to Fig. 5-27.

5-27   Circuit for problems 5-9 and 5-10.

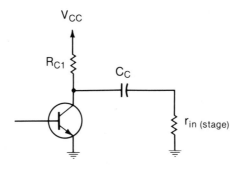

5-9. If R_{C1} is 2 KΩ and $r_{in(\text{stage})}$ is 1 KΩ, find a value for the coupling capacitor to make the lower 3-dB cut-off frequency 50 Hz.

$C_c =$

5-10. If R_{C1} is 1.5 KΩ and $C_c = 2$ μF, what must be the minimum value of $r_{in(\text{stage})}$ to make $f_1 = 40$ Hz?

$r_{in(\text{stage})} =$

5-11. When viewing the output of an amplifier at high frequencies, the shunting capacitance of the scope lead can affect the output amplitude. For this reason, you should use an oscilloscope probe (lead) with a—(a)high, (b)low—capacitance.

5-12. In order to get as wide a bandwidth as possible, individual stages should be designed with—(a)moderately low, (b)very high—gain. (*Hint:* Consider whether high-gain circuits need large or small values for R_C. Then study equation 5-3 to see the effect of R_C on high-frequency response.)

5-13. Negative feedback—(a)increases, (b)decreases—gain and also—(c)increases, (d)decreases—distortion.

5-14. Negative feedback makes an amplifier—(a)more, (b)less—temperature stable and also—(c)more, (d)less—dependent on transistor characteristics.

For the next four problems, refer to Fig. 5-22.

5-15. Equation 5-4 states that the gain of an amplifier using feedback across two stages is $A_v = R_F/R_E + 1$. Which two resistors represent R_F and R_E in the AF AMP circuit (X_{103} and X_{104})?

5-16. What is the low-frequency gain of the AF AMP? (Neglect the effect of bypass capacitors.)

$A_v =$

5-17. What is the approximate gain of the preamp (X_{101} and X_{102}) at lower frequencies when S_1 is in the record position? (Neglect the effect of bypass capacitors.)

$A_v =$

5-18. Considering the effect of C_{107} in the PREAMP, the gain of the circuit—(a)increases, (b)decreases—at higher frequencies.

For the next four problems, refer to Fig. 5-28.

5-28   Circuit for problems 5-19 through 5-22.

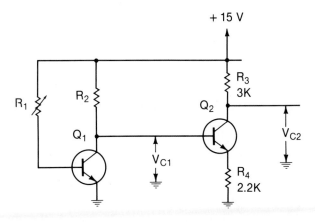

5-19. If R_1 is adjusted so that $V_{C1} = 4$ V, what is the collector current of Q_2?

$I_C =$

5-20. Referring to problem 5-19, what is the collector voltage of Q_2?

$V_c =$

5-21. If R_1 is made smaller, I_{C2} will (a)increase. (b)decrease.

5-22. If Q_1 opens, V_{C2} will (a)go up. (b)go down. (c)remain the same.

For the next four problems, refer to Fig. 5-22.

5-23. If R_{121} in the AF AMP opens, X_{103} will—(a)turn on harder, (b)turn off—causing its collector voltage to go (c)up. (d)down.

5-24. If R_{121} opens, the d-c collector voltage of X_{104} will (a)go up. (b)go down. (c)not be affected.

5-25. If C_{120} opens, the gain of the AF AMP will be (a)higher. (b)lower.

5-26. If C_{118} in the AF AMP becomes disconnected, the d-c collector voltage of X_{103} will—(a)go up, (b)go down, (c)remain the same—but the high-frequency gain will (d)go up. (e)go down. (f)remain the same.

For the next four problems, refer to Fig. 5-29.

5-29 Circuit for problems 5-27 through 5-30.

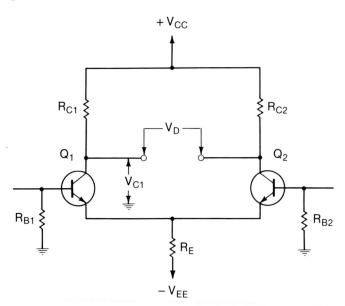

5-27. If $V_{EE} = -12$ V and $R_E = 3$ KΩ, what is the approximate value of the emitter current through Q_1? (Assume that $R_{B1} = R_{B2}$ and the two transistors are identical.)

$I_E =$

5-28.   Referring to problem 5-27, what is the value of V_{C1}, assuming $R_{C1} = R_{C2}$ = 4 KΩ and V_{CC} = +12 V? What is the value of V_D?

$V_{C1} =$

$V_D =$

5-29.   Assume $R_{C1} = R_{C2} = 4$ KΩ and the d-c emitter currents are each set at 1.5 mA. If a difference signal of 0.5 mVDC is applied between the two bases, what is the value of the differential output signal V_D?

$V_D =$

5-30.   Referring to problem 5-29, if the temperature rises, the differential output signal V_D will (a)go up. (b)go down. (c)remain the same.

For the next two problems, refer to Fig. 5-30.

5-30   Circuit for problems 5-31 and 5-32.

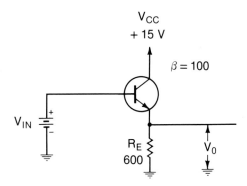

5-31.   What is the resistance seen by the signal source in V_{in} looking into the base of the transistor?

$r_{in} =$

5-32.   Assuming $V_{BE} = 0.7$ V, if $V_{in} = 6$ V, what is the value of V_0? If V_{in} goes up to 8 V, what is the value of V_0? If a 0.25-VAC signal is superimposed on (placed in series with) V_{in}, what will be the a-c output signal voltage?

$V_0 =$

$V_0 =$

$V_0 =$

For the next ten problems, refer to Fig. 5-22.

5-33.   If transistor X_{105} has a β of 100, what is the approximate value of input resistance seen looking into its base (to the right of R_{127})? Ignore the meter circuit in the emitter lead.

$r_{in} =$

5-34. What is the value of the dc flowing through X_{105}? (*Hint:* Use the voltage readings given on the schematic.)

$I_E =$

5-35. If C_{122} shorts, the current through X_{105} will (a)increase. (b)decrease. (c)remain the same.

5-36. What is the value of the emitter current through the record amp X_{106}?

$I_E =$

5-37. If R_{139} opens, the collector voltage of X_{106} will (a)go up. (b)go down. (c)remain the same.

5-38. With an audio signal injected into TP 12, no output is seen on an oscilloscope connected to TP 47, but the signal at TP 28 looks OK. R_{115} and R_{119} are set at midrange. Which of the following would be a good next test? (a)Connect an oscilloscope to the collector of X_{102}. (b)Connect an oscilloscope to the collector of X_{104}. (c)Check the resistance of the volume control. (d)Check if C_{121} is shorted.

5-39. Suppose no signal is seen at TP 28, but there is a signal at TP 21. When the record/playback switch is changed to the record position, a signal is seen at TP 28. Which of the following is the most likely cause? (a)Defective record amp stage X_{106}. (b)Defective preamp stage X_{101} or X_{102}. (c)Open R_{115}. (d)Open R_{119}.

5-40. Suppose the signal at TP 21 is weaker than normal, and d-c voltage at the collector of X_{102} reads about 12 V. Which of the following is the most likely cause? (a)Open or disconnected C_{112}. (b)Shorted R_{114}. (c)Shorted C_{103}. (d)Open R_{105}.

5-41. Referring to problem 5-40, which of the following could also be a cause? (a)Open C_{110}. (b)Open R_{112}. (c)Shorted C_{105}. (d)Open R_{104}.

5-42. In the record mode, a normal signal is seen at TP 35, but the signal at TP 43 is weaker than normal. The collector voltage of X_{106} reads about 12 V. Which of the following is the most likely cause? (a)R_{140} changed value. (b)Open L_{101}. (c)Open R_{138}. (d)Defective X_{106}.

EXPERIMENT 5-1   CASCADED AMPLIFIER

In this experiment, you will study a two-stage capacitively coupled amplifier. You will measure its gain with and without feedback. The same circuit will be used for experiments 5-2 and 5-3, so do not disassemble it when you finish this experiment.

EQUIPMENT
- 12-VDC power supply
- audio signal generator
- oscilloscope
- volt–ohmmeter
- (2) NPN silicon transistors (low-power general purpose)

- (2) 75-KΩ, ½-W ± 5% resistors
- (2) 12-KΩ, ½-W ± 5% resistors
- (1) 10-KΩ, ½-W ± 5% resistor
- (2) 4.7-KΩ, ½-W ± 5% resistors
- (3) 1-KΩ, ½-W ± 5% resistors
- (1) 10-Ω, ½-W ± 5% resistor
- (1) 25-KΩ pot
- (2) 20-μF, 25-WVDC capacitors
- (2) 5-μF, 25-WVDC capacitors
- (1) 0.1-μF, 25-WVDC capacitor

E5-1 Circuit for experiments 5-1 through 5-3.

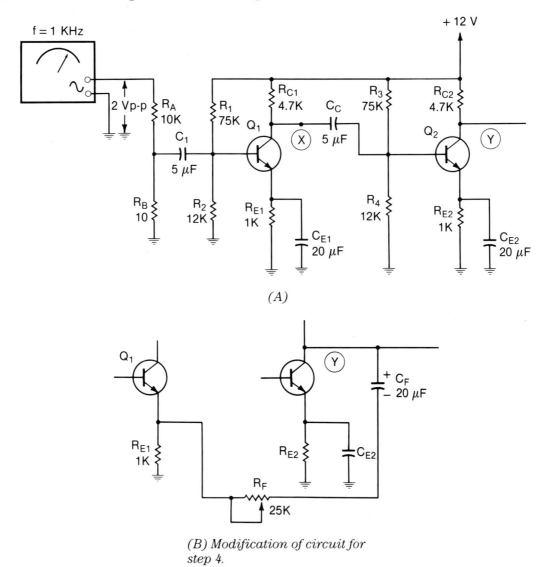

(A)

(B) Modification of circuit for step 4.

PROCEDURE

1. Build the circuit of Fig. E5-1A. Apply d-c power, but do not apply an a-c signal yet. Measure the d-c collector voltage of each transistor to make sure that both stages are operating normally.

$V_{C1} =$

$V_{C2} =$

If both collector voltages measure between 6 and 8 V from ground, the circuit is probably working normally, so go on to the next step. Otherwise recheck your wiring.

2. Now connect the signal generator to the input. Set the amplitude of V_s to 2 Vp-p at a frequency of about 1 KHz. This will make $v_{in} = 2$ mV as a result of the attenuator R_A–R_B.

3. Measure the a-c signal at each collector.

$v_{01} =$

$v_{02} =$

If your circuit is working correctly, v_{01} should be much larger than v_{in}, and v_{02} should be much larger than v_{01}. The output signal v_{02} should not be clipped. Determine the gain at the first stage.

$A_{v1} = v_{01}/v_{in} =$

Determine the gain at the second stage.

$A_{v2} = v_{02}/v_{01} =$

Now find the overall gain.

$A_{v(tot)} = v_{02}/v_{in} =$

4. Now test the amplifier with feedback. Shut off power and remove the 20-µF capacitor C_{E1} from the emitter of Q_1. Then connect the capacitor in series with the feedback resistor (25-KΩ pot) from the collector of Q_2 to the emitter of Q_1. See Fig. E5-1B. Adjust the pot to its maximum resistance, and apply power.

5. Remove the input signal attenuator R_A–R_B from the circuit, and connect the signal generator directly to the input capacitor. Adjust the input signal amplitude to 0.1 Vp-p. Measure the output signal at point Y.

$v_{02} =$

Calculate the gain of the circuit.

$A_v = v_{02}/v_{in} =$

6. Adjust the feedback pot to its lowest value and measure v_0.

$v_0 =$

What is the gain now?

$A_v =$

By adjusting R_F higher and lower, the output signal amplitude increases and decreases. R_F makes an excellent gain control.

Go on to the next experiment.

EXPERIMENT 5-2 TROUBLESHOOTING A TWO-STAGE AMPLIFIER

Use the circuit in Fig. E5-1 to try some troubleshooting techniques. Start by making measurements to the amplifier while it is working normally. Then make changes to a few components to see how various waveforms and voltage readings are affected.

EQUIPMENT

Use the same equipment as in experiment 5-1.

PROCEDURE

1. Apply power to the circuit of Fig. E5-1, and inject a small signal. Adjust v_s to 2 V as in experiment 5-1. (Feedback pot R_F should not be connected.)
2. Record the amplitudes of the a-c waveforms at points X and Y, v_{b2}, and v_{b1} in row 1 of Table E5-2.

Table E5-2

Condition	X v_{c1}	Y v_{c2}	V_{c1}	V_{B1}	v_{b1}	V_{C2}	V_{B2}	v_{b2}
1 normal								
2 R_1 open								
3 R_4 open								
4 C_{E1} shorted								
5 Q_2 open								

3. Next measure the d-c voltage at the collectors and bases, and record these in row 1 of the table for V_{C2}, V_{C1}, V_{B2}, V_{B1}. The readings just recorded in row 1 are the *normal* values for your amplifier. Compare these readings to abnormal readings, taken in later steps.
4. Disconnect R_1 from the circuit, and repeat all measurements. Record your measured values in row 2 of Table E5-2.
5. Reconnect R_1 back into the circuit, and remove R_4 from the circuit. Repeat all measurements, recording your values in row 3 of the table.
6. Replace R_4 in the circuit, and connect a clip lead from the emitter of Q_1 to ground. This simulates a shorted capacitor C_{E1}. Repeat all measurements, recording your values in row 4 of the table.
7. Remove the short from the circuit. Disconnect the collector of Q_2 from the circuit. This simulates an open transistor. Repeat all measurements, recording the values in row 5 of the table.

OPTIONAL

To see how well you can troubleshoot the amplifier, wrap each resistor and each capacitor with masking tape so you cannot tell the value. Then have a lab partner or your instructor substitute one resistor or capacitor with one of a different value, say half or two times as large. Have this substitute resistor masked also. Then inject a signal into the amp, and see if you can locate the bogus part with a minimum number of checks. (Use the flowchart. Do not just measure every resistor with an ohmmeter!) You can also have them substitute known defective transistors or capacitors in the circuit to see if you can locate these.

1.   A change in resistance value in the first stage affects the d-c readings in (a)the first stage only. (b)the second stage only. (c)both stages.

2.   A change in resistance in the first stage affects the a-c readings in (a)the first stage only. (b)the second stage only. (c)both stages.

3.   A change in resistance in the second stage affects the d-c readings in (a)the first stage only. (b)the second stage only. (c)both stages.

4.   Shorting C_{E1} causes Q_1 to—(a)turn off, (b)turn on hard—as can be seen by the fact that V_{C1} went—(c)up, (d)down—due to—(e)more, (f)less—current flow through R_{C1}.

5.   Opening R_4 causes Q_2 to (a)turn off. (b)turn on hard.

6.   Opening R_1 causes Q_1 to (a)turn off. (b)turn on hard.

7.   Refer to the flowchart of Fig. 5-3 and consider the readings in row 5 of Table E5-2, the open transistor Q_2 would have been found by taking the—(a)YES, (b)NO— route from the box labeled "WAVEFORM X OK?"

8.   Referring to question 7, when reaching the block labeled "V_{C2} OK?" the— (a)YES, (b)NO—route would be followed.

9.   Referring to question 8, a good indication of an open transistor is when (a)V_C is normal. (b)V_C is much lower than normal. (c)V_C is equal to V_{CC}.

10.   If Q_2 were shorted, V_{C2} would have been (a)normal. (b)much lower than normal. (c)equal to V_{CC}.

EXPERIMENT 5-3   FREQUENCY RESPONSE

Now you will measure the frequency response of a two-stage capacitively coupled amplifier. You will make changes to the circuit to see what factors affect the high- and low-frequency response.

EQUIPMENT

Use the same equipment as in Experiment 5-1.

PROCEDURE

1.   Using the circuit of Fig. E5-1, connect an oscilloscope to the amplifier output at point Y. Apply power, and adjust input signal v_s so that v_{02} measures 1 Vp-p at a frequency of 1 KHz. Measure v_s.

$v_s =$

Be sure to keep v_s at this same value for all measurements. You might have to readjust the signal generator amplitude control when you change to different frequencies.

2.   Change the generator frequency to 40 Hz, but keep v_s at the same value as measured in step 1. Record the value of output voltage in the space provided in Table E5-3 in row 1.

Table E5-3

Frequency, Hz

Run	40	100	200	400	1K	2K	4K	10K	20K	40K	100K	
1												$V_{02},$ Vp-p
2												
3												

3. Change the generator frequency to 100 Hz, and measure v_{02} again. Record it in the space provided in row 1.

4. Make output voltage measurements at each frequency listed in Table E5-3. Be sure to keep the amplitude of v_s constant. Fill in row 1 only.

E5-3 Semilog graph.

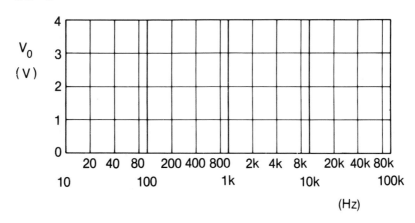

5. On the axes of Fig. E5-3, mark each point from Table E5-3 row 1 with a dot. Then connect these dots with a smooth curve. This is the frequency response curve of your amp, and it should look similar to that of Fig. 5-7.

6. Using the curve, find the 3-dB half-power points.

$f_1 =$

$f_2 =$

(It's possible that f_2 is beyond 100 KHz and won't appear on your curve.)

7. Replace capacitor C_c in your circuit with a 0.1-μF capacitor.

8. Measure v_{02} at all points listed in Table E5-3 row 2. Be sure to keep v_s at the same value as in step 1.

$v_{02} =$

9. Plot each of the points from Table E5-3 row 2 onto the axes of Fig. E5-3 marking each point with a small x. Then connect these points with a smooth curve using a different colored pen than in step 5. Using this curve, find f_1 and f_2.

$f_1 =$

$f_2 =$

10.   Replace C_c with the original value of 5 μF, and connect a 0.1-μF capacitor from point X to ground.

11.   Measure v_{02} at each point listed in Table E5-3, and record these values in row 3.

12.   Plot the response curve again, this time using a small triangle to locate each point from row 3. Then connect these points with a smooth curve using a third color. Using this curve, find f_1 and f_2.

$f_1 =$

$f_2 =$

QUIZ

1.   What is the bandwidth of your circuit using the measurements of row 1?

2.   What is the effect on the low-frequency response when the coupling capacitor is made smaller?

3.   What is the effect on the high-frequency response when the coupling capacitor is made smaller?

4.   What is the effect on the low-frequency response when a small capacitance is shunted from the collector of Q_1 to ground, as in step 10?

5.   Referring to question 4, what is the effect on the high-frequency response?

EXPERIMENT 5-4   DIRECT-COUPLED AMPLIFIER

In capacitively coupled amplifiers, we saw that changes in d-c values in one stage do not affect the d-c readings in other stages. This is not the case in direct-coupled amplifiers. You will now build and test a commonly used direct-coupled amplifier circuit. Then you will add an emitter follower to the circuit to see how it prevents loading.

EQUIPMENT

- 12-VDC power supply
- oscilloscope
- audio signal generator
- (3) NPN silicon general purpose transistors
- (1) 33-KΩ, ½-W ± 5% resistor
- (3) 10-KΩ, ½-W ± 5% resistors
- (1) 7.5-KΩ, ½-W ± 5% resistor
- (1) 2-KΩ, ½-W ± 5% resistor
- (1) 1-KΩ, ½-W ± 5% resistor
- (1) 510-Ω, ½-W ± 5% resistor
- (1) 10-Ω, ½-W ± 5% resistor
- (1) 100-μF capacitor
- (1) 5-μF capacitor

E5-4 Circuit for experiment 5-4.

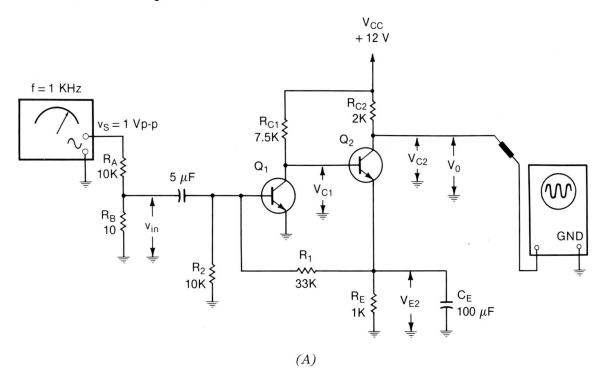

(A)

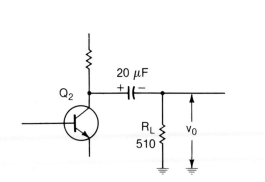

(B) R_L connected to Q_2.

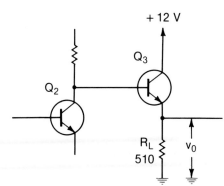

(C) Emitter follower connected to Q_2.

PROCEDURE

1. Build the circuit of Fig. E5-4A. Apply power and measure the d-c voltages at the two collectors and at the emitter of Q_2.

$V_{C1} =$

$V_{C2} =$

$V_{E2} =$

If your circuit is operating normally, V_{C1} should read between 3 and 4 V, and V_{C2} should read near 6 V.

2. To see how d-c values of Q_1 affect Q_2, shunt R_2 with another 10 KΩ, and measure the voltages again.

$V_{C1} =$

$V_{C2} =$

$V_{E2} =$

Remove the extra 10-KΩ resistor.

3. Next, adjust v_s to 0.5 Vp-p at a frequency of about 1 KHz. This makes $v_{in} = 0.5$ mV. Measure v_0.

$v_0 =$

Calculate the voltage gain of the amplifier.

$A_{v(\text{tot})} = v_0/v_{in} =$

4. Now let's see how loading affects the output amplitude. Connect a 510-Ω resistor to the collector of Q_2 through a 20-μF capacitor, as shown in Fig. E5-4B. Measure v_0.

$v_0 =$

5. Now disconnect the load, and replace the coupling capacitor with emitter follower Q_3, as shown in Fig. E5-4C. Measure v_0.

$v_0 =$

QUIZ

1. Refer to Fig. E5-4. The emitter voltage of Q_2 is—(a)independent of, (b)controlled by—the collector voltage of Q_1.

2. If Q_1 is made to conduct less, V_{C1} (a)goes up. (b)goes down. (c)remains the same.

3. Referring to question 2, this causes V_{E2} to (a)go up. (b)go down. (c)remain the same.

4. When V_{E2} goes up, Q_1 is forced to conduct—(a)more, (b)less—due to the feedback path R_1–R_2.

5. If the temperature rises, both transistors will conduct—(a)much more, (b)much less, (c)about the same—because of the action of the feedback network.

6. Connecting a relatively low-resistance load across the output, as in step 4—(a)increased, (b)decreased, (c)did not affect—the gain.

7. Using the emitter follower in step 5—(a)increased, (b)decreased—the output signal from that of step 4 because the emitter follower (c)has high gain. (d)prevented loading of Q_2.

8. Comparing the output amplitude in step 5 with that in step 3, what is the approximate gain of the emitter follower?

Power Amplifiers

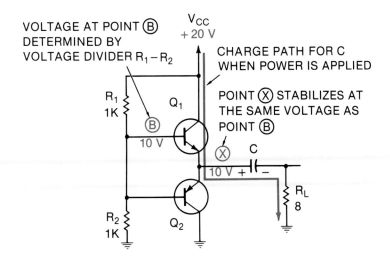

Whenever an amplifier must work into a low-resistance load, such as a speaker, or has to deliver more than a few milliwatts to a load, it is considered to be a *power amplifier*. In this chapter, we will look at a few common power amplifier circuits.

COMPLEMENTARY SYMMETRY OUTPUT STAGE

The *complementary symmetry* amplifier gets its name from the fact that it uses complementary transistors in the output stage. That is, the output stage is built with one NPN and one PNP-type of transistor.

Figure 6-1 shows a simplified version of a complementary symmetry amp. Notice that the PNP-type (Q_2) is drawn upside down. The transistor is biased normally, because the collector for Q_2 is made more negative than its emitter.

Because of the voltage divider R_1–R_2, the base of Q_1 is held at a d-c voltage of ½ V_{CC}, or 10 V. When power is first turned on, there is no charge on capacitor C. Therefore, the emitter of Q_1 is effectively returned to ground through the load resistor R_L. (R_L represents the speaker in an audio power amp.) So Q_1 turns on hard, drawing current up through R_L and charging C, as shown. The voltage at point X rises as C charges, until eventually point X reaches 10 V. Point X cannot go above

6-1 **Simplified complementary symmetry power output stage.**

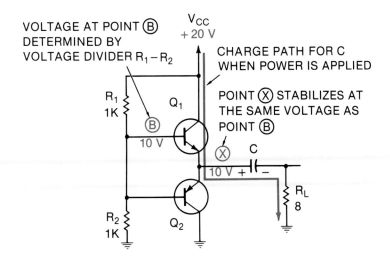

10 V. If it did Q_1 would become back biased and stop conducting. The whole point is that shortly after power is applied, point X reaches a voltage approximately equal to that at point B. Once C is charged, both transistors remain cut off.

Now let's see what happens when we apply an input signal. As shown in Fig. 6-2A, Q_1 conducts on the positive half of the input cycle. This happens because its base is being driven more positive than its emitter. Notice that Q_1 acts like an *emitter follower* driving load resistor R_L. That is, as the base of Q_1 goes positive, the emitter of Q_1 follows the base, pulling the left side of C more positive. This causes a charging current to flow through Q_1, C, and R_L as shown. Since Q_1 acts like an emitter follower, the voltage developed across R_L during the positive half cycle looks the same, and has the same amplitude, as the input signal at point B.

When a transistor is biased so that it conducts only half (180°) of the input cycle, that transistor is said to be operated *Class B*. The amplifiers in earlier chapters were all biased *Class A*, because they conducted the entire 360° of the input cycle.

On the negative alternation of the input signal, Q_1 turns off and Q_2 turns on, because point B is driven *negative* with respect to point X. Current flow through C and R_L is in the opposite direction from that on the positive alternation, as shown in Fig. 6-2B. In this way, current flow through R_L is an *alternating* current whose waveform looks like that of Fig. 6-2C.

The peak amplitude of voltage across R_L can be equal to ½ V_{CC} or 10 V. In other words, the peak-to-peak voltage across R_L can be equal to V_{CC}, if the input signal amplitude is large enough to drive each transistor alternately into saturation.

6-2 Complementary transistors conduct on alternate half cycles.

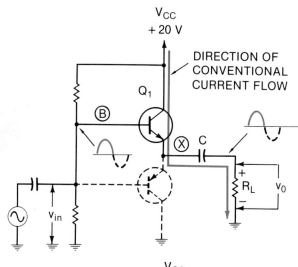

(A) Q_1 conducts on positive half cycle.

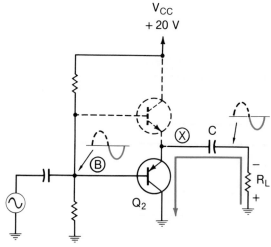

(B) Q_2 conducts on negative half cycle.

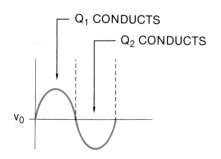

(C) Composite output signal across R_L.

The peak amplitude of current I_C through either transistor can be found by using the peak voltage divided by R_L. That is,

$$I_{C(\text{peak})} = \frac{\frac{1}{2} V_{CC}}{R_L} \qquad \text{6-1}$$

Similarly, we find the power delivered to the load by using the rms value of voltage developed across the load.

$$V_{\text{rms}} = 0.707 \, V_P = 0.707 \times \frac{1}{2} \, V_{CC} \qquad \text{6-2}$$

Then

$$P_{\text{load}} = \frac{V_{\text{rms}}^2}{R_L} \qquad \text{6-3}$$

Since these power transistors are operated Class B, the power dissipated by each transistor is actually less than $\frac{1}{4}$ of the load power, when delivering full power to the load. That is,

$$P_D = 0.25 \, P_{\text{load}} \qquad \text{6-4}$$

where P_D is the power dissipated in each transistor.

EXAMPLE 6-1 In the circuit of Fig. 6-2, if $V_{CC} = 20$ V and $R_L = 8 \, \Omega$, what is the peak current through Q_1? How much power can be delivered to R_L? What should be the minimum power rating of the transistors? (Assume that the input signal is large enough to saturate each transistor.)

SOLUTION Since the peak voltage across R_L is 10 V,

$$I_{\text{peak}} = \frac{V_p}{R_L} = \frac{10}{8} = 1.25 \text{ A}$$

To calculate power, we use the rms voltage across R_L.

$$V_{\text{rms}} = 0.707 \times V_{\text{peak}} = 0.707 \times 10 \text{ V} = 7.07 \text{ V}$$

Then

$$P_L = \frac{V^2}{R_L} = \frac{7.07^2}{8} = 6.25 \text{ W}$$

The minimum rating should be

$$P_D = 0.25 \times 6.25 = 1.56 \text{ W}$$

Coupling capacitor C is usually very large, hundreds, or even thousands, of microfarads. Its reactance must be small compared to R_L at low audio frequencies.

REDUCING CROSSOVER DISTORTION

Although the signal developed across R_L is an a-c signal, it is somewhat distorted. Remember that a silicon transistor needs about 0.7 V of forward bias before it begins to conduct. Fig. 6-3 shows the result of keeping both transistors cut off with no signal applied. Notice that Q_1 does not begin to conduct until V_{in} reaches about 0.7 V on the positive alternation. Likewise, Q_2 does not begin to conduct until V_{in} reaches about -0.7 V on the negative alternation. The resulting composite current waveform through R_L looks like that shown in Fig. 6-3B. The wrinkle in the output waveform as it crosses zero is called *crossover distortion*. It adds odd harmonics to the output that were not present at the input.

6-3   Crossover distortion caused by transistors being normally cut off.

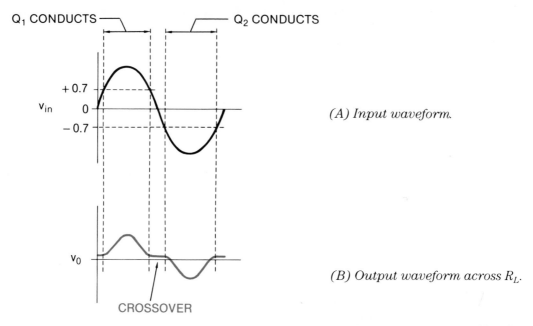

(A) Input waveform.

(B) Output waveform across R_L.

One way to minimize crossover distortion is to forward bias each transistor *slightly*. Fig. 6-4 shows one simple way to do this. Diode D_1 is placed in series with voltage divider R_1–R_2, in such a way that it is forward biased. Since the bases of Q_1 and Q_2 are connected to opposite ends of the diode, there is approximately 0.7-V difference between the bases. Effectively each transistor acts as though it has about 0.35 V of forward bias. This is not enough forward bias to really turn the transistors on very hard, but the input signal does not have to go quite as far in either direction to make Q_1 or Q_2 conduct. The net effect is that the distortion is reduced. Typically, this "idle" current through each output transistor is about 5 mA.

6-4   Diode D_1 keeps transistors slightly forward biased to reduce crossover distortion.

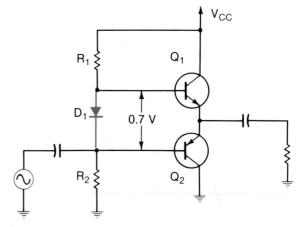

Since the transistors in this new circuit conduct for slightly more than 180°, but less than 360°, they are said to biased *Class AB*.

The input signal is applied to the lower end of D_1, but D_1 does not rectify the input. D_1 *remains forward biased all the time*. It simply acts as if it were a 0.7-V battery between the two bases. D_1 does not decrease the amplitude of the a-c signal applied to Q_1, because being forward biased makes it look like a short circuit to ac.

ADDING A DRIVER TO THE COMPLEMENTARY SYMMETRY STAGE

We can add a driver stage to the complementary symmetry amp so that it does not need quite as large an input signal to develop full output. Fig. 6-5 shows transistor Q_3 replacing voltage–divider resistor R_2 of Fig. 6-4. Q_3 amplifies input signal v_{in} and applies it to the bases of Q_1 and Q_2.

6-5   **Adding a driver to the complementary symmetry stage.**

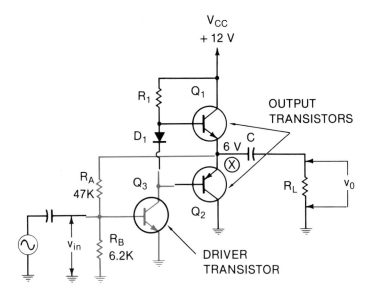

Notice that Q_3 is directly coupled to the power transistors, which gives good low-frequency response. However, as you know, direct-coupled amplifiers are susceptable to temperature instability drift. But this circuit makes use of a clever feedback biasing arrangement. Resistors R_A and R_B provide Q_3 with a form of voltage–divider bias, but R_A does not connect directly to V_{CC}. Instead, R_A connects to the junction of the emitters of the power transistors.

Here's how it works. When power is first applied, Q_3 is cut off. Therefore R_1 pulls the base of Q_1 toward V_{CC}. Of course the emitter of Q_1 (point X) follows its base positive. As point X goes positive, forward-biasing voltage is applied to Q_3, through R_A. As a result, Q_3 begins to turn on. The ratio of R_A to R_B is such that approximately 0.7 V will appear across R_B when point X reaches ½ V_{CC}. If point X tries to go too high, Q_3 will turn on harder. This, in turn, will pull the bases of Q_1 and Q_2 lower and bring point X back down again. This circuit maintains temperature stability in much the same manner as did the direct-coupled amplifier of Fig. 5-14.

Fig. 6-6 is a General Electric Model M8710A tape recorder. The highlighted section is a complementary symmetry amp, like that of Fig. 6-5. The circuit is drawn slightly differently, so study it until you recognize the similar components.

Notice the 1-Ω resistors in the emitter leads of the power transistors Q_4 and Q_5. These serve two purposes. First, they help equalize the characteristics of the two transistors, making each of them conduct approximately the same peak current even if the β of the two differ. Second, the resistors act as safety devices, limiting the transistor currents, in the event that the two overheat and tend to thermally run away. The 1-Ω resistors might even burn open, like fuses, thus protecting Q_4 and Q_5 if the transistors turn on and stay on.

One other component to notice in Fig. 6-6 is capacitor C_{10} from a tap (TP 16) on the biasing resistors R_{14}–R_{15} to the output at TP 17. This capacitor increases the input impedance seen looking into the base of Q_4 by a technique known as "*Bootstrapping*." Here's how bootstrapping works. As Q_4 is driven on harder by a positive-going signal applied to its base, its emitter follows positive also. Capacitor C_{10} is so large that it cannot discharge significantly at the frequency being amplified. Therefore, C_{10} remains charged to the voltage difference that was between points TP 16 and 17 before the input signal was applied. In other words, C_{10} acts like a small d-c source (battery) between TP 16 and 17. As TP 17 goes positive, it forces TP 16 positive also. This action keeps the voltage across R_{15} essentially constant.

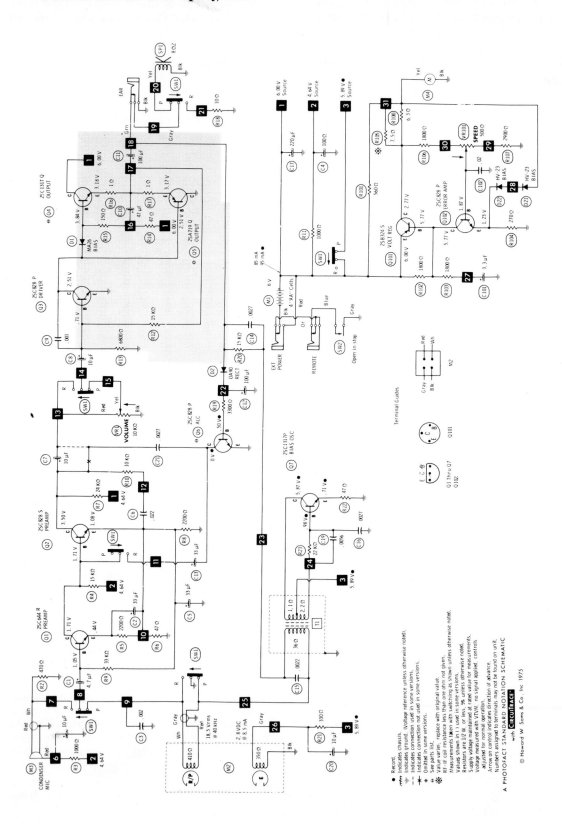

Now if the voltage across R_{15} is constant (no a-c component), there is no a-c current flow through R_{15}. This means that as far as the signal source is concerned, R_{15} looks like an open circuit to ac.

Remember from Chapter 5 that the impedance seen looking into a stage is the combination of $r_{in(\text{base})}$ in parallel with any biasing resistors. Since R_{15} is a rather low-resistance value, it would make the a-c input resistance to the final stage quite low, if it were not for C_{10}. The a-c resistance, seen by the collector of Q_3, affects the voltage gain of Q_3. So by making R_{15} look like an open circuit to ac, the a-c load seen by Q_3 is higher. Therefore, the gain of Q_3 is higher. If C_{10} is removed, the gain of Q_3 drops substantially.

A variation of the output stage, sometimes found in high-power amplifiers, is shown in Fig. 6-7. Notice that *two separate power supplies* are used, one positive and the other negative with respect to ground. The transistors are biased such that the output (point X) is held at 0 VDC, rather than at half the power supply voltage. In this way, no coupling capacitor is needed between point X and the speaker. Point X swings above and below ground when a signal is amplified.

6-7 Output stage of complementary symmetry amp using two power supplies.

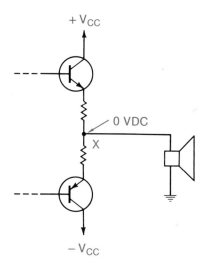

QUASICOMPLEMENTARY AMPS

Circuits like that of Fig. 6-5 are used for moderate amounts of output power, say up to a few watts. We will now take a look at a circuit that can deliver high power to a load.

Fig. 6-8 shows a typical quasicomplementary (similar to complementary) amplifier that is often used in high fidelity or guitar amplifiers. To best study its operation, first disregard the two highlighted transistors Q_4 and Q_5. Notice that the remaining circuit is almost identical to that of the complementary symmetry amps of Figs. 6-5 and 6-6. Two diodes (D_1 and D_2), rather than just one, are used to minimize crossover distortion. This is necessary to make Q_2 and Q_3 conduct slightly more, and thereby reduce crossover even better.

Transistor Q_1, called the predriver in this circuit, is biased by means of the voltage divider R_1–R_2 from output to ground, as before. Resistor R_5 in the emitter of Q_1 gives Q_1 better temperature stability.

With power turned on, but no signal applied, point X will stabilize at about ½ V_{CC}, as before. Transistors Q_2 and Q_3 are biased near cutoff, with just an idling current of a couple of milliamps flowing through them. Capacitor C_2 is used for bootstrapping, as was C_{10} in Fig. 6-6. Study this part of the circuit thoroughly until you see the similarity to Figs. 6-5 and 6-6.

Now let's look at the power output stage, transistors Q_4 and Q_5. Notice that they are not complementary types, but rather are both NPNs. They are stacked, one on top of the other, in a "totem pole" fashion.

With Q_2 and Q_3 idling near cutoff, very little current flows through R_6 and R_7. So the voltage developed across these two resistors is too small to forward bias Q_4 and Q_5. In other words, Q_4 and

6-8 Quasicomplementary power amp.

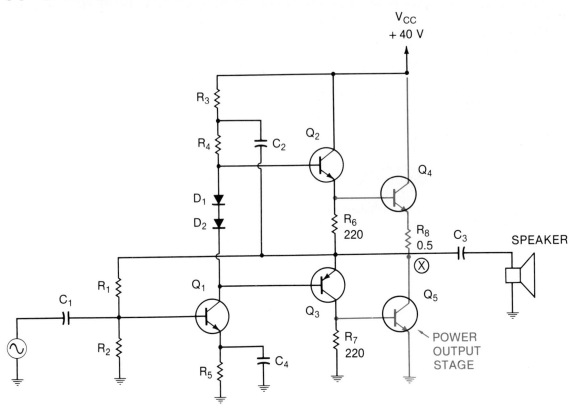

Q_5 are nearly cut off, as long as no signal is applied. When an input signal is applied, Q_2 and Q_3 conduct on alternate half cycles, as before.

Fig. 6-9A shows how power transistor Q_4 conducts on the positive half cycle. When Q_2 conducts, it develops a voltage across its emitter resistor R_6. This signal voltage forward biases Q_4, making it conduct. So Q_4 pulls the left side of C_3 toward V_{CC}, causing current to flow through the speaker.

6-9 Power transistors conduct on alternate half cycles.

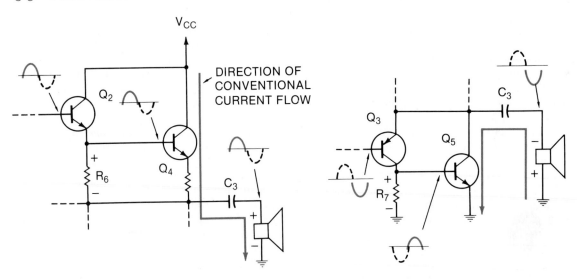

(A) Q_4 conducts on positive half cycle.

(B) Q_5 conducts on negative half cycle.

Then on the negative half cylce, Q_3 gets turned on as shown in Fig. 6-9B. Notice that resistor R_7 is in the *collector* circuit of Q_3, rather than in the emitter. Therefore, the voltage developed across R_7 is 180° out of phase with the input signal. In other words, when a negative-going signal is applied to the base of Q_3, a *positive-going* voltage is developed across its collector resistor R_7. This positive-going signal across R_7 is applied between base and emitter of Q_5, causing it to turn on. So on the negative half of the input cycle, Q_5 conducts. This pulls the left side of the C_3 toward ground, causing current to flow through the speaker as shown. In this circuit, Q_2 and Q_3 act as drivers for the high-power transistors Q_4 and Q_5. Q_2 and Q_3 do not deliver much power to the load.

Fig. 6-10 is a Kay Model 771 guitar amplifier. You should recognize the highlighted section as being a quasicomplementary amplifier. Study this circuit. You will see that it is almost identical to that of Fig. 6-8.

After studying the power amp section, look at the rest of Fig. 6-10. Notice the power supply shown directly above the power amp stage. Notice the small 0.01-μF capacitor is used for high-frequency filtering. Aluminum electrolytic capacitors do a good job of filtering out line frequency variations, but they do not work well at high frequencies, about 10 KHz or higher. This small capacitor is usually a ceramic disc or plastic dielectric type. It is very effective in filtering out high-frequency noise.

You should also be able to trace your way through the preamp section from the inputs. Notice that three input jacks sum into the first preamp stage. This allows mixing and amplification of three musical instruments simultaneously.

TRANSFORMER-COUPLED PUSH–PULL CIRCUIT

Another type of power amplifier, which has been around since early vacuum-tube days, is the transformer-coupled push–pull circuit, shown in Fig. 6-11.

Notice that the input signal is applied to the two power transistors through transformer T_1. Also notice that the center tap of the secondary of T_1 is grounded, keeping the two bases at 0 VDC. In other words, the two transistors are both normally cut off with no input signal.

Since the bases of Q_2 and Q_3 are connected to opposite ends of the secondary of T_1, the signals applied to the bases are 180° out of phase. Therefore, Q_2 and Q_3 are *alternately* turned on for each half of the input cycle. Fig. 6-12A shows how Q_2 conducts when its base is driven positive. Likewise, Fig. 6-12B shows that Q_3 conducts on the other half cycle. The collectors of the two power transistors are connected to opposite ends of the primary winding of output transformer T_2. The speaker, or other load, is connected across the secondary of T_2.

Since collector current pulses flow in opposite directions through the primary of T_2 on alternate half cycles, the net effect is the same as if there were *alternating* current flowing in the primary. Therefore, an alternating voltage is developed across the speaker.

The transformer turns ratios are chosen for impedance matching. That is, by having more turns on the primary of T_2 than on the secondary, the impedance seen by each collector is much higher than the load impedance. For this reason, it is important to try to get an exact replacement for any defective transformer. Also, to have a good low-frequency response, these transformers must be physically large and have a lot of iron. This makes them very costly and heavy. So designers want to do away with transformer coupling and favor other types, such as the complementary symmetry or quasicomplementary. However, some transformer coupled circuits are still in use.

Figure 6-13 shows an output stage of a Channel Master Model AE-6802 automobile tape player. Notice that the driver stage is coupled to the output stage with a transformer. But the output transistors are connected as a quasicomplementary pair, thus eliminating the bulky, expensive output transformer.

In this circuit, the driver stage sees a high impedance (for high gain) when looking into the primary of T_1. This is a big advantage. In addition, drift problems are less severe in the final stage because the circuit does not use direct coupling over several stages. You can appreciate that electronic equipment built for mobile use must be able to withstand wider ambient temperature variations than most equipment built for home use.

As usual, the two power transistors are biased near cutoff by a voltage divider network R_{10} through R_{13}. Diodes D_1 and D_2 provide further temperature stabilization, because the forward drop across these diodes decreases with an increase in temperature. This action reduces the bias on the transistors to prevent thermal runaway.

6-10 **Kay Model 771 guitar amp. From Jack Darr, *Electronic Guitar Amplifier Hand-book*, fourth edition, Howard W. Sams & Company, Inc.**

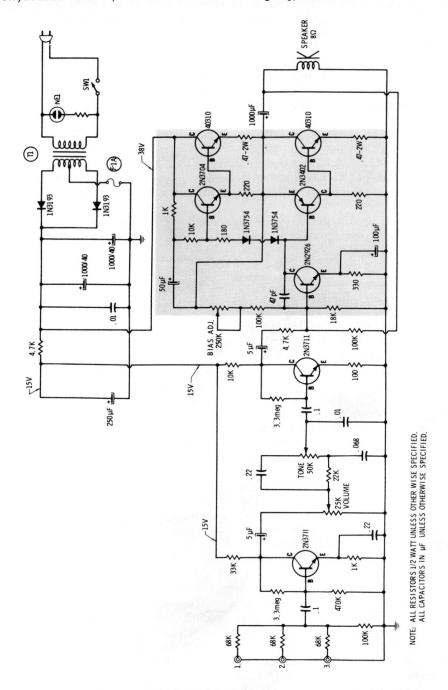

HIGH-POWER MOSFET AMPLIFIER

As stated in chapter 4, high-power field effect transistors are being used more and more in applications where previously only bipolar transistors were used. This is true in the area of high-power audio amplifiers too. Until recently, only N-channel (NMOS) transistors were available with good power-handling capabilities. But in recent years, P-channel (PMOS) transistors have become readily available at a reasonable price. PMOS transistors have made possible the design of complementary amplifiers with even better characteristics than were obtainable with bipolar types. For example, complementary MOS (CMOS) amplifiers generally have a wider bandwidth and lower

6-11 Simplified transformer-coupled push–pull circuit.

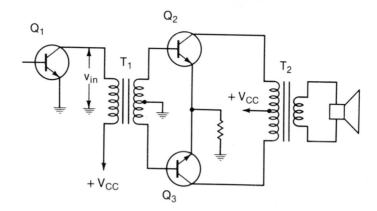

6-12 Push–pull transistors cause alternating current to flow in transformer secondary.

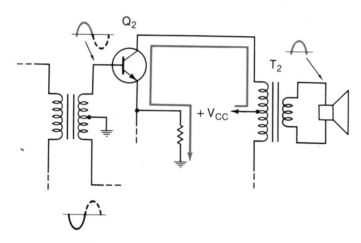

(A) Q₂ conducts on one half cycle.

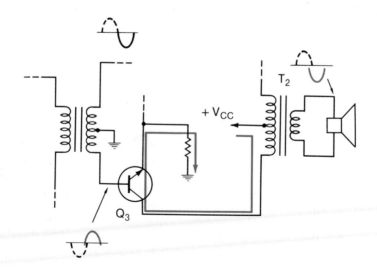

(B) Q₃ conducts on other half cycle.

6-13   Output stage of Channel Master Model AE-6802 using transformer coupling between stages and transformerless output. From *Auto Radio Series No. 236*, Howard W. Sams & Company, Inc.

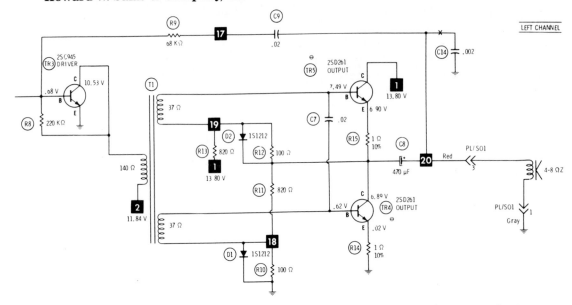

distortion level than do bipolar types. A good-quality CMOS power amp may have an output power of up to 250 W/channel, a frequency response of 5 Hz to 1 MHz, and less than 0.05% distortion. In addition, CMOS amps are easier to design, because the transistors are voltage-controlled devices, and they are not susceptible to thermal runaway, as bipolar types are.

Fig. 6-14 shows the output stage of a typical CMOS power amplifier. The biasing and voltage amplifier stages are omitted for simplicity. Note that the output stage consists of PMOS transistors Q_3 and Q_4 in parallel connected to the positive power supply, while NMOS Q_5 and Q_6 in parallel are connected to the negative rail. The speaker is connected to the drains of all MOS transistors. The point where the speaker connects should normally rest at 0 V in order to keep dc out of the speaker. Note also that it is perfectly OK to connect MOS transistors in parallel in order to get more output power than can be obtained from a single transistor. In some designs, three or even four transistors are connected in parallel to increase the total current capability.

The PMOS transistors are driven by emitter follower Q_1, while the NMOS transistors are driven by emitter follower Q_2. However, the MOS transistors are not acting as followers, but are connected in the common source configuration so that they also have some voltage gain. In this way, drivers Q_1 and Q_2 do not have to swing the full supply voltage in order to get the maximum output across the speaker. Incidentally, gate resistors R_3 through R_6 do not limit gate current, since there isn't any, but they suppress parasitic oscillations which could occur at very high frequencies.

In order to make the circuit efficient, the output transistors are normally biased at cutoff, like the transistors in previous circuits. A look at the characteristic curves of the IRF630 transistor, Fig. 6-15, shows that for gate-to-source voltages less than about $+3.5$ V, the transistor is cut off. (The IRF9630 PMOS transistor has identical curves but opposite polarities.) Normally, Q_2 is made to conduct just enough so that the voltage across it is 3.5 V. That makes the gate-to-source voltage for Q_5 and Q_6 exactly at cutoff. Similarly, Q_1 conducts just enough so that there are 3.5 V across it, thus keeping Q_3 and Q_4 at cutoff. With all of the MOS transistors cut off, the voltage at point X is zero, as it should be.

When a positive-going signal is applied to the base of Q_1, its emitter (point Y) follows its base in the positive direction. The voltage applied between the gate and source of Q_3 and Q_4 gets even smaller, so Q_3 and Q_4 remain cut off. However, this same positive-going signal applied to the base of Q_2 (shifted down a few volts) drives its emitter (point Z) in the positive direction, thus turning Q_5 and Q_6 on. This pulls point X in the negative direction. Similarly, a negative-going signal applied to the bases of Q_1 and Q_2 will turn Q_3 and Q_4 on, pulling point X in the positive direction. Thus, as with complementary symmetry amplifiers, either the pull-up transistors or the pull-down transistors are conducting with signal, but not both at the same time.

6-14 High-power FET amplifier output stage.

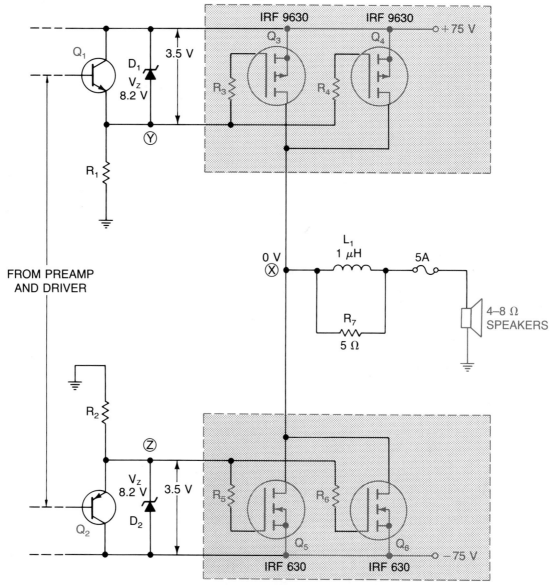

Zener diodes D_1 and D_2 are used to prevent the power transistors from being overdriven. Normally, D_1 and D_2 are both nonconducting. However, if the voltage across Q_1 ever exceeds 8.2 V, D_1 conducts and prevents the voltage at point Y from ever getting lower than 8.2 V less than the positive power supply. In this way, the gate-to-source signal applied to the PMOS transistors can never exceed 8.2 V. Similarly, D_2 prevents the voltage applied to the NMOS transistors from ever getting larger than 8.2 V and thus prevents damage to the transistors.

Coil L_1 and resistor R_7 help to minimize the capacitive effects of the load at high frequencies, and the fuse protects the speaker if any power transistor should short or be driven on continuously.

When troubleshooting this circuit, you would first make sure that the voltage at point X is zero volts. Also make sure that you measure 3.5 V across Q_1 and across Q_2. If you measure more than 3.5 V across Q_1, it could mean that the bias voltage applied to the base of Q_1 is incorrect, or it could be that Q_1 is open. If Q_1 is open, D_1 should prevent the voltage across it from exceeding 8.2 V. So if you measure 8.2 V across Q_1, you look for an open Q_1 or improper bias on it. If the fuse blows and the voltage at point X is not zero, do not reconnect the speaker until you find out why the voltage is incorrect. Can you tell what the symptom would be if the bias voltage across Q_1 was much less than 3.5 V? That is, what would you observe if you looked at the signal at point X with a scope?

6-15 Characteristics of IRF630 MOSFET.

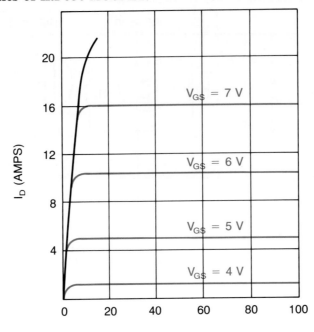

I_D (AMPS)

V_{GS} = 7 V

V_{GS} = 6 V

V_{GS} = 5 V

V_{GS} = 4 V

V_{DS} (VOLTS)

TROUBLESHOOTING POWER AMPLIFIERS

Now we will discuss some points to keep in mind while troubleshooting power amplifiers. Assume that the amp has either no output or severe distortion.

As usual, the first thing to do is look for obvious symptoms, such as signs of smoke or overheating. If the fuse in the power supply blows, even with no input signal, check for possible power supply problems. As mentioned in chapter 5, if possible, disconnect the power amp from the supply. Often the amp will have a jumper wire located at the power supply input terminal, which can be easily removed for this purpose. See Fig. 6-16A.

If the fuse does not blow, but there is no output signal, measure the d-c voltage at the amp output terminal. As shown in Fig. 6-16A, point X should measure approxiamtely ½ V_{CC} if the complementary symmetry, or quasicomplementary stage, if biased properly. If point X measures near ½ V_{CC}, assume that all of the direct-coupled states are operating normally. However, if *any one* of the direct-coupled transistors is bad, possibly all bias voltages will be incorrect. When testing a power amp with dual-power supplies, the amplifier output terminal (point X in Fig. 6-16B) should measure exactly 0 V.

Fig. 6-17 shows a simplified flowchart for troubleshooting a direct-coupled amplifier, such as a complementary symmetry or quasicomplementary type. Trace through it to see the general procedure you should follow.

The most common cause of trouble in power amps is failure of one or more of the transistors. So if you read an erroneous bias voltage, check the transistors.

The easiest way to find a bad transistor is to use an in-circuit transistor checker, such as the one shown in Fig. 6-18. All you do is connect three clips to the three-transistor terminals. The tester shows whether the transistor is good or not. Of course, you must follow the manufacturer's directions in using the instrument. It is extremely easy to use, and you do not have to unsolder the component from the circuit. In fact, the tester will tell you whether the transistor is an NPN or PNP, in case you don't know. This type of tester pays for itself in no time if you do a lot of servicing.

If you do not have an in-circuit tester, you can still check transistors in the circuit by using the ohmmeter technique discussed in chapter 3. If you measure a low resistance in one direction across the base–emitter diode and a high resistance when the leads are reversed, the junction is probably OK. Repeat the same measurement on the collector–base junction. You should measure a reasonably high resistance in both directions between collector and emitter.

6-16 **Measuring d-c voltage at output of power amps.**

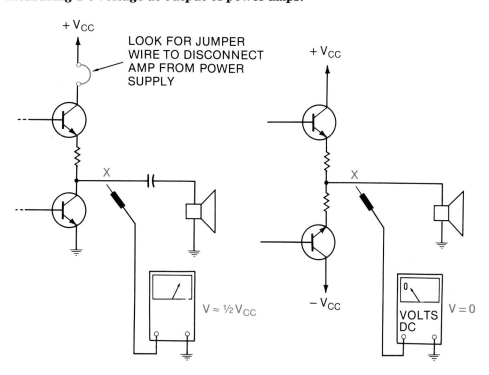

(A) Voltage at point X should measure near ½ V_{CC} for single supply.

(B) Voltage at point X should measure zero for dual-supply amp.

Whether you use the in-circuit transistor tester or the ohmmeter, be sure to test a suspected defective transistor again, after removing it from the circuit. Parallel components can give false indications.

When replacing a defective transistor, try to use an exact replacement if it is readily available. However, many times it is not available. Several manufacturers publish cross-reference lists of equivalent replacement types. They should at least equal or exceed the specifications of the original part.

If you don't know the specs of the original part, you can roughly determine values for a suitable replacement. The most important specs are maximum values of collector-to-emitter voltage $V_{CE(max)}$, collector current $I_{C(max)}$, and power dissipation P_D. As long as $V_{CE(max)}$ is equal to or greater than the power supply voltage, it should be OK. In a dual-supply amp, make sure $V_{CE(max)}$ is at least equal to the *sum* of the two supply voltages.

You can determine the peak-collector current for a single supply amp by using equation 6-1. For a dual amp, find $I_{C(max)}$ by dividing V_{CC} by the load impedance.

As mentioned before, transistors in the output stage of a class B amplifier dissipate only about 20 to 25% of the power delivered to the load. Just make sure that the transistor can handle about 25% of the load power, and it will work fine.

EXAMPLE 6-2 Suppose you are servicing the amplifier of Fig. 6-10 and you find that one of the 40310 power transistors is defective. You cannot get an exact replacement. What values of $V_{CE(max)}$, $I_{C(max)}$, and P_D would you look for in a substitute part?

SOLUTION The power supply voltage is $+38$ V, so $V_{CE(max)}$ should be at least 38 V. The peak collector current might be as high as

6-17   Simplified flowchart for troubleshooting d-c power amplifiers.

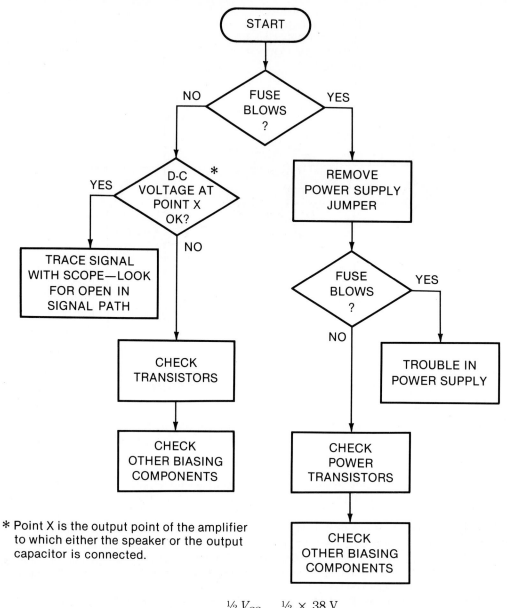

* Point X is the output point of the amplifier to which either the speaker or the output capacitor is connected.

$$I_{C(\text{max})} = I_{L(\text{max})} = \frac{\frac{1}{2} V_{CC}}{R_L} = \frac{\frac{1}{2} \times 38\,\text{V}}{8\,\Omega} = 2.375\,\text{A}$$

And finally, the maximum load power is

$$P_L = \frac{V_{\text{rms}}^2}{R_L} = \frac{(0.707 \times 19\,\text{V})^2}{8\,\Omega} = 22.55\,\text{W}$$

So the transistors must be capable of about 25% of P_L, or $0.25 \times 22.55 \approx 5.6$ W.

Your substitute transistor should have ratings of $V_{CE(\text{max})} \geqslant 38$ V, $I_{CE(\text{max})} \geqslant 2.4$ A, and $P_D \geqslant 5.6$ W.

> **NOTE**   These values are probably higher than necessary because most amplifiers are not driven hard enough to saturate the power transistors.

6-18   Dynascan Model 510 portable transistor tester. Courtesy of B&K Precision.

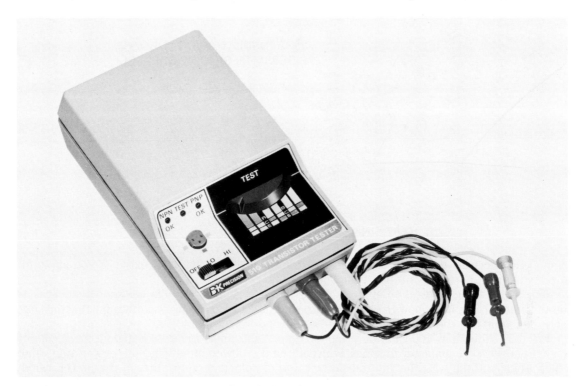

Often, manufacturers recommend replacing the output transistors as a matched pair. The pair should have as close to identical characteristics as possible. Normally, this is only true in very high-quality amps, in which distortion is to be kept extremely low. For most low- to moderately priced amps, matched pairs are not required.

Although the value β of a power transistor is usually not critical, darlington pairs are sometimes used. The darlington pair is really two transistors, often in a single package, which act like one "super-β" transistor. See Fig. 6-19.

The emitter of the first transistor is directly connected to the base of the second. Base current flowing into the first transistor thus controls the collector current of the second. And when we consider the two transistors acting as one device, they act like a transistor whose β (I_C/I_B) is the *product* of the β of the two transistors. Obviously, when you replace a darlington pair, you must do it with another darlington pair.

Since power transistors develop considerable heat, they must be mounted to either a metal chassis or some sort of heat sink to get rid of the heat. Fig. 6-20 shows a typical mounting arrangement. Notice the insulating mica washer used between the transistor and metal heat sink. Both sides of the washer should be coated liberally with silicone grease to obtain good heat flow. Insulating bushings and washers are also needed for the mounting hardware, unless nylon screws are used.

Suppose you find a shorted power transistor in an amplifier. First find a suitable replacement, and install it properly on the heat sink. Now you are ready to test it. Should you just flip on the power switch and hope for the best? Chances are that there may still be some other defective

6-19   Darlington pair transistor.

6-20 **Mounting a power transistor to a heat sink.**

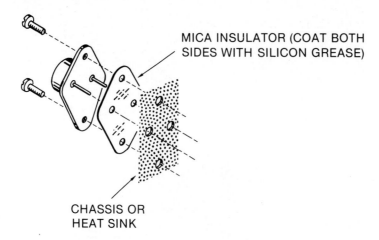

MICA INSULATOR (COAT BOTH
SIDES WITH SILICON GREASE)

CHASSIS OR
HEAT SINK

components. Perhaps the cause of the original transistor's going bad still exists. If you apply full power, you might burn out your new component again.

The safest thing to do is to apply power *gradually* using a variable a-c supply, such as an autotransformer. Fig. 6-21 shows a typical autotransformer, called a powerstat. By adjusting the large control knob on top, the output voltage can be adjusted from 0 to 140 VAC.

Fig. 6-22 shows how you can safely apply power to the amp without blowing out more components. Connect a d-c ammeter in series, between the d-c power supply and the power amp's final stages. Then, starting with the autotransformer output voltage at zero, turn on the power switch. Gradually increase the a-c voltage applied to the amplifier power supply. Watch the ammeter while you increase the a-c voltage.

6-21 **Power stat variable transformer Model 1168. Courtesy of The Superior Electric Co.**

As you increase the a-c input, the d-c output voltage of the power supply will gradually rise. If you see the idle current climb too high (with no input signal), turn down the a-c line voltage and look for additional troubles in the amplifier. The idling current through the power transistors should be only a small fraction of the current drawn at full power. You have problems if the d-c current is more than about 1 A when the a-c line voltage is about a third its normal value, even for high-power amps. But if the idle current is low, keep cranking up the line voltage all the way to 120 VAC. Then let it "cook" for a few minutes and watch for signs of overheating or drift. If no trouble appears, you are ready to inject a signal.

> **CAUTION** Do not apply a signal to a transformer-coupled circuit without a load on the secondary. An unloaded secondary will cause an extremely high impedance to be reflected into the primary. This can cause an excessive voltage to develop across the primary when a signal is applied, which can ruin the power transistors. Also do not short circuit the load when testing a transformerless amp. This can blow the transistors.

So with either type of amp, put a normal 8- or 16-Ω load across the output terminals. Then apply power, and inject a signal. Crank up the signal amplitude until near full power is being delivered to the load. It is usually more convenient to use a dummy load (resistive load) rather than to drive speakers. This allows to "burn in" the amp for several minutes without the deafening noise. Put an oscilloscope across the load, and watch for signs of distortion or clipping.

Of course amplifier current drain from its power supply will be quite high with signal applied. Therefore, either adjust the ammeter accordingly or remove it from the circuit for final testing.

6-22 Measuring d-c drain of amplifier while gradually increasing a-c line voltage.

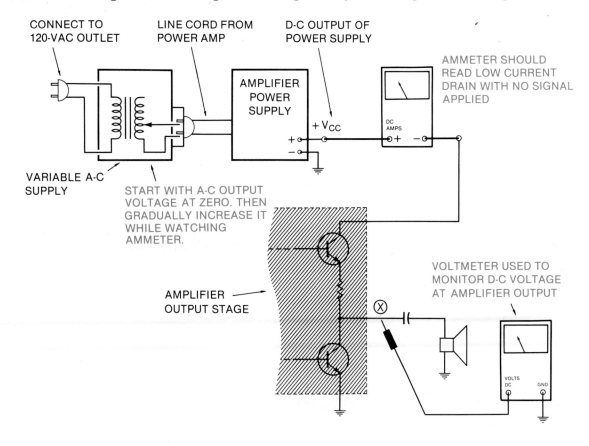

MISCELLANEOUS TROUBLESHOOTING TIPS

INTERMITTENT OPERATION

Intermittent problems are about the most frustrating and difficult to service. You usually have to apply power and wait until the symptom shows up before you can service it. Two helpful devices are a heater (like a hot-air hair dryer) and a can of coolant spray.

If the trouble only shows up after prolonged use, or in hot weather, try to isolate the defective part by heating localized areas a few degrees above ambient to see if the problem appears.

The opposite approach is also very useful in localizing a defective part. When an intermittant problem *does* show up, cool various parts, one at a time, using a can of coolant spray. This often pinpoints the bad part quickly.

D-C VOLTAGE ACROSS SPEAKER

Across a speaker, d-c voltage can cause distortion or, in severe cases, can ruin the speaker. If you measure any d-c voltage across the speaker, here's what to do. If the amp has a capacitor in the output, it is probably leaking. Check it, and replace it. If the amp is directly coupled to the speaker (dual-supply type), the idle bias needs adjusting.

NO SOUND OUTPUT

If you get a normal signal across a dummy load, but no sound when you connect the speaker system, the trouble is either in the speakers or in a crossover network. A simplified crossover network is shown in Fig. 6-23. Here are two speakers, a large woofer for handling low frequencies and a small tweeter for higher frequencies. Coil *L* allows low frequencies to pass to the woofer but offers a high impedance to higher frequencies. Similarly, capacitor *C* couples the highs to the tweeter and blocks the lows. Before replacing either speaker, check the crossover components.

6-23 Speaker system with crossover network.

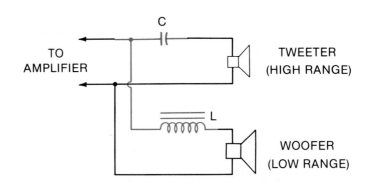

Speakers can be checked for continuity with an ohmmeter. Usually the d-c resistance is a little lower than the rated impedance. You will usually hear a click or crackling sound when you connect an ohmmeter across a good speaker.

Most older amps work reasonably well into speakers of either slightly higher or lower than recommended impedances. But some of the newer amps must work into a specific value of speaker impedance to get good results.

When replacing speakers, keep in mind that more expensive speakers can usually be rebuilt for less than a replacement would cost.

Noisy Controls

Pots can usually be cleaned sufficiently by spraying the control with a suitable contact cleaner and rotating the pot shaft back and forth several times.

PROBLEMS

For the next four problems, refer to Fig. 6-24. They will serve to review your understanding of direct-coupled amplifiers as used in power amps. Assume ideal transistors, that is $V_{BE} = 0$.

6-24 **Circuit for problems 6-1**
through 6-4.

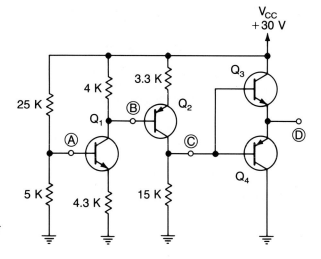

Note: Assume V_{BE} = 0.7 V
for each conducting transistor

6-1.   Find the voltage at point A and the current through Q_1.

$V_A =$

$I_{C1} =$

6-2.   Find the voltage at point B and the current through Q_2.

$V_B =$

$I_{C2} =$

6-3.   Find the voltage at point C and the voltage at point D.

$V_C =$

$V_D =$

6-4.   If Q_1 shorts, collector to emitter, the voltage at point D will (a)go up. (b)go down. (c)remain the same.

For the next four problems, refer to Fig. 6-25.

6-25 **Circuit for problems 6-5**
through 6-8.

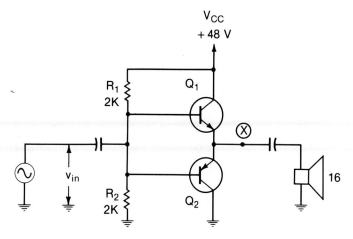

6-5.   What is the quiescent voltage at point X?

$V_X =$

6-6.   If v_{in} is large enough to drive Q_1 into saturation on the positive half cycle, what is the peak collector current of Q_1?

$I_{C(max)} =$

6-7.   What is the rms voltage across the load, if the circuit is driven as hard as possible without distortion? What is the power delivered to the load?

$V_{rms} =$

$P_L =$

6-8.   What should be the minimum power ratings at Q_1 and Q_2?

$P_D =$

$P_D =$

For the next four problems, refer to Fig. 6-5.

6-9.   If R_A is increased to 62 KΩ, the voltage at point X will (a)go up. (b)go down. (c)remain the same.

6-10.   If D_1 shorts, which of the following will happen? (a)Q_1 will go into saturation. (b)Crossover distortion will increase. (c)Q_3 will turn on too hard. (d)There will be no output signal.

6-11.   What is the class of operation of Q_3? (a)Class A. (b)Class B. (c)Class AB.

6-12.   If $R_L = 8\ \Omega$ and β of $Q_1 = 100$, what is the impedance seen looking into the base of Q_1, when it is delivering a signal to the load?

$r_{in} =$

For the next four problems, refer to Fig. 6-6.

6-13.   Assume that the voltage at the collector of Q_3 is much lower than normal. Which of the following could be the cause? (a)R_{13} open. (b)D_1 shorted. (c)C_{11} open. (d)Open Q_4.

6-14.   If D_1 opens, the emitter voltage of Q_4 (a)will go up. (b)go down. (c)remain the same.

6-15.   If R_{16} opens, the d-c voltage at TP 17 will (a)go up. (b)go down. (c)remain the same.

6-16. All d-c voltages measure normal, but there is no a-c output signal at TP 17. Which of the following is the most likely cause? (a)Defective transistor, either Q_3, Q_4, or Q_5. (b)Open R_{12}. (c)Shorted C_{10}. (d)Open C_8.

For the next four problems, refer to Fig. 6-8. Assume that V_{CC} is changed to 30 V.

6-17. If the circuit is operating normally, what voltage would you expect to read at point X?

$V_X =$

6-18. If the circuit is operating normally, what voltage would you expect to read between the collector of Q_1 and the base of Q_2?

$V =$

6-19. Output transistors Q_4 and Q_5, (a)normally are both biased near cutoff. (b)conduct on alternate half cycles. (c)must be capable of dissipating approximately 25% of the load power. (d)All of these.

6-20. If the resistance of R_4 is increased by 20%, the voltage gain of Q_1 will (a)go up. (b)go down. (c)remain about the same.

For the next four problems, refer to Fig. 6-10.

6-21. If the 2N3704 driver transistor shorts from collector to emitter, (a)both power transistors will turn off. (b)both power transistors will turn on hard, possibly blowing the fuse. (c)the amplifier will continue to operate, but with a distorted output signal. (d)there would be no noticeable effect on a-c output.

6-22. What is the probable purpose of the 47-pF capacitor from base to collector of the 2N2926 predriver?

6-23. If one of the 1N3754 biasing diodes opened, could it result in blowing the fuse? (a)Yes. (b)No.
Explain.

6-24. Suppose the amplifier output signal is distorted. The d-c voltage measurements on the power amp section read near normal, but the collector voltages of both of the 2N3711 preamp transistors are low. Which of the following is the most likely cause? (a)Open diode in power supply. (b)Bias adjust pot set incorrectly. (c)Leaky 250-μF capacitor in power supply. (d)Both preamp transistors are conducting too heavily.

For the next four problems, refer to Fig. 6-13.

6-25.   The secondaries of T_1 are connected so that two output transistors conduct (a)at the same time. (b)on alternate half cycles.

6-26.   If D_1 opens, (a)TR_4 will tend to conduct more as the temperature goes up. (b)both transistors will turn on hard. (c)the signal will not be rectified on one half cycle. (d)TR_5 will tend to conduct more.

6-27.   What d-c voltage would you expect to read at TP 18?

$V =$

6-28.   What d-c voltage would you expect to read at TP 20?

$V =$

EXPERIMENT 6-1   COMPLEMENTARY-SYMMETRY AMP

You will now test and troubleshoot a complementary-symmetry amp, like that in the GE Model M8710A tape recorder.

EQUIPMENT

- 6-VDC power supply
- oscilloscope
- voltmeter
- audio signal generator
- (1) TIP 31 or equivalent NPN power transistor
- (1) TIP 32 or equivalent PNP power transistor
- (1) 2N2222 NPN transistor
- (1) silicon diode
- (1) 15-KΩ, ½-W ± 5% resistor
- (1) 6.8-KΩ, ½-W ± 5% resistor
- (1) 150-Ω, ½-W ± 5% resistor
- (1) 47-Ω, ½-W ± 5% resistor
- (2) 1-Ω ½-W ± 5% resistors
- (1) 8-Ω, 1-W resistor or 8-Ω speaker
- (1) 200-μF, 6-WVDC capacitor
- (1) 50-μF, 6-WVDC capacitor
- (1) 10-μF, 6-WVDC capacitor

PROCEDURE

1.   Build the circuit of Fig. E6-1. Apply power and measure the d-c voltage at point X.

$V_X =$

If the voltage at point X is near 3 V, go on to the next part. Otherwise recheck your wiring.

2.   Connect a signal generator to the input, with the frequency set at about 1 KHz. Observe the output with an oscilloscope, and adjust the signal generator amplitude until V_0 measures about 4 Vp-p. Measure v_{in}.

E6-1 Circuit for experiment 6-1.

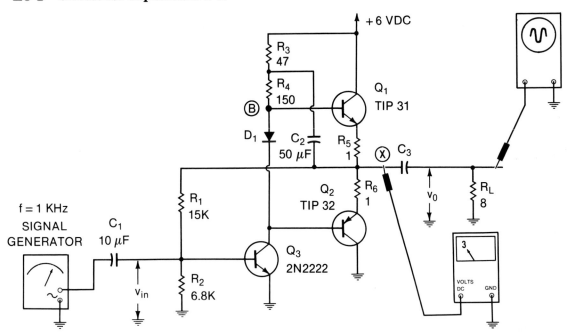

$v_{in} =$

Calculate the voltage gain of the overall circuit.

$A_{v\,(tot)} = v_o/v_{in} =$

3. Measure the amplitude of the a-c signal at point B, and caclulate the voltage gain of the output stage.

$A_v =$

4. Next, we will make some changes to the circuit to simulate malfunctions. For each of the changes listed in Table E6-1, make the change, then measure and record the d-c voltage at point X, and the a-c output voltage. If any distortion of the output signal appears, draw two cycles of the distorted waveform in the space provided.

QUIZ

1. Since the output stage has a gain of near unity, most of the overall voltage gain can be attributed to which transistor? (a)Q_1. (b)Q_2. (c)Q_3.

2. Removing C_2 from the circuit—(a)increased, (b)decreased—the voltage gain of the circuit. This is because the a-c load impedance seen by Q_3—(c)increased, (d)decreased—when C_2 was removed.

3. Shorting D_1—(a)increased, (b)decreased—crossover distortion by biasing both transistors (c)on harder. (d)completely off.

4. Reducing R_1 caused Q_3 to conduct—(a)more, (b)less—which brought the d-c voltage at points B and X (c)up. (d)down.

	Malfunction	V_X	v_0	Distortion?
1	Open D_1			
2	Shorted D_1			
3	Open C_2			
4	R_1 reduced 50%			
5	Open R_2			
6	Open R_3			
7	Open Q_3			
8	Open Q_2			
9	Open Q_1			
10	Open R_6			

*To simulate an open transistor, simply remove it from the circuit.

5. Opening R_2 caused Q_3 to conduct (a)more. (b)less.

6. Opening R_3 caused Q_1 to conduct (a)more. (b)less.

7. An open Q_3 caused Q_1 to conduct (a)more. (b)less.

8. When Q_2 was opened, the a-c output signal went—(a)up, (b)down—because (c)there was less load on Q_1. (d)the discharge path for C_3 was removed.

9. An open R_6 has the same symptoms as an open (a)Q_1. (b)Q_2. (c)Q_3.

10. Before replacing an open resistor like R_6 in an actual amplifier, you should first look for the reason it burned open. One possible cause could be a shorted (a)Q_3. (b)D_1. (c)Q_2.

EXPERIMENT 6-2   QUASICOMPLEMENTARY AMPLIFIER

The amplifier in this experiment can deliver more power to a load than the amp in experiment 6-1. This is primarily because we used a higher voltage power supply. To prevent the possibililty of ruining the power transistors, make all circuit changes with power off. Build your circuit neatly, without a lot of loose wiring, and check it carefully before applying power.

It is important that you understand what to expect in each step of the experiment before actually performing it. So read each step thoroughly before doing it.

NOTE This experiment requires a considerable amount of time, because we use a large number of components.

EQUIPMENT

- 20-VDC adjustable power supply at 1-A or 20-VDC* fixed output power supply and autotransformer to vary line voltage
- oscilloscope
- voltmeter
- ammeter
- audio signal generator
- (2) TIP 31 NPN power transistors
- (1) 2N5447 or equivalent PNP transistors
- (2) 2N2222 or equivalent NPN transistors
- (2) silicon diodes
- (1) 100-KΩ pot
- (1) 47-KΩ, ½-W ± 5% resistor
- (1) 10-KΩ, ½-W ± 5% resistor
- (1) 5-KΩ, ½-W ± 5% resistor
- (1) 1-KΩ, ½-W ± 5% resistor
- (1) 330-Ω, ½-W ± 5% resistor
- (2) 220-Ω, ½-W ± 5% resistors
- (2) 1-Ω, 1-W ± 5% resistors
- (1) 500-μF capacitor at 25 WVDC
- (2) 50-μF capacitors at 25 WVDC
- (1) 10-μF capacitor at 25 WVDC
- (1) 8-Ω, 10-W resistor (for dummy load)
- (1) 8-Ω speaker capable of handling 6-W heat sinks or metal plate to mount power transistors miscellaneous mounting hardware, including Mica washers, silicon grease, etc.

*If you have a power supply capable of anywhere between about 16 and 24 V at 1 A, it should work fine. The 20-V value shown here is just a nominal value.

E6-2 **Set-up for experiment 6-2.**

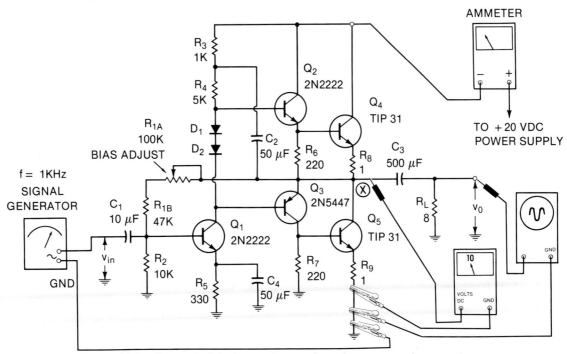

Note: Connect all instrument grounds and power supply ground to same point to minimize tendency for oscillation or distortion.

PROCEDURE

1. Build the circuit of Fig. E6-2, but leave the power transistors, Q_4 and Q_5 out of the circuit. Also leave C_3 disconnected from point X.

> **NOTE** It is very important when building this circuit to keep ground leads short. If possible, connect your power supply ground as close as possible to output transistor Q_5. Similarly, connect the V_{CC} lead as close as possible to the collector of Q_4. Also connect all instrument grounds to the same point as the power supply ground. If you don't, you will have problems with oscillation and/or distortion due to ground loops.

2. Apply power, and adjust the BIAS ADJUST pot until the voltage at point X is equal to ½ V_{CC}. This adjustable bias compensates for variations in transistor characteristics, and even allows you to use different transistors from those specified.

3. With C_3 disconnected (no load), connect a 1-KHz signal generator to the input. Watch point X with an oscilloscope. Adjust the signal generator amplitude so that the signal at point X measures about 10 Vp-p.

Step 3 demonstrates a possible test procedure when troubleshooting a defective amplifier. It allows you to check the entire amplifier up to the final stages, without the danger of damaging the output transistors, in the event something else is bad.

If you see a good undistorted signal at point X, go on to the next step, otherwise recheck your wiring. Show this setup to your instructor.

Instructor's initials_____

4. Shut off the power, and connect C_3 and R_L to point X (the bottom end of R_6). Apply power. Do not readjust the input signal, but measure the output signal at point X again.

$v_0 =$

You should notice a drastic reduction in v_0. Next we will see the effect of adding the power transistors.

5. Disconnect the input signal v_{in}. Shut off power and connect Q_4 and Q_5 into the circuit. These power transistors should be mounted on heat sinks, because they will each be dissipating about 1 W when signal is applied.

6. If you have an adjustable d-c supply, decrease the supply voltage to near zero, or at least to just a few volts. If your supply is not adjustable, connect an autotransformer in series with the a-c line cord of the power supply, as shown in Fig. 6-20. Have the line voltage being applied to the power supply initially set near zero.

7. With an ammeter connected in series between the amplifier and the d-c power supply, turn on the power. Gradually increase the d-c supply voltage until $V_{CC} = 20$ V. Record the idling current.

$I_{dc} =$

This current should be quite low with no signal applied.

8. Reduce the d-c supply voltage to zero again, and connect a clip lead from collector to emitter of Q_2 (simulating a shorted transistor).

9. Slowly start cranking up the d-c voltage again while watching the ammeter.

Does the idle current rise more rapidly? (a) Yes. (b) No.

Don't let the current exceed about ½ A.

10. Shut off power, and remove the short from across Q_2. Then apply power and crank up the d-c supply voltage again. You should read the same idling current as in step 7. Now connect the signal generator to the input. Starting with v_{in} near zero, gradually increase the amplitude of v_{in}. Watch the output on an oscilloscope connected across R_L. Also watch the ammeter reading. What happens to the d-c reading as the input signal amplitude is increased?

(a) Increases. (b) Decreases. (c) Remains about the same.

11. Adjust v_{in} so that $v_0 = 8$ V peak (16 Vp-p). (You may have to either set the ammeter on a high-current scale or remove it.) Measure v_{in}.

$v_{in} =$

Calculate the voltage gain.

$A_v = v_0/v_{in} =$

Occasionally check to see if any parts are getting too hot.

12. Connect the oscilloscope from ground to the emitter of Q_5. Draw two cycles of the waveform you see.

13. *Optional.* Replace R_L with a speaker. Drive the input with either the signal generator or the output from a tape recorder or radio to see what your amp sounds like.

QUIZ

1. In step 3, since Q_4 and Q_5 were not in the circuit, the amplifier was basically a—(a)class A, (b)complementary-symmetry, (c)quasicomplementary—type.

2. In step 4, the output signal dropped drastically when the load was connected. This was because (a)complementary-symmetry amps cannot deliver any power. (b)most of the output signal was dropped across the 220-Ω resistors.

3. In step 7, the current drain from the supply was—(a)low, (b)high—with no signal applied. This indicated that Q_4 and Q_5 are biased (c)class A. (d)near cutoff.

4. In step 9, even though no signal was applied, the current drawn from the supply was—(a)higher than, (b)lower than, (c)the same as—in step 7. This was because Q_4 was turned—(d)on, (e)off—hard.

5. In step 10, we saw that the dc drawn by the amplifier—(a)increases, (b)decreases, (c)remains the same—when signal is applied.

6. Step 12 showed that Q_5 conducts for—(a)one half of, (b)the entire—input cycle.

7. Q_5 must be operating class (a)A. (b)B.

8. Based on the observed operation, we can say that class—(a)A, (b)B—amplifiers draw—(c)more, (d)less—current when input signal is applied.

Transformer-Coupled Circuits

Most of the amplifiers that we have studied so far were either direct or capacitively coupled. But sometimes transformers are used to transfer signals from one stage to the next. This is called *transformer coupling*. There are times when transformer coupling offers advantages. The first is when it is necessary to make a low value of impedance look like a high value, or vice versa. Another is when it is desirable to amplify only a narrow band of frequencies. We will look at both of these applications in this chapter.

UNTUNED INTERSTAGE TRANSFORMER COUPLING

In chapter 2, you worked with power supply transformers. Transformers designed for coupling between amplifier stages, or between the final stage and the load, are similar to power supply types. But, in general, coupling transformers are designed to have lower losses at higher frequencies, and they often can operate over a wider frequency range. When designed to operate over a wide range of frequencies, such as the entire audio spectrum, the transformers are said to be *untuned*.

7-1   **Untuned transformer with 4:1 step down turns ratio.**

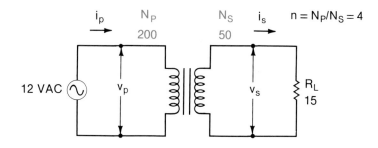

The transformer shown in Fig. 7-1 has 200 turns on the primary and 50 turns on the secondary. As explained in chapter 2, the secondary voltage is related to the primary voltage by the ratio of the number of turns on each winding. That is,

$$n = \frac{N_p}{N_s} = \frac{v_p}{v_s}$$

7-1

where

$$n = \text{turns ratio}$$
$$N_p = \text{number of primary turns}$$
$$N_s = \text{number of secondary turns}$$
$$v_p = \text{primary voltage}$$
$$v_s = \text{secondary voltage}$$

Since there are less turns on the secondary than there are on the primary, the transformer is said to be a *step-down* transformer.

EXAMPLE 7-1   For the transformer of Fig. 7-1, what is the turns ratio? What is the secondary voltage?

SOLUTION

$$n = \frac{N_p}{N_s} = \frac{200}{50} = 4$$

so

$$v_s = \frac{v_p}{n} = \frac{12}{4} = 3 \text{ V}$$

Once we know the secondary voltage, we can find the current that flows through a load connected across the secondary by the equation,

$$i_s = \frac{v_s}{R_L} \qquad\qquad 7\text{-}2$$

Using the values shown in Fig. 7-1, we see that

$$i_s = \frac{3 \text{ V}}{15 \text{ }\Omega} = 200 \text{ mA}$$

Another question now arises. What is the value of primary current? Well, the transformer is basically an energy transfer device. It takes energy from the generator connected to its primary and transfers that energy to the load connected across its secondary. If the transformer is 100% efficient, *all* of the energy deliverd to the primary is transferred to the secondary. Or putting it another way, the primary power is equal to the secondary power.

$$P_P = P_S \qquad\qquad 7\text{-}3\text{A}$$

Since $P = V \times I$, we can substitute to get

$$v_p\, i_p = v_s\, i_s \qquad\qquad 7\text{-}3\text{B}$$

From which we get

$$i_p = i_s \times \frac{v_s}{v_p} \qquad\qquad 7\text{-}3\text{C}$$

And since

$$\frac{v_s}{v_p} = \frac{1}{n}$$

We see that

$$i_p = \frac{i_s}{n} \qquad\qquad 7\text{-}4$$

So for the circuit of Fig. 7-1,

$$I_P = \frac{200 \text{ mA}}{4} = 50 \text{ mA}$$

Since the primary voltage is 12 V and the primary current is 50 mA, we can say that the generator looking into the primary sees an impedance of

$$Z = \frac{v_p}{i_p} = \frac{12 \text{ V}}{50 \text{ mA}} = 240 \ \Omega$$

We call this impedance the *reflected resistance*. That is, because of the 15 Ω resistance connected across the secondary, there is a reflected resistance of 240 Ω in the primary.

Substituting for i_p in equation 7-4, we get

$$r_{ref} = \frac{v_p}{i_p} = \frac{v_p}{i_s} \times n$$

Then since $i_s = v_s/R_L$ we see that

$$r_{ref} = \frac{v_p}{v_s} \times n R_L$$

But since $v_p/v_s = n$, we can say that

$$r_{ref} = n^2 R_L \qquad\qquad 7\text{-}5$$

EXAMPLE 7-2 If a transformer has 300 turns on the primary and 100 turns on the secondary, what will be the reflected resistance seen looking into the primary, when a load of 8 Ω is connected across the secondary?

SOLUTION

$$n = N_p/N_s = 300/100 = 3$$

$$r_{ref} = n^2 R_L = 3^2 \times 8 = 72 \ \Omega$$

So we see that the transformer can make a small resistance connected across its secondary look like a much larger resistance when viewed looking into its primary. Keep in mind that this reflected resistance seen looking into the primary is an *a-c* resistance. It results from the need for the generator to supply power to the primary whenever power is drawn from the secondary. The *d-c* resistance of the windings is usually very low, often less than 1 Ω.

What do you suppose would be the value of the primary current if the secondary were left open circuited? Looking at equation 7-4, if i_s were zero, i_p would be zero also.

This is essentially true in a good transformer. But if the d-c resistance of the primary is very low, what limits the primary current when the secondary load is disconnected? The answer is the *inductive reactance* of the primary.

Remember that the primary is constructed of many turns of wire wrapped around a highly permeable iron core. So it has a large amount of inductive reactance. In fact, transformers are usually designed so that the inductive reactance of the primary is at least ten times the reflected resistance at the lowest frequency to be handled. The reflected resistance is effectively seen in parallel with the reactance, so we see essentially a pure resistance looking into the primary, when there is a load across the secondary. However, when the secondary load is disconnected, primary current drops to a very low value, because it is limited by the inductive reactance.

You can see that if a transformer is to be used at low audio frequencies, the core must have a lot of iron so that the inductive reactance will be high. This makes the transformer heavy and expensive, particularly if it is to be used in high-power applications where core saturation might occur as a result of high currents flowing in the windings.

On the other hand, if the core has a lot of iron, and/or many turns of wire to increase the inductance, it also has a considerable amount of distributed capacitance. That is, between each turn of wire and other turns, and between the turns and the core, capacitance exists, which tends to shunt high frequencies to ground. So if the transformer is designed to have good low-frequency response, it will not have good high-frequency response, and vice versa. All in all, transformers usually are suitable only over a limited frequency range.

TRANSFORMER-COUPLED AMPLIFIER

In chapter 3 we discussed how the gain of a transistor amplifier varies directly with the value of the a-c load seen by its collector. Or, in equation form, $A_V = r_L/r_e$. If it is necessary for the transistor to work into a small value of load resistance, the gain of the amp will be very small. In fact, severe distortion of the output signal can occur if the amp works into too small a load resistance. The transformer offers a way to make a low value of resistance look like a large value. This then allows the transistors to see a high-impedance load, while actually delivering a signal to a small resistance.

The circuit of Fig. 7-2 shows how a transformer is used to make a low-resistance load look like a higher value of resistance when seen by the transistor. Just like the generator of Fig. 7-1, the transistor sees the reflected resistance when looking into the primary. It is this reflected resistance that determines the gain of the amp. Resistors R_1, R_2, and R_3 are used to bias the stage as before.

7-2   Transformer-coupled amplifier stage.

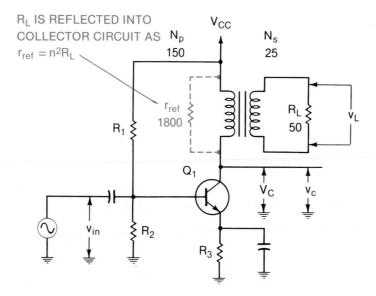

EXAMPLE 7-3 Referring to the circuit of Fig. 7-2, find the turns ratio n, the reflected resistance r_{ref}, the gain of the amplifier (v_c/v_{in}), and the amplitude of the signal v_L developed across the load. Assume $r_e = 12\ \Omega$ and $v_{in} = 5\ \text{mV}$.

SOLUTION

$$n = N_p/N_s = 150/25 = 6$$

$$r_{ref} = n^2 R_L = 6^2 \times 50 = 1800$$

$$A_v = \frac{r_L}{r_e} = \frac{r_{ref}}{r_e} = \frac{1800}{12} = 150$$

$$v_c = v_{in} \times A_v = 5\ \text{mV} \times 150 = 750\ \text{mV}$$

Then from equation 7-1, since $v_L = V_{sec}$ and $v_c = V_{pri}$,

$$v_L = v_c \frac{N_s}{N_p} = 750\ \text{mV} \times \frac{1}{6} = 125\ \text{mV}$$

Even though the voltage developed across the load is lower than the signal developed at the collector, the load voltage is still larger than it would be if the 50-Ω load had been connected directly to the collector of the transistor without using the transformer. If we used the same value of r_e and the 50-Ω load, the gain of the stage would have been $A_v = 50/12 = 4.16$. This gain would give us a load voltage of only about 20 mV. So we see that the transformer helps considerably.

TESTING AND TROUBLESHOOTING TRANSFORMER-COUPLED AMPLIFIERS

As stated earlier, the d-c resistance of transformer windings is usually very low. If you measure the d-c collector voltage of a transformer-coupled amplifier stage, such as that of Fig. 7-2, *you will read a collector voltage nearly equal to the V_{CC} supply voltage.* This does not indicate that the transistor is cut off but merely that there is very little d-c drop across the low-resistance primary winding.

If you look at the collector of the transistor with an oscilloscope when the stage is amplifying a signal, you will see a pattern like that of Fig. 7-3. This happens since the d-c collector voltage is near V_{CC}. Notice that the collector voltage actually goes *higher* than V_{CC} on the positive half of the cycle. This never happens in a transistor stage that works into resistive load. But when working into a

7-3 **Signal at collector of circuit of Fig. 7-2.**

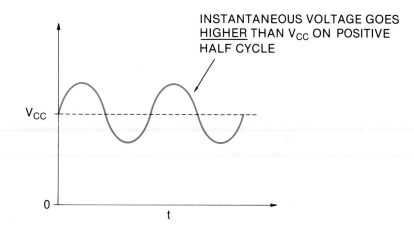

INSTANTANEOUS VOLTAGE GOES HIGHER THAN V_{CC} ON POSITIVE HALF CYCLE

transformer, the collector current flows through an inductor. When the input signal v_{in} goes in the negative direction, the collector current decreases. The collapsing magnetic field around the inductor induces a voltage in the coil in such a direction as to try to keep the same amplitude of current flowing. As a result, the polarity of voltage developed across the inductor is such that the bottom end (collector end) actually goes positive with respect to V_{CC}. You may have heard this phenomenon referred to as *inductive kickback*, or some other similar term, in your study of a-c circuits.

The fact that the collector voltage goes above V_{CC} is important to keep in mind for two reasons. First of all, when you are testing this circuit, you will know what to expect. Second, it is apparent that *the voltage rating of the transistor must be higher than the power supply voltage.*

In chapter 6, we noted that you should never operate a transformer-coupled stage (push–pull output) without a load connected to the secondary. Now you can appreciate the reason for this. If the load is removed from the secondary, the reflected impedance will be almost infinite (or at least extremely high). Therefore, the a-c gain of the amplifier stage will be extremely high, causing the a-c signal at the collector to be very large. The kickback voltage developed across the transformer primary can be so large that it can break down the transistors. In some cases, it might cause arcing between transformer windings, thus ruining the transformer. Consequently, you should *never drive a transformer-coupled stage without a load connected across the secondary!*

Like power supply transformers, coupling transformers can develop open windings, or occasionally, shorted turns. An open primary winding in the circuit of Fig. 7-2 would obviously show up when you tried to measure the collector voltage of the transistor. An open secondary would give 0 V across the load, but plenty of signal across the primary. A short across either the primary or secondary will cause the reflected impedance to be near zero, so almost no a-c signal will appear at the collector when the stage is driven.

Refer back to the power amp of Fig. 6-13 to see where an interstage coupling transformer is used in a commerical circuit. Transformer T_1 couples the driver transistor TR_3 to the two power transistors. The transformer serves two puposes in this circuit. First, its two secondaries feed out-of-phase signals to the push–pull output transistors. Second, it serves to make the low input impedance of the power transistors look much higher as seen by the collector of the driver. The driver, therefore, has more gain.

In the circuit of Fig. 6-13, notice that the primary has a d-c resistance of 140 Ω. So it does have a measurable d-c drop of slightly more than 1 V across it. Even so, the collector voltage of TR_3 is nearly equal to V_{CC}.

When replacing a bad transformer, it is almost always best to get an exact replacement, since there are many characteristics of the transformer that must be considered. Besides the turns ratio, which must be exactly the same as the original part, other important factors are 1) current rating of the two windings, 2) frequency responses of the transformer, which is related to the physical dimensions, amount of iron, etc., 3) physical size (for mounting purposes), and 4) temperature characteristics. However, in a pinch, if an exact replacement is not available, at least make sure that the turns ratio of the replacement transformer is the same as that of the original part. Also, look at the physical size of the replacement. Most often, if the physical size of the new part is about the same as the old one, it probably will have similar frequency response characteristics, similar wire size, etc.

TUNED TRANSFORMERS

The circuit of Fig. 7-4A shows a generator driving a coil and a capacitor in parallel. Such a parallel L–C circuit is often called a *tank* circuit. At a frequency where $X_L = X_C$, the circuit is said to be *resonant*. The impedance of the parallel L–C circuit reaches a *maximum value at resonance*, as shown in Fig. 7-4B. The value of the impedance at resonance can be shown to be equal to

$$Z_t = Q\,X_L \qquad\qquad 7\text{-}6$$

Q is the ratio of X_L/R_c, where R_c is the series resistance of the coil itself, and X_L, the inductive reactance of the coil, is equal to $2\pi f L$ (See Fig. 7-4C.)

If the series resistance of the coil is small compared to X_L ($Q \geqslant 10$) the frequency at which resonance occurs is

$$f_r = \frac{1}{2\pi\sqrt{LC}}$$

where

f_r = resonant frequency
L = inductance of coil in henries
C = capacitance in farads

7-4 Driving a parallel *L–C* circuit near resonance.

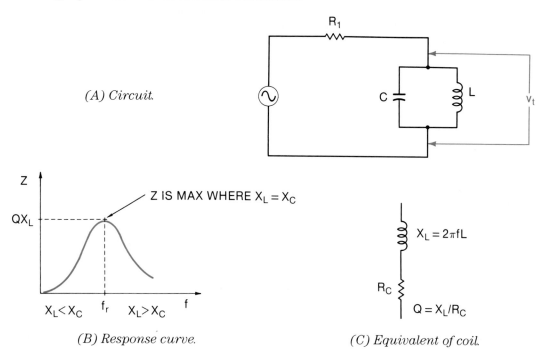

(A) Circuit.

(B) Response curve.   *(C) Equivalent of coil.*

EXAMPLE 7-4 If in the circuit of Fig. 7-4A, $L = 2$ mH, $C = 0.003$ μF, and the resistance of the coil is 20 Ω, find the resonant frequency f_r, value of Q for the circuit, and tank impedance.

SOLUTION

$$f_r = \frac{1}{2\pi\sqrt{LC}} = \frac{1}{6.28 \times \sqrt{2 \times 10^{-3} \times 3 \times 10^{-9}}} = 64.9 \text{ KHz}$$

$$X_L = 2\pi fL = 6.28 \times 64.9 \times 10^3 \times 2 \times 10^{-3} = 817 \text{ } \Omega$$

$$Q = \frac{X_L}{R_C} = \frac{817}{20} = 40.8$$

$$Z_t = QX_L = 40.8 \times 817 = 33.4 \text{ K}\Omega$$

At resonance, since $X_L = X_C$, the reactive components of the circuit current cancel each other out, so that *the circuit looks purely resistive.* Above resonance, more current flows through the capacitor than does through the coil; therefore, the circuit looks capacitive. On the other hand, more current flows through the coil below resonance. So the circuit looks inductive.

The circuit impedance does reach a maximum at resonance. Therefore, if you connect an oscilloscope across the tank and observe the voltage v_t while you vary the generator frequency from some value below resonance to some value above resonance, you will see the voltage v_t peak up as you pass through resonance. This is because, at all frequencies, the tank and R_1 in series form a voltage divider across the signal generator output. As resonance is approached, the tank impedance goes up, causing more of the generator output voltage to appear across the tank. The tank voltage, as a function of frequency, has the same shape as the curve of Fig. 7-4B.

We have just reviewed parallel resonance briefly. For a more thorough discussion of resonance, refer to a good text on a-c circuit theory.

Many coils have movable iron (ferrite) slugs (threaded cylinders) that make the value of inductance variable over a limited range. Fig. 7-5 shows a typical symbol for a slug-tuned coil. When adjusting these coils, a nonmagnetic tuning wand (tool) is needed. An iron-based tool would change the tuning when brought near the coil. Some tuning slugs require a flat screwdriver tip for adjustments, while others need an allen wrench tip. So a variety of tools is available. Do not adjust coils unless necessary, since some can be adjusted only a couple of times before damaging the screw threads of the slug.

7-5 Slug-tuned coil.

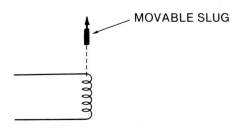

MOVABLE SLUG

AMPLIFIERS WORKING INTO PARALLEL-TUNED CIRCUITS

Fig. 7-6 shows a transistor amplifier working into a parallel L–C circuit. You can probably guess the sort of response this amplifier has. Since the tank circuit has an extremely high impedance at resonance, the gain of the amplifier is very high at resonance. As the frequency of the amplifier input signal is varied above or below resonance, the amplifier's gain decreases. In fact, the gain drops more and more, the farther away from resonance the signal is varied. If we plot the gain versus frequency response curve of the amplifier, we get a curve very similar to the impedance curve of Fig. 7-4B.

7-6 Tuned amplifier.

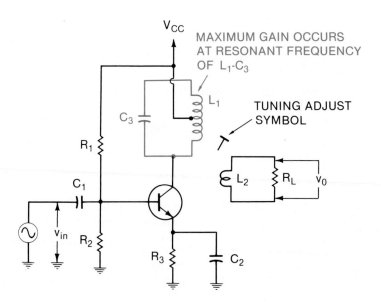

Looking once again at the circuit of Fig. 7-6, notice that the power supply connection is actually made to a tap on the primary. This is often done to improve the sharpness of tuning. Also note that the load R_L is connected to a secondary coil L_2. Just as with untuned-transformer coupling, by having less turns on the secondary than on the primary, the secondary load resistance is reflected back into the primary as a much higher value. We will not go into the exact calculations involving the value of

7-7   Portion of Courier Model Cruiser citizen's band transceiver. From *CB Radio Series No. 79*, Howard W. Sams & Company, Inc.

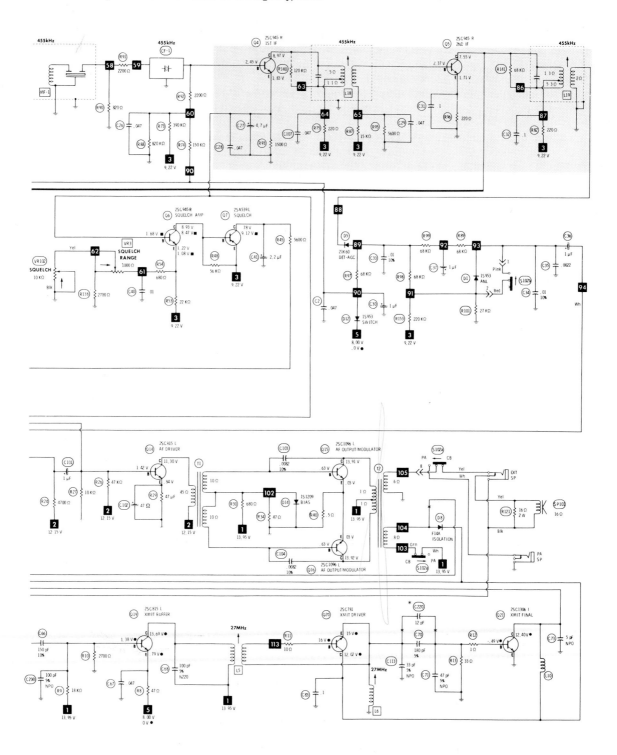

reflected resistance, since it is more complex for the tuned circuit. The idea is, however, the same as for an untuned transformer.

The primary function of a tuned amplifier is to amplify a *narrow band* of frequencies around some center frequency. The tuned circuit is adjusted so that resonance occurs at the center frequency, thereby giving maximum gain at this point.

A typical example of the use of a tuned amplifier is that of an *intermediate frequency* (IF) amplifier, as used in radio receivers. The IF amplifiers, used in AM broadcast radios, are usually designed with a center frequency of 455 KHz. The bandwidth is quite narrow, perhaps only 10 KHz or so. (Remember that the bandwidth of an amplifier is defined as the span of frequencies over which the gain of the amplifier is at least 0.707 its maximum value.)

The midband gain of such an amplifier can be very high, perhaps over 1000 for a single stage. This is caused by the extermely high impedance presented by the parallel resonant circuit. It is impractical to get a gain that high with a resistor in the collector lead. This is because of the large dc drop across a large collector resistor as a result of I_C flow. For example, if you use a 100-KΩ collector load resistor, 100 V will drop across it with just $I_C = 1$ mA flowing through it. But by using a tuned transformer in the collector lead, the a-c impedance seen can be extremely high, while the d-c drop across the coil is negligible.

The parallel-resonant transformers used in IF amplifiers are usually mounted inside aluminum enclosures, called *cans*, for shielding. Both windings and the tuning capacitor are mounted inside the can, which has an access hole for reaching the tuning slug.

Fig. 7-7 shows a portion of a Courier Model Cruiser citizen's band transceiver. Notice in the highlighted portion that the output of the first IF stage is transformer coupled to the input of the second IF. The dashed lines indicate the IF can, which is grounded for shielding. The output of the second IF is then coupled to the detector stage. Typical voltage-divider bias is used for both stages, and the coupling transformers are tunable, as indicated by the arrows between the coils.

The process of tuning the coils is called *alignment*. To tune the coils, first, set a signal generator to 455 KHz, and connect it to the base of the first IF amp. With an oscilloscope connected across the secondary of the first IF transformer (base of Q_5), adjust the tuning slug of the transformer for maximum signal delivered to the base of Q_5. Next, connect the oscilloscope across the secondary of the second IF can. Adjust the second IF transformer tuning slug for a maximum reading on the oscilloscope.

Manufacturers sometimes describe slightly different ways to align the stages, such as tuning for a maximum reading on a d-c voltmeter connected to the automatic gain control (agc) point after the last IF amp. But the idea is the same. The IF transformers are simply tuned to resonance at 455 KHz.

One precaution must be taken when aligning an IF amplifier. Any test instruments, even oscilloscopes, that are connected across a tank must have very high input impedance, otherwise they will upset the circuit under test. Oscilloscopes should use high-impedance (attenuator) probes, with an impedance of 10 MΩ or so. Any resistive loading on a tuned circuit tends to *broaden* the tuning curve.

Another thing to watch out for is capacitive loading on the tank. If the oscilloscope probe had a significant amount of capacitance, do not connect it across the tank circuit. If you do, you will detune the tank severely. A low-capacitance probe must be used. Connecting the oscilloscope across the secondary has much less effect than connecting it across the resonant primary circuit.

Other than the loading and/or detuning of the resonant circuits, IF amplifiers can develop much the same troubles as any other amplifier stage. Transistors can open or short, biasing can change, or transformer coils can open or become shorted.

PROBLEMS

7-1.   If a transformer has 180 turns on the primary and 20 turns on the secondary, what is the turns ratio?

$n =$ *9*

7-2. If a transformer has 60 turns on the primary and the turns ratio is 0.2, how many turns are on the secondary?

$N_S =$ 300

7-3. If 120 VAC is applied to the primary of the transformer in problem 7-1, what is the secondary voltage?

$v_s =$

7-4. If 20 VAC is applied to the transformer primary of problem 7-2, what is the secondary voltage?

$v_s =$

7-5. Referring to problems 7-1 and 7-3, how much current flows through a 10-Ω resistor connected across the secondary?

$i_s =$

7-6. Referring to problems 7-2 and 7-4, how much current flows through a 50-Ω resistor connected across the secondary?

$i_s =$

7-7. Referring to problem 7-5, what is the value of the primary current?

$i_p =$

7-8. Referring to problem 7-6, what is the value of the primary current?

$i_p =$

7-9. A certain transformer has a turns ratio N_p/N_s of 8. What is the reflected resistance if a 16-Ω load is connected across its secondary?

$r_{ref} =$

7-10. What primary to secondary turns ratio is necessary to make a 30-Ω load across the secondary appear as a 2.4-KΩ reflected resistance in the primary?

$n =$

7-11. In the circuit of Fig. 7-8, what reflected resistance does the transistor see looking into the primary?

$r_{ref} =$ 800

7-8   Circuit for problems 7-11 through 7-18.

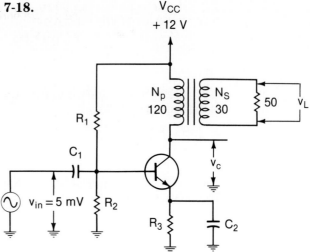

7-12.   In the circuit of Fig. 7-8, if R_L is changed to 8 Ω, what reflected resistance is seen by the transistor looking into the primary?

$r_{ref} =$

7-13.   Referring to problem 7-11, if $r_e = 25\ \Omega$, what is the voltage gain of the amplifier? What a-c voltage would you measure at the collector? What would you expect to measure across the load?

$v_C/v_{in} =$

$v_C =$

$v_L =$

7-14.   Referring to problem 7-12, if $r_e = 8\ \Omega$, what is the voltage gain of the circuit? What a-c voltage would you expect to measure at the collector? What would you expect to measure across the load?

$v_C/v_{in} =$

$v_C =$

$v_L =$

For the next four problems, suppose you are troubleshooting the circuit of Fig. 7-8.

7-15.   If you measure 0 VDC at the collector, which of the following is most likely? (a)Open R_1. (b)Open R_3. (c)Open primary. (d)Open secondary.

7-16.   If no d-c voltage appears across the load, which of the following is true? (a)This is normal. (b)Open secondary. (c)Open primary. (d)Open R_L.

7-17.   If you see a higher than normal a-c signal at the collector, but no signal across the load, the trouble might be (a)an open primary. (b)an open secondary. (c)improper bias. (d)a short across secondary.

E7-1 Circuits for experiment 7-1.

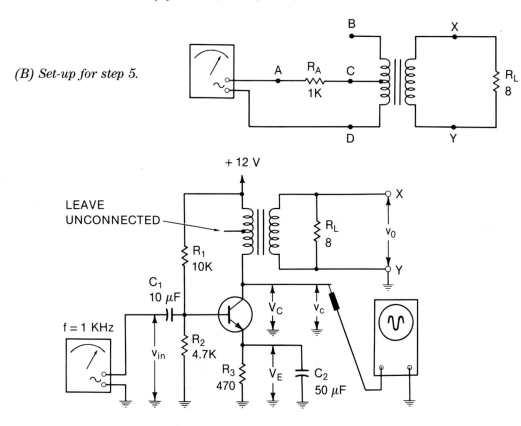

(A) Set-up for steps 2, 3, and 4.

(B) Set-up for step 5.

(C) Set-up for steps 8 through 12.

2.   Next connect the transformer primary to the signal generator through a 1-KΩ resistor, as shown in Fig. E7-1A. Set the generator frequency at 1 KHz, and adjust the amplitude so that the voltage across points A–D measures 4 Vp-p. Measure the primary voltage from point B to D.

$V_{BD} =$

Now measure the secondary voltage from point X to Y.

$V_{XY} =$

Calculate the turns ratio by taking the ratio of the primary voltage to the secondary voltage.

$$n = v_p/v_s =$$

3. Since the current flowing in the primary also flows through R_A, you can determine the impedance reflected into the primary of T_1 by comparing the voltage drop across R_A to that across the primary. In other words

$$\frac{r_{ref}}{R_A} = \frac{v_{BD}}{v_{AB}}$$

where v_{AB} is the voltage across R_A and v_{BD} is the voltage across the primary. Substitute for the known values, and solve for r_{ref}.

$$r_{ref} = R_A \times V_{BD}/V_{AB} =$$

4. Now compare this experimental value of r_{ref} to the theoretical value.

$$r_{ref(theoretical)} = n^2 R_L$$

Use your experimental value for n and solve for the reflected impedance.

$$r_{ref(theoretical)} =$$

5. Next, reconnect your circuit as shown in Fig. E7-1B. Readjust the generator amplitude until you read 4 Vp-p across points A to D again. Measure the voltage from C to D.

$$V_{CD} =$$

Measure the secondary voltage again.

$$V_{XY} =$$

Calculate the turns ratio now.

$$n = v_{pri}/v_{sec} =$$

Determine the reflected impedance as before.

$$r_{ref} = R_A \times v_{CD}/v_{AC} =$$

Calculate the theoretical value of reflected impedance.

$$r_{ref(theoretical)} = n^2 R_L =$$

6. Connect the right end of R_A to point B again, as in Fig. E7-1A. Now disconnect the 8-Ω load from the secondary. Leave the secondary open circuited. Adjust the generator amplitude to get 4 Vp-p from point A to point D. Now measure the voltage across the transformer primary.

$$V_{BD} =$$

7. Finally, short out the secondary from point X to point Y with a clip lead. Readjust the generator voltage to 4 Vp-p again. Now measure the voltage across the primary.

$$V_{BD} =$$

8. Build the circuit of Fig. E7-1 C. Now we will see how the transformer is used in an amplifier. Measure d-c voltages V_E and V_C.

$V_E =$

$V_C =$

9. Connect an oscilloscope between collector and ground, and adjust v_{in} until v_C reads 4 Vp-p. Measure v_{in}.

$v_{in} =$

What is the voltage gain of the circuit?

$A_v = v_C/v_{in} =$

Also measure the output signal across R_L.

$v_0 =$

10. Now change the connections to the transformer so that the collector sees only half the primary. That is, connect the collector to the center tap of the primary, but leave the power supply connected to one end as before. Leave the other end of the primary disconnected. Again measure V_E and V_C.

$V_E =$

$V_C =$

11. Measure the a-c signal at the collector gain.

$v_C =$

Now calculate the voltage again.

$A_v = v_C/v_{in} =$

Measure v_0 across R_L again.

$v_0 =$

12. Leave the transformer primary connected as in step 10, but change R_L to 22 Ω. Do not change v_{in}. Measure v_C and v_0.

$v_C =$

$v_0 =$

How do these values compare to those of Step 11?

QUIZ

1. Based on your calculated value of turns ratio n in step 2, what voltage would you expect to measure across the load if $v_p = 6$ Vp-p?

$v_s =$

2.   If the generator in step 2 were adjusted to 6 Vp-p and if $R_L = 20\ \Omega$, what voltage should appear across points B and D?

$v_{BD} =$

3.   In step 5, connecting the generator to the primary center tap—(a)doubled, (b)halved—the turns ratio and caused the reflected impedence to be—(c)half, (d)one quarter, (e)twice, (f)four times—as great as in step 1.

4.   Step 6 showed that an open secondary reflects a very—(a)high, (b)low—impedance back into the primary.

5.   In step 7, short circuiting the secondary reflects a very—(a)high, (b)low—impedance into the primary.

6.   The d-c measurements in Step 8 showed that (a)the transistor was cut off. (b) the transistor was saturated. (c)there was very little d-c across the transformer primary.

7.   The fact that an a-c voltage could be measured at the collector in step 9 showed that (a)the transistor sees a high a-c resistance, even though the d-c resistance is low. (b)the a-c signal is developed across the low d-c resistance of the transformer.

8.   When viewing the signal at the collector, we see that the peak voltage appearing across the transistor—(a)does, (b)does not—exceed the power supply voltage.

9.   In step 11, the voltage gain—(a)went up, (b)went down, (c)remained the same—compared to step 9, because the reflected impedence was (d)higher. (e)lower. (f)the same.

10.   Step 12 showed that the voltage gain—(a)increases, (b)decreases, (c)remains the same—if the secondary load is increased.

EXPERIMENT 7-2   TUNED-TRANSFORMER COUPLING

This experiment demonstrates the use of a tuned transformer for coupling a narrow band of frequencies from one amplifier stage to another. It also shows the effect of a parallel LC circuit near resonance.

The tuned transformer (IF can) used in this experiment is taken from a portable transistor radio kit manufactured by Graymark International, Inc. Since IF cans are usually designed for specific radios, universal replacement types may be difficult to find. Therefore, if you cannot obtain the part called for in the parts list, you can substitute an IF transformer from an old broken portable radio. Just make sure that the can is designed for operation at 455 KHz and that it has a tuned primary and untuned secondary, as shown in Fig. E7-2A. Your measured values may differ from those obtained with the specified part, but the general operation should be about the same.

EQUIPMENT
- 9-VDC power supply
- oscilloscope and low-capacitance probe
- voltmeter

- signal generator (variable around 455 KHz)
- transformer, IF (455 KHz) Graymark International, Inc., 62718 or equivalent
- (1) NPN transistor
- (2) 0.1-μF disc capacitor
- (1) 47-KΩ, ½-W ± 5% resistor
- (1) 10-KΩ, ½-W ± 5% resistor
- (1) 4.7-KΩ, ½-W ± 5% resistors
- (2) 1-KΩ, ½-W ± 5% resistors

E7-2   Set-ups for experiment 7-2.

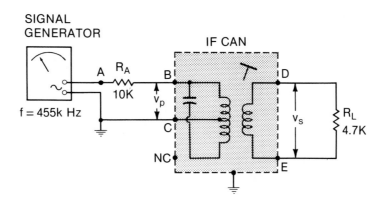

(A) Circuits for steps 1 through 7.

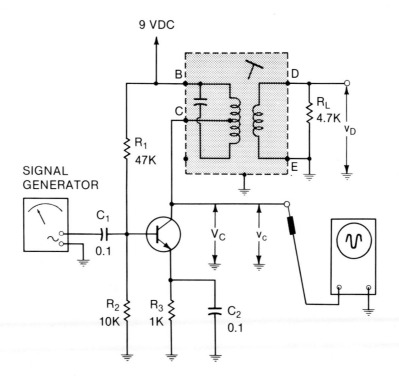

(B) Circuits for steps 8 through 12.

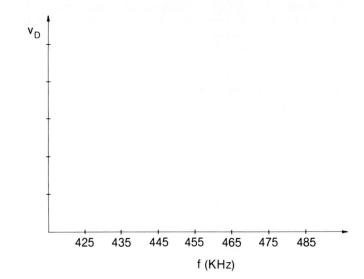

(C) Response curve.

PROCEDURE

1. Assemble the circuit of Fig. E7-2A. Using a low-capacitance attenuator probe, connect an oscilloscope input to point B and the oscilloscope ground to point C. Rock the signal generator frequency control back and forth from about 400 to 500 KHz. You should notice a peak (maximum signal) as the generator frequency passes through 455 KHz.

If the maximum amplitude does not occur at $f = 455$ KHz, adjust the tuning slug on the coil. Check with your lab instructor before adjusting the tuning slug, because some coils are easily damaged if adjusted more than a couple of times.

If it is OK to tune your coil, adjust your signal generator to exactly 455 KHz. Now with a nonmagnetic (plastic) tool, screw the slug in or out as needed, while watching the oscilloscope until the observed signal peaks (reaches maximum amplitude). Leave the adjustment at this setting for the remainder of this experiment.

2. With the signal generator still at 455 KHz, move the oscilloscope probe to point A, and adjust the generator amplitude to exactly 1 Vp-p. Move the probe back to point B, and measure the amplitude of signal across the primary winding.

$v_p =$

Using the technique from Experiment 7-1, determine the impedance of the primary at this resonant point.

$Z_{p(\text{at resonance})} =$

3. Next, connect the oscilloscope probe to point A again, and change the generator frequency to 425 KHz (30 KHz below resonance). Adjust the generator amplitude to 1 Vp-p again, then measure the voltage across the primary.

$v_p =$

Calculate the tank impedance at this frequency.

$Z_{p\ (425\ \text{KHz})} =$

4. Now readjust the generator frequency to 485 KHz. Adjust the generator amplitude to 1 Vp-p at point A again, and measure v_p.

$v_p =$

Calculate the tank impedance.

$Z_{p\,(485\,\text{KHz})} =$

5. Next, adjust the generator amplitude to 1 Vp-p at f = 455 KHz, as you did in step 2. Connect the oscilloscope across the secondary (points D and E), and measure the secondary voltage.

$v_{s(\text{at resonance})} =$

6. Leave the oscilloscope across the secondary. Do not change the generator amplitude, but rotate the frequency output control to f = 425 KHz. Measure the secondary output voltage.

$v_{s\,(425\,\text{KHz})} =$

7. Repeat the measurement of step 6 at f = 485 KHz.

$v_{s\,(485\,\text{KHz})} =$

8. Build the circuit of Fig. E7-2B. Apply power, and measure the d-c voltage at the collector.

$V_C =$

9. Set the generator amplitude to 5 mVp-p. Connect the oscilloscope across the transformer secondary (points D–E). Now, starting with f = 485 KHz, measure the a-c voltage at the collector (point C) and also across the secondary (point D). Enter the values in the space provided below. Repeat for each frequency shown. Use a frequency counter across the generator output, if one is available; otherwise try to read the generator frequency dial carefully. Be sure to check that v_{in} = 5 mVp-p at each new setting.

f (KHz)	v_C	v_D
425	_____	_____
435	_____	_____
445	_____	_____
455	_____	_____
465	_____	_____
475	_____	_____
485	_____	_____

10. Calculate the voltage gain v_C/v_{in} of the amplifier at the following frequencies:

f (KHz)	A_v
425	_____
455	_____
485	_____

11.   Using the measured values of secondary voltage v_D, plot the frequency response curve for the tuned amplifier in Fig. E7-2C.

12.   Connect a 1-KΩ resistor across the secondary, in place of the 4.7 KΩ. Rock the frequency control back and forth through resonance a few times while watching the secondary waveform with an oscilloscope. What do you notice about the sharpness of the tuning?

QUIZ

1.   The impedance of a parallel L–C circuit is—(a)maximum, (b)minimum—at resonance.

2.   When a parallel L–C circuit is fed from a high-resistance source, the voltage developed across the tank reaches a—(a)maximum, (b)minimum—value when the applied frequency passes through resonance.

3.   When a secondary coil is coupled magnetically to a parallel L–C circuit, the voltage induced in the secondary reaches a—(a)maximum, (b)minimum—at resonance.

4.   The d-c collector voltage of a transistor driving a tank circuit usually measures near V_{CC}. (a)True. (b)False.

5.   The voltage gain of a parallel tuned amplifier is—(a)maximum, (b)minimum—at resonance.

6.   The voltage gain of a tuned amplifier—(a)increases, (b)decreases—above and below resonance because the load seen by the collector is—(c)higher, (d)lower—than at resonance.

7.   If the primary coil should open between the center tap and the collector, the d-c collector voltage would—(a)go up, (b)go down, (c)remain the same—and the secondary a-c voltage would (d)go up, (e)go down, (f)remain the same.

8.   If the tank capacitor should short, the d-c collector voltage would—(a)go up, (b)go down, (c)remain the same—and the secondary a-c voltage would (d)go up. (e)go down. (f)remain the same.

9.   When the load on the secondary of a tuned transformer is made smaller, the tuning (a)sharpens. (b)broadens. (c)remains the same.

10.   The bypass and coupling capacitors C_1 and C_2 used in this experiment are much smaller than those used in Experiment 7-1. Why?

Operational Amplifiers 8

In chapter 5, you learned that multistage amplifiers are needed for high gain. Problems of drift and mismatch become very difficult to overcome when it is necessary to obtain high gain of small d-c signals, say 1000 or more. Yet such amplifiers are needed in instrumentation and analog computer circuits. Direct-coupled amplifiers with exceptionally high gain and good stability were developed to handle mathematical operations in analog computers. These special circuits became known as *operational amplifiers*, or simply *op amps*. Although we now use these amplifiers for a wide variety of applications, the name op amp has stuck.

OP AMP CIRCUITRY

What gives the op amp its stability is its differential input stage. The discrete component differential stage was discussed in chapter 5. Recall that the output signal is an amplified version of the *difference* signal applied between the bases of the two input transistors. Practical op amps have just such an input stage, followed by additional amplifier stages to increase the gain well up into the thousands.

Discrete component op amps were very expensive because of the number of parts required and the necessity of carefully matching their characteristics. But in the late 1960s, a giant step forward was made. Manufacturers began building single-chip (monolithic), IC op amps. All of the transistors, resistors, and diodes necessary for the entire circuit were combined on a single silicon crystal chip. Since all components are made on the same crystal and in close proximity, their characteristics are very similar. And, because so little material is used and clever fabrication processes have been found, these single chip op amps cost little to produce. Today, many IC op amps can be purchased for less than a dollar each. A similar quality op amp made from discrete components would probably sell for well over a hundred dollars.

Figure 8-1 shows a simplified version of a typical IC op amp. Note that the input differential stage is quite similar to the circuit of Fig. 5-18.

Additional stages, which further amplify the difference signal and shift the d-c level of the output signal down toward ground, follow the input stage. (One characteristic of an op amp is that the output signal should be at *0 VDC* when the *input difference* signal is zero.) Also notice that the op amp has a single-ended output. That is, only one lead is brought out, and the output signal is measured from ground to the output terminal. If the output is at 0 V when the input difference is zero, the amplifier is said to be *nulled*. Many amplifiers have balance terminals, which are used to null the output to zero. This null control compensates for any slight variation in component characteristics.

Once nulled with no input signal, the output can swing in both the positive and negative directions when signal is applied. This is similar to the power amp with two supplies in chapter 6. In

8-1 Simplified typical IC op amp.

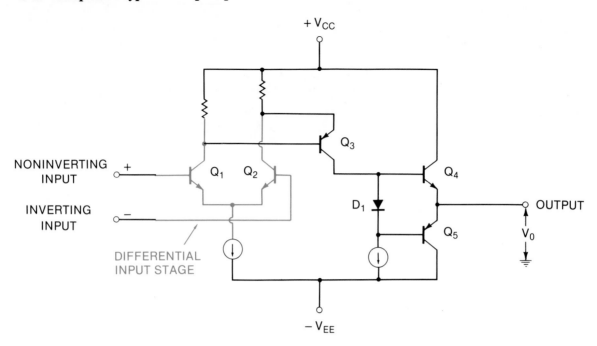

fact, if you examine the complementary symmetry output stage of Fig. 8-1, you should see the similarity between it and the power amp final stages that you are familiar with. This power amp output stage gives the op amp low-output impedance.

Fig. 8-2 shows the 8-pin dual inline package (DIP) and the pinouts of a 741 op amp. The 741 is one of the most popular op amps in use because it is inexpensive and reliable. National Semiconductor calls it LM741, and the Fairchild number is μA741, but the circuits are the same.

8-2 741 op amp in 8-pin DIP.

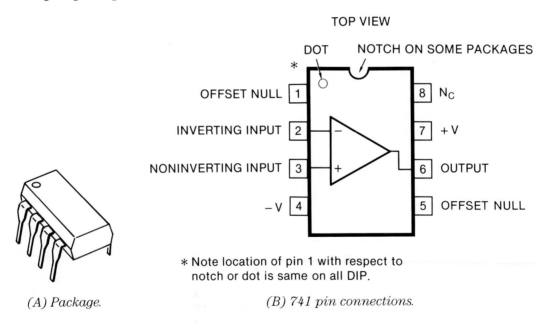

(A) Package.   *(B) 741 pin connections.*

The 741 is made with all bipolar transistors; therefore, the input transistors need some bias current. A newer type of op amp, called a *BIFET* (bipolar-FET), uses field effect transistors for the two input transistors, while the remainder of the amplifier is bipolar. As a result, the input impedance

8-3   Simplified schematic of LF351 BIFET op amp.

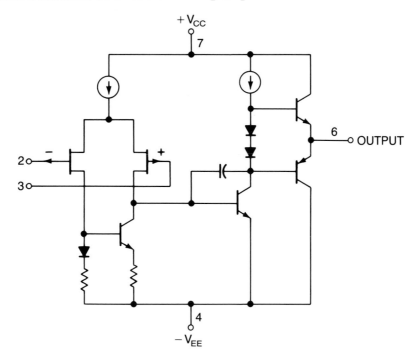

of the BIFET is extremely high, which is desirable for instrumentation circuits. The LF351 BIFET is a pin-compatible replacement for the 741 and is being used in upgraded circuits. The LF351 has an input impedance of 1 tΩ (teraohm) (10^{12}). In case you're wondering how large a resistance of 1 tΩ is, be assured that it is large. You would need more than *50 miles* of 1-MΩ resistors, soldered end to end, to make 1 tΩ! Fig. 8-3 shows a simplified version of the LF351.

BASIC CONFIGURATIONS OF OP AMP CIRCUITS

Op amps are used in a wide variety of circuit arrangements, but there are a few more or less standard configurations that are commonly used. Before we study these standard arrangements, let's take a look at what characteristics an ideal amplifier would have.

First of all, remember that the op amp has several stages of amplification on the single chip, so that its open-loop gain is extremely high. The open-loop gain is defined as the ratio of the output voltage to the input differential voltage with no feedback components. See Fig. 8-4. In equation form,

$$A_{VOL} = \frac{V_0}{V_{id}}$$

8-1

8-4   Op amp without feedback.

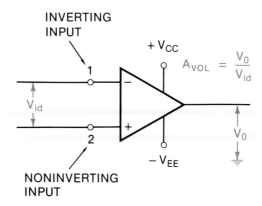

Ideally, the *open-loop gain should be infinite.* Of course in practical amplifiers, the gain is not infinite, but it is extremely high, on the order of several hundred thousand to over a million. An ideal op amp should also have *infinite input impedance,* seen looking into its input terminals, as well as *zero output impedance.*

Once again, practical op amps are not exactly ideal, but input impedances are often in the hundreds of megaohms, and output impedances less than 1 Ω. So for most applications, we can say that the characteristics of real op amps are very close to ideal.

Three of the most commonly used op amp circuit configurations are the *inverting* amp, the *noninverting* amp, and the *voltage follower.* We will look at these circuits as used to amplify d-c signals. (The same equations apply to a-c signal amplification as well.)

Fig. 8-5 shows the circuit of an *inverting* amplifier with its feedback and input resistors. As in the circuit of Fig. 8-4, the ratio of V_0 to V_{id} is A_{VOL}. Now if V_0 has some finite value less than the power supply voltage (say ± 15 V) and if the open loop gain is infinite (or at least extremely high), we see that

$$V_{id} = \frac{V_0}{\infty} \approx 0$$

In other words, for any finite value of V_0, the input differential voltage is so small that we can say it is *virtually* equal to zero. We will use this approximation later.

8-5   Inverting amplifier.

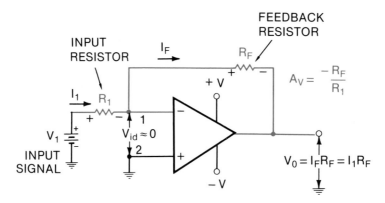

If an input signal is applied to the left end of resistor R_1, as shown in Fig. 8-5, it will cause a current to flow through the resistor. Also, since $V_{id} \approx 0$, terminal 1 on the amplifier is at about the same potential as terminal 2, namely at ground. We say that terminal is at *virtual* ground, because it is not actually shorted to ground, but is at zero potential. The current through R_1 is then

$$I_1 = \frac{V_1}{R_1} \qquad\qquad 8\text{-}2$$

From this we see that *the input impedance seen by the source is just resistance R_1.*

Now we can determine the output voltage. If the op amp has nearly an infinite input impedance, all of the current that flows through R_1 must flow through R_F. In other words, $I_F = I_1$. When I_F flows through R_F, it causes a voltage drop across it equal to

$$V_0 = I_F R_F = I_1 R_F \qquad\qquad 8\text{-}3$$

The voltage drop across R_F is the same as the output voltage measured from ground because the left end of R_F is at virtual ground. Dividing equation 8-3 by equation 8-2, we get the *closed-loop voltage gain A_v,*

$$A_v = \frac{V_0}{V_1} = \frac{-I_1 R_F}{I_1 R_1} = \frac{-R_F}{R_1} \qquad\qquad 8\text{-}4$$

The negative sign indicates that the polarity of the output voltage is opposite from that of the input voltage.

EXAMPLE 8-1   In the circuit of Fig. 8-5, if $R_F = 100$ KΩ and $R_1 = 2$ KΩ, what is the voltage gain of the circuit, and what is the input impedance seen by the driving source?

SOLUTION

$$A_v = \frac{-R_F}{R_1} = \frac{-100 \text{ K}\Omega}{2 \text{ K}\Omega} = -50$$

$$Z_{in} = R_1 = 2 \text{ K}\Omega$$

The diagram of Fig. 8-6 is that of a *noninverting* amplifier. Note that input voltage of V_2 is applied between ground and terminal 2. Since the input signal is applied to the noninverting input, the output polarity is the same as that of the input. Again, assume voltage $V_{id} \approx 0$, which means that a voltage equal to V_2 also appears across R_1. This causes current I_1 to flow through R_1. Since no current flows into or out of the amplifier input terminals themselves, the same current that flows through R_1 also flows through R_F. Since the left end of R_1 is grounded, the output voltage V_0 is the sum of the voltage drops across R_F and R_1. That is,

$$V_0 = I_1 R_F + I_1 R_1 = I_1(R_F + R_1)$$

Since $V_{id} = 0$, the voltage across R_1 is equal to V_2 or

$$V_2 = I_1 R_1$$

Solving for closed loop voltage gain,

$$A_v = \frac{V_0}{V_2} = \frac{I_1 (R_F + R_1)}{I_1 R_1}$$

or   8-5

$$A_v = \frac{R_F}{R_1} + 1$$

The input impedance of a noninverting amplifier is much higher than for an inverting amp, since almost no current flows into terminal 2. Practical values run into many megaohms.

8-6   Noninverting amplifier.

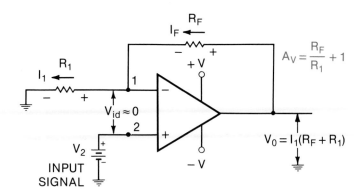

EXAMPLE 8-2 In the circuit of Fig. 8-6, if $R_F = 20$ KΩ and $R_I = 1$ KΩ, what is the voltage gain?

SOLUTION

$$A_v = \frac{R_F}{R_1} + 1 = \frac{20 \text{ KΩ}}{1 \text{ KΩ}} + 1 = 21$$

Chapter 5 discussed the emitter follower, which is used to match a low-impedance load to a high-impedance source. The op amp equivalent of an emitter follower is a *voltage follower*. Its circuit is shown in Fig. 8-7. The input signal is applied to terminal 2, which produces an output voltage that is in phase with the input. Notice that the output is directly tied to input terminal 1. So since $V_{id} \approx 0$, V_0 is always the same amplitude and the same phase as the input signal voltage. The input impedance of a typical voltage follower is on the order of hundreds of megaohms, and its output impedance is usually less than 1 Ω.

8-7 Voltage follower.

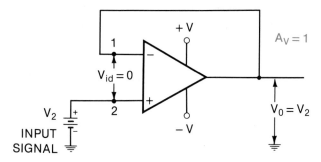

AMPLIFYING A-C SIGNALS WITH OP AMPS

Besides being able to handle d-c signals, op amps can be used to amplify small a-c signals. Because of its high gain, a single op amp chip can replace transistor circuits containing many discrete components.

The circuit of Fig. 8-8 show an inverting amplifier being used to amplify the signal for generator source v_g. Coupling capacitor C_c is used for the same purpose as it is used in discrete component circuits. It blocks any d-c level the source might have, and it couples the a-c signal to the amplifier input.

8-8 Op amp used to amplify a-c signals.

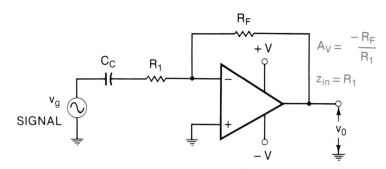

The gain of the a-c amplifier is equal to the ratio of R_F/R_1, just as in the d-c amp circuit. And once again, the input impedance of the amplifier is equal to R_1, making it easy to match to any driving source.

You might wonder why discrete components are used at all anymore, if op amps are so easy to use. Frequency response is one reason. Typical op amps, like the 741, cannot have high gain at high frequencies. Fig. 8-9 shows the open loop response curve for the 741, which is published by the manufacturer. Notice that the open-loop gain is about 200,000 at dc (or near 0 Hz), and falls off as the frequency increases. Observe that the gain drops to 1000 at 1 KHz, down to 10 at 100 KHz, and is unity at 1 MHz.

8-9   Open loop voltage gain versus frequency for 741 op amp.

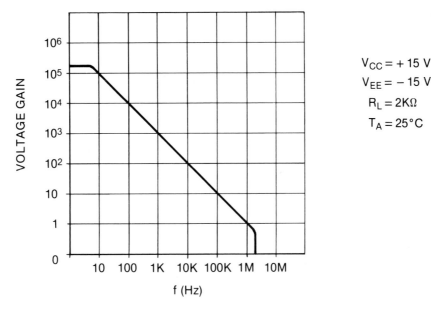

To find the bandwidth of an amplifier, we simply draw a horizontal line on the graph indicating its closed-loop gain. This line intersects the horizontal axis at the maximum frequency that the amplifier will handle.

EXAMPLE 8-3 In the circuit of Fig. 8-8, suppose $R_F = 200$ KΩ, $R_1 = 2$ KΩ, and the amplifier is a type 741. What is the gain of the circuit at low frequencies, and what is the highest frequency it can amplify?

SOLUTION

$$A_v = \frac{-R_F}{R_1} = \frac{-200 \text{ K}\Omega}{2 \text{ K}\Omega} = -100$$

Next, plot a horizontal line on the response curve at Voltage Gain = 100, as shown in Fig. 8-10. Then drop a vertical line from the intersection. The gain will then fall off beyond 10 KHz.

In other words, the amplifier in this example would not be suitable over the entire audio spectrum. On the other hand, if we could get by with less gain, the bandwidth will increase.

The low frequency cut-off point for the circuit of Fig. 8-8 is determined by the values of C_c and the input resistance by R_1. The 3-dB cutoff occurs when $X_C = R_1 + R_{\text{(generator)}}$, as in the case of discrete component circuits.

The 741 is said to be *internally compensated*. This means that it has internal components that are used to roll off the response as indicated in Fig. 8-9. There are some operational amplifiers available that need to be *externally* compensated to prevent the amplifier from going into oscillation. The 709 op amp shown in Fig. 8-11 is such a circuit.

Components C_1, R_2, and C_2 are chosen by the designer to give the amplifier its desired bandwidth. We will not go into this selection here. Just keep in mind that they control the gain of the amp at high frequencies. Without these components, the circuit becomes unstable.

Sometimes a voltage follower is needed to act as a high input impedance to a-c signals. At first thought, it seems like the circuit of Fig. 8-12A is all that is needed. But notice that there is no d-c

8-10 Finding upper cut-off frequency.

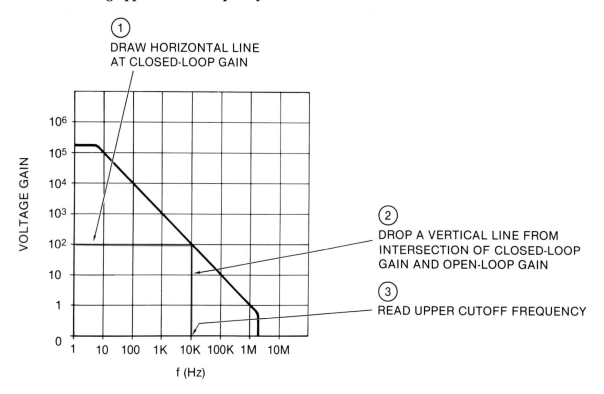

8-11 External components used to compensate op amp.

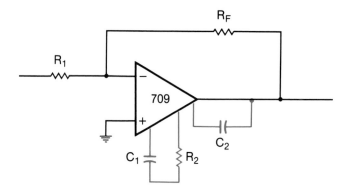

return to ground on the noninverting input. Remember that the input terminals connect to the bases of two transistors in a differential arrangement. If we do not provide a d-c path for bias current to flow into the base of the noninverting transistor, it will be cut off.

What if we connect a resistor R to ground, as shown in Fig. 8-12B? It would provide a d-c path, but then the generator sees R as the input impedance to the amplifier. If $R = 100$ KΩ, or say $R = 1$ MΩ is high enough, the circuit of Fig. 8-12B works fine. But if the generator must work into many megaohms, say 100 MΩ, it is impractical to use a resistor that large. The solution is to use a bootstrapping technique, similar to that method introduced in chapter 6. Fig. 8-12C employs bootstrapping.

In Fig. 8-12C, the output signal is coupled back to the junction of R_A and R_B via capacitor C_F. Since the circuit is a voltage follower, the output signal is of the same amplitude and same phase as that of the input signal. Since points X and Y have exactly the same amplitude and phase of signal, *there is no a-c drop across R_A*. Therefore, no alternating current flows through R_A. In other words, R_A *looks like an open circuit* as seen by the generator. So the generator sees only the input impedance of the amplifier chip itself, which may be hundreds of megaohms.

8-12 Voltage follower in a-c circuit.

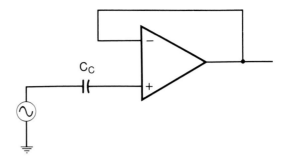

(A) No d-c return to ground on noninverting input. This circuit will not work.

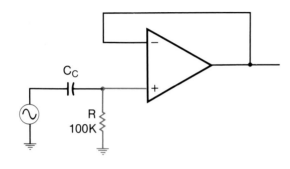

(B) Resistor R provides bias current path.

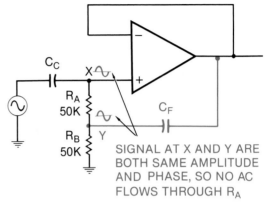

SIGNAL AT X AND Y ARE BOTH SAME AMPLITUDE AND PHASE, SO NO AC FLOWS THROUGH R$_A$

(C) Bootstrapping increases input impedance.

OP AMPS IN THE DIFFERENTIAL MODE

The circuit arrangement shown in Fig. 8-13 is called the differential mode or simply a difference amplifier. In this configuration, the circuit amplifies the *difference* between the two input voltages V_1 and V_2. The differential gain is equal to the ratio of R_F/R_1, just like the inverting amplifier. The differential gain refers to the ratio of V_0 to the *difference* of $V_1 - V_2$.

For example, if $V_1 = V_2 = 0$, then of course $v_{id} = 0$, so $V_0 = 0$. Now if we increase both V_1 and V_2, say to 3 V each, v_{id} is still equal to zero, as long as V_1 and V_2 are exactly the same. Therefore, V_0 is still zero.

Since V_1 and V_2 are both the same amplitude and same polarity, we say that the voltage is *common* to both inputs. It is, therefore, called a *common-mode* signal. The ability of a differential

8-13 Difference amplifier.

$$A_V = V_0/(V_1 - V_2) = -R_F/R_1$$

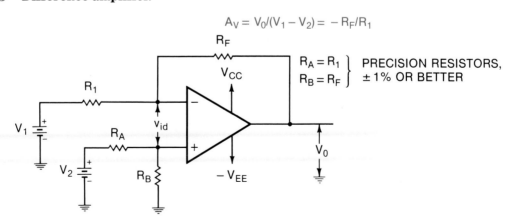

$$\left.\begin{array}{l} R_A = R_1 \\ R_B = R_F \end{array}\right\} \text{ PRECISION RESISTORS,} \pm 1\% \text{ OR BETTER}$$

amp to ignore, or reject, this common-mode signal is called its *common-mode rejection ratio* (CMRR). Ideally, the CMRR is infinite, but in practical amplifiers it is very high, perhaps on the order of 20,000 or 30,000. Manufacturers often specify the common-mode rejection in *decibels* rather than as a ratio, where

$$\text{CMR (in dB)} = 20 \log_{(10)} \text{CMRR} \qquad\qquad 8\text{-}6$$

EXAMPLE 8-4 Express a CMRR of 30,000 in decibels.

SOLUTION

$$\text{CMR} = 20 \log_{(10)} 3 \times 10^4 = 20 \times 4.48 = 89.5 \approx 90 \text{ dB}$$

The ratio of the common-mode output voltage V_{cmo} to the common-mode input voltage V_{cm} is called the common-mode gain, A_{cm}. We use the CMRR to find the common-mode gain by the equation

$$A_{cm} = \frac{V_{cmo}}{V_{cm}} = \frac{A_v}{\text{CMRR}} \qquad\qquad 8\text{-}7$$

where

$$\begin{aligned} A_{cm} &= \text{common mode gain} \\ A_v &= \text{closed-loop gain} \\ \text{CMRR} &= \text{common-mode rejection ratio specified by the manufacturer} \end{aligned}$$

EXAMPLE 8-5 In the circuit of Fig. 8-13, suppose $V_1 = V_2 = 3$ V, $R_F = R_B = 100$ KΩ, and $R_1 = R_A = 10$ KΩ. What is the value of the common-mode gain and the common-mode output signal? Assume CMRR = 20,000.

SOLUTION

$$A_v = \frac{R_F}{R_1} = \frac{100 \text{ KΩ}}{10 \text{ KΩ}} = 10$$

$$A_{cm} = \frac{A_v}{\text{CMRR}} = \frac{10}{20,000} = 0.5 \times 10^{-3}$$

then

$$V_{cmo} = V_{cm} \times A_{cm} = 3 \times 0.5 \times 10^{-3} = 1.5 \text{ mV}$$

Note that this common-mode output signal is so small that it can be considered negligible. In order for the common-mode signal to be rejected, resistors R_1, R_F, R_A, and R_B should be precision resistors.

The difference amplifier is often used to amplify signals from a bridge circuit. Fig. 8-14 shows a difference amplifier connected to a bridge circuit in which one of the resistances is a thermistor. The thermistor is a heat-sensitive element whose resistance value decreases as the temperature increases. (The thermistor has a negative temperature coefficient.)

Suppose that at some starting temperature like 0°C (32°F) the resistance of the thermistor is near 100 Ω, as shown. The slider on pot R_W is adjusted to *balance* the bridge. That is, R_W is adjusted so that its resistance is made exactly equal to the resistance of R_T. This makes the voltage from point Y to ground exactly the same as the voltage from point X to ground, namely 2 V. Since the difference voltage applied to the amplifier is zero, the output voltage is zero *even though both inputs have +2 VDC applied!*

8-14   Difference amplifier connected to thermistor bridge.

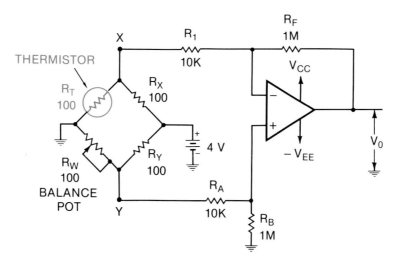

Now when the temperature of the thermistor increases, its resistance decreases. This causes the voltage at point X to become less positive than the voltage at point Y. The *difference* voltage is amplified by the op amp circuit and shows up as a positive output voltage V_0. Because of the closed loop voltage gain of the circuit (100), any slight variation in temperature causes a substantial voltage change at the output. The output voltage can be used to indicate temperature or to control some device, such as a heater or cooler.

Occasionally, difference amplifiers are operated open loop, that is, with no feedback components, as shown in Fig. 8-15A. This circuit is called a *zero crossing detector*, because its output switches to V_{CC} or $-V_{EE}$ whenever the input crosses zero. Let's see what happens when a signal is applied to one of its terminals.

Fig. 8-15B shows a sine-wave a-c signal being applied to the noninverting input, while the inverting input is connected to ground. As soon as the applied signal goes slightly positive, by just a few millivolts, the op amp tries to amplify this signal by the open loop gain A_{VOL}. Suppose the open loop gain is 1,000,000. The op amp tries to amplify this input signal 1,000,000 times. Of course, the output cannot go thousands of volts positive, so it is driven as far positive as it *can* go. In other words, V_0 quickly goes approximately to V_{CC}. Then as the input signal goes negative by a few millivolts, the output is driven quickly to $-V_{EE}$.

INTEGRATOR

Another special circuit that does not use a feedback resistor is the *integrator*. As shown in Fig. 8-16, capacitor C replaces feedback resistor R_F. When voltage V_1 is applied as shown, current flows through R_1, whose magnitude is $I_1 = V_1/R_1$. Since essentially no current flows into the inverting input, all of I_1 flows into capacitor C. That is, $I_C = I_1$. Under the condition where V_1 is constant, I_C remains constant. Now if a capacitor is charged with a constant current, a linearly changing voltage, or *linear ramp*, is developed across the capacitor. The rate of change of output voltage with respect to time can be found using the equation:

$$\frac{\Delta V_0}{\Delta t} = \frac{-V_1}{R_1 C}$$

8-8

EXAMPLE 8-8   Suppose in the circuit of Fig. 8-16, $R_1 = 1$ MΩ, $C = 2$ μF, and $V_1 = +3$ V. Find the rate of change of output voltage.

SOLUTION

$$\frac{\Delta V_0}{\Delta t} = \frac{-V_1}{R_1 C} = \frac{-3}{1 \times 10^6 \times 2 \times 10^{-6}} = -1.5 \text{ V/s}$$

8-15   Zero crossing detector.

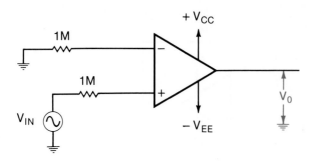

(A) Basic circuit.

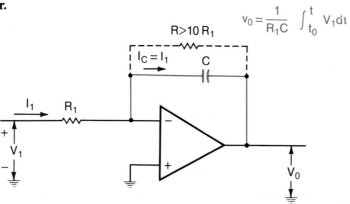

WHEN V_{IN} GOES
SLIGHTLY NEGATIVE,
V_0 SWITCHES TO $-V_{EE}$

WHEN V_{IN} GOES
SLIGHTLY POSITIVE,
V_0 SWITCHES TO $+V_{CC}$

(B) Input signal.

(C) Output signal.

8-16   Integrator.

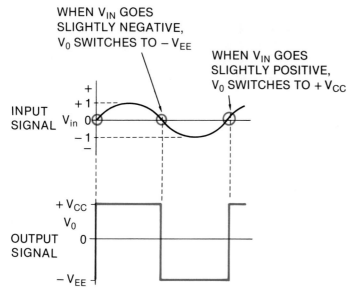

$$R > 10 R_1 \qquad V_0 = \frac{1}{R_1 C} \int_{t_0}^{t} V_1 dt$$

See Fig. 8-17. The output voltage will continue to ramp down until it reaches the $-V_{EE}$ power supply value.

In general, the output of the integrator is equal to the integral of the input voltage, or

$$V_0 = \frac{1}{R_1 C} \int_{t_0}^{t} V_1 dt \qquad\qquad 8\text{-}9$$

where

V_0 = output voltage
V_1 = input voltage (may be time varying)
R_1 = input resistance
C = feedback capacitor
t_0 = starting time
t = total elapsed time

8-17 V_0 versus time for integrator.

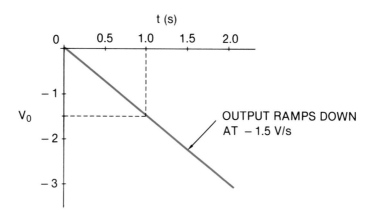

Practical op amp integrators usually have a large resistor, 10 MΩ or so, across capacitor C as shown by the dashed lines in Fig. 8-16. The reason for the resistor is to prevent capacitor C from charging up as a result of slight bias currents in the amplifier.

COMPARATOR

A special-purpose op amp, which is used extensively in switching and interfacing circuits, is the *comparator*. The comparator is designed to have its output switched to either V_{CC} or ground, depending on its inputs. In most amplifiers we have studied thus far, the amplifier was supposed to have an output signal which looked like its input signal, except perhaps larger in value. We normally avoided cutoff and saturation. The comparator, on the other hand, always has its output either saturated or cut off, never in between.

Fig. 8-18A shows the pinouts of an LM339 comparator chip. The LM339 has four comparators on the single chip and normally uses only a single power supply rather than dual supplies, as does the 741. Each comparator has an open-collector output. *Open collector* means that there is no internal collector resistor; the user must supply one. See Fig. 8-18B. Inside the package is a high-gain, direct-coupled amplifier, like in the 741. But the output of the chip is a single transistor which is driven either to cutoff or saturation, depending on the input signal. Whenever the (+) input terminal is driven positive with respect to the (−) input, the transistor is cut off. The external pull-up resistor pulls the output up to V_{CC}. The external pull-up resistor can actually be tied to a voltage which is higher or lower than the V_{CC} supply of the chip, thus making the chip more versatile. Likewise, when the (+) input is less positive than the (−) input (even by a few millivolts), the transistor is saturated, thus pulling the output to ground. So you can see that the (+) terminal acts like the noninverting input of the ordinary op amp. The LM339 can operate with V_{CC} supply voltages between 2 and 18 V.

A couple of examples will help to show how the comparator is used. Fig. 8-19A shows the diagram of a circuit that can be used to change small-amplitude sine waves into large-amplitude square waves of the same frequency. Notice that the (−) input is grounded, and the input signal is applied to the (+) input of the chip. Whenever the input signal, shown in part B, goes slightly positive with respect to ground, the output of the comparator gets pulled high to + 5 V by the pull-up resistor because the output transistor is switched off. But when the input signal goes slightly

8-18 LM339 comparator.

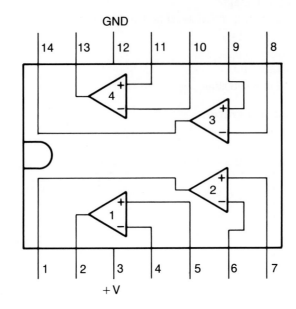

(A) Dual in-line package (top view).

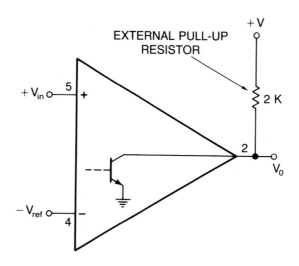

(B) Open-collector output.

below ground, the output goes low (to ground potential) because the output transistor switches on.

The amplitude of the input signal to the circuit could be very small (a few millivolts) or large (several volts), and the output will still be a square wave of 5 V peak amplitude. This type of squaring circuit is very useful in digital or computer circuits. The digital circuits cannot process sine waves or any small-amplitude signals. So the comparator is used to make the input signals compatible with the digital circuits.

8-19 Comparator squaring circuit.

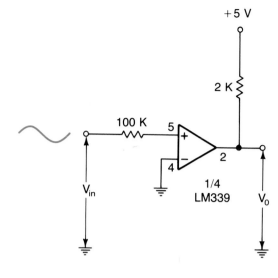

(A) Circuit

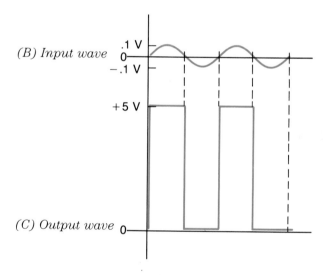

(B) Input wave

(C) Output wave

Another example using the comparator is shown in Fig. 8-20, the LO-BATTERY indicator. In this application, we want an alarm buzzer to sound when the terminal voltage of the 6-V battery drops too low. (The 6-V battery could possibly be the power supply battery of some system.) Notice that the ($-$) input to the comparator is connected to a separate 1.5-V mercury reference battery. The ($+$) input is connected to a voltage divider connected across the 6-V battery. If the terminal voltage of the 6-V battery is near $+6$ V, the voltage at the ($+$) input of the comparator is more positive than that of the ($-$) input, so the output transistor of the comparator is cut off. But as the 6-V battery gets weaker, its terminal voltage drops, thus lowering the voltage to the ($+$) input of the comparator. With the values of R_1 and R_2 shown, when the terminal voltage of the 6-V battery drops below about $+5$ V, the voltage at the ($+$) input of the comparator will drop below 1.5 V, causing the output transistor to turn on. This completes the circuit for the alarm buzzer, causing it to sound. The buzzer chosen for this application is a piezoelectric type, commonly available these days, which draws only a few milliamps when buzzing.

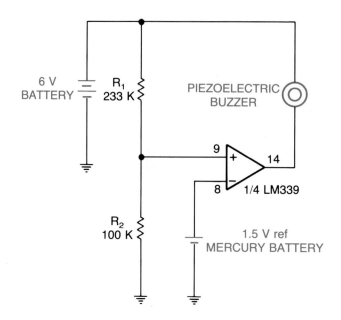

There are other comparators besides the LM339, but it is one of the most commonly used. Another interesting chip is the TLC374 comparator. It is the MOS version of the LM339, which means that it draws less power supply current and has an extremely high input impedance, about 1 TΩ. The TLC374 is also pin-compatible with the LM339, which means that you can unplug an LM339 from its socket and replace it with a TLC374 and the system will work exactly the same—no circuit changes are needed.

Comparator circuits are very easy to test and troubleshoot. Let's suppose you want to test the comparator in the LO-BATTERY indicator circuit. All you have to do is bridge resistor R_2 with another resistor so that the voltage at the (+) input drops below the voltage at the (−) input. When you do, the alarm should sound. If it doesn't, you might still want to check whether the alarm itself is OK. All you have to do is short the output of the comparator (pin 14) to ground with a clip lead. This will sound the alarm. No damage will be done to the comparator by shorting its output to ground.

LOCATING FAULTS IN OP AMP CIRCUITS

Now that you know how several op amp circuits work, let's discuss some of the problems you may encounter when working with them. Although there is nothing inside an op amp chip package to "wear out," occasionally a chip does go bad, from fatigue, or whatever, and has to be replaced. However, many times the failure of the device is the result of misuse or improper handling. Some of the reasons the chips are damaged follow.

POWER SUPPLY VOLTAGE TOO HIGH

Manufacturers specify a maximum safe power supply voltage. This is often in the neighborhood of ± 18 V, although sometimes higher for some units. If this voltage is exceeded, the op amp will be permanently damaged.

POWER SUPPLY CONNECTIONS REVERSED

If an op amp is inadvertently connected into the circuit backwards, so that the V_{CC} and $-V_{EE}$ connections are reversed, the amp will be ruined. One way to help prevent this from occurring, in breadboard or test circuits, is to connect diodes in series with the power supply leads, as shown in Fig. 8-21.

8-21   Protecting an op amp from reverse supply voltages.

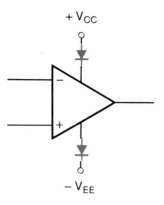

INPUT VOLTAGE SWING TOO LARGE

If the input leads are driven with signals larger than the maximum recommended value, the inputs to the amplifier will be damaged. The maximum input signal is usually less than the power supply voltage being used. The circuit of Fig. 8-22 can be used to help protect the inputs. One or the other of the diodes will become forward biased if the input signal gets too large.

Incidentally, whether the op amp uses protective diodes or not, under normal operating conditions the inverting and noninverting inputs of an op amp should be at approximately the same potential. In other words, we said before that V_{id} is approximately equal to zero. So one quick test you can make on an op amp is to connect a voltmeter across its inputs, from the $(-)$ input to the $(+)$ input. If the op amp is good, you should always measure near zero volts. If you measure any appreciable voltage, the op amp is probably bad and should be replaced.

8-22   Diodes protect inputs for excessive signals.

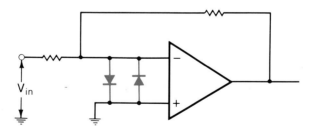

SHORT CIRCUIT ON OUTPUT

Many of the newer op amps have built-in short-circuit protection. But older op amps do not. By placing a resistor in series with the output, as shown in Fig. 8-23, the output is protected from being shorted directly to ground. By connecting the feedback lead to the output end (right side) of the protection resistor, the gain equations for the circuit are still valid. That is, the value of the protection resistor does not affect the gain of the circuit and does not have to be taken into account. If may, however, decrease the maximum possible output signal capability.

8-23 **Resistor protects output from short circuit.**

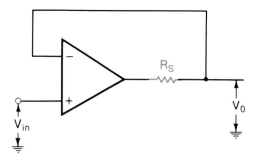

OUTPUT SATURATED

Occasionally when testing an op amp circuit, you will find the output saturated at either the positive or the negative power supply voltage. Does this mean that the amp is bad? Not necessarily. Some things that may cause this symptom follow.

Latch Up

In some op amps, if an input is overdriven so that the output is driven into saturation (to V_{CC} or $-V_{EE}$), the output will remain in saturation even when the input signal is removed. The cure is simple. Just shut off the power supply, then turn it back on. If the problem goes away, it was latch up. You may then want to take some precautions to see that the input cannot be overdriven. The problem of latch up has been eliminated in most newer op amps.

Open Input Leads

If either of the input leads from the op amp become disconnected, the output will probably slam into one power supply or the other. A break in printed circuit foil, a cold solder joint, or a disconnected lead could cause this symptom.

Open Feedback Resistor

As you know, without the feedback resistor, the amplifier is operating open loop. Any slight input voltage will then drive the output into saturation.

Null Control Open or Misadjusted

As mentioned earlier, we would normally like the output of the op amp to be at 0 V when the input signal is zero. However, because of slight differences in transistors or other components inside the op amp, it is practically impossible to expect the output to be *exactly* zero when the input is zero. This is especially true when the amplifier closed-loop gain is high. Any slight differences are amplified.

The circuit of Fig. 8-24 shows how a null control is used. Pot R_N is connected between the $+V$ and $-V$ power supplies. By adjusting the pot slider one way or the other, the voltage at the noninverting input can be made slightly positive or negative with respect to ground. This small voltage causes the transistor connected to the noninverting input to conduct more or less as necessary to bring the output exactly to zero.

8-24   Null circuit used to make $V_0 = 0$ when $V_{in} = 0$.

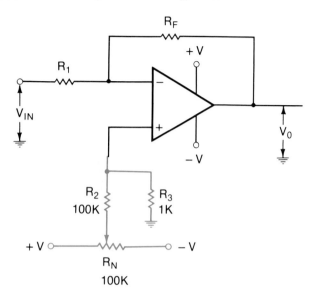

As you can see, if the pot is badly misadjusted or if one side of it becomes open, the output may be driven quite far off ground. There are a couple more problems that are somewhat more subtle.

SOLDER FLUX BETWEEN HIGH-IMPEDANCE FOILS

You know that the input impedance seen looking into the amplifier terminals is very high. For FET-input op amps, it may be as high as 10^{12} Ω. When these high-impedance circuits are soldered into a printed circuit board, any dirt, solder flux, etc., between adjacent foils will decrease the resistance between foils drastically. It may be enough to stop the circuit from operating completely, or it may just cause crosstalk between signals. In either case, you have problems. The solution, of course, is to clean the circuit board properly. Use a good grade of solder flux remover or degreaser that will not leave a residue.

GUARD BAND DIRTY OR NOT USED

Particularly when FET-input op amps are used, the signal input terminals must be isolated electrically from surrounding voltages. See Fig. 8-25. Notice the ring of copper foil around the two signal input leads. The ring, called a *guard-band*, provides electrical isolation between the input terminals and nearby printed circuit foils, which might carry signals or high voltages. The guard band prevents leakage current from flowing across the surface of the circuit board from a nearby foil.

8-25 Using a guard band around high-impedance inputs.

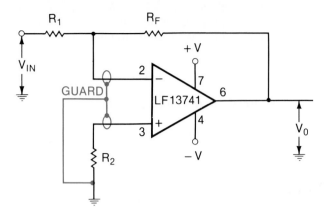

(A) Schematic.

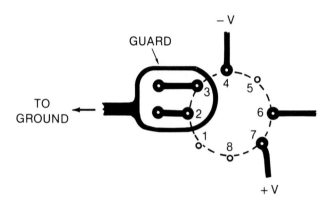

(B) PC layout.

PROBLEMS

8-1.   Referring to the circuit of Fig. 8-1, if a positive-going signal is applied to the noninverting input, Q_1 will conduct (a) more. (b) less. This will cause Q_3 to conduct—(c)more, (d)less—thereby pulling the base of Q_4 in the (e)positive direction. (f)negative direction. The net result is that V_0 will go more (g)positive. (h)negative.

8-2.   Diode D_1 in Fig. 8-1 is probably used to (a)rectify the signal. (b)minimize crossover distortion. (c)protect Q_5 from wrong polarity voltages.

8-3.   The advantage of a BIFET op amp over a bipolar type is its (a)high input impedance. (b)much higher gain. (c)higher output impedance.

8-4.   The LF351, shown in Fig. 8-3, uses—(a)JFET, (b)MOSFET—as input transistors and, therefore, do not require special handling precautions, as compared to bipolar types.

For the next three problems, refer to Fig. 8-26.

8-5.   If $V_1 = 2.5$ V, $R_1 = 20$ KΩ, and $R_F = 60$ KΩ, find the closed loop voltage gain, the output voltage, and the input impedance seen by voltage source.

8-26 **Circuit for problems 8-5 through 8-7.**

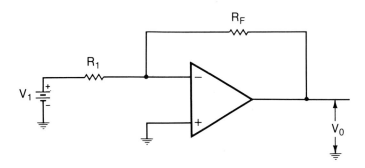

$A_v =$

$v_O =$

$r_{in} =$

8-6.   Choose values for R_1 and R_F so that the circuit has an input impedance of 600 Ω and a closed-loop voltage gain of 80.

$R_1 =$

$R_F =$

8-7.   Suppose resistor R_F is replaced by a pot whose resistance varies between 1 and 25 KΩ. If $V_1 = 0.2$ V and $R_1 = 1$ KΩ, what will be the maximum and minimum values of V_0 as R_F is varied over its limit?

V_0, max $=$

V_0, min $=$

8-27 **Circuit for problem 8-8.**

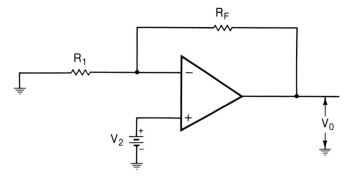

8-8.   In the circuit of Fig. 8-27, if $R_1 = 5$ KΩ, $R_F = 20$ KΩ, and $V_2 = 1.5$ V, what is the value of V_0?

$V_0 =$

What is the input impedance seen by source V_2 looking into the amplifier?

(a)Very low. (b)About 5 KΩ. (c)Probably many megaohms.

8-28 Circuit for problem 8-9.

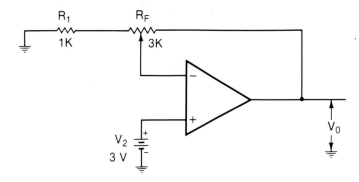

8-9. Refer to Fig. 8-28. What are the maximum and minimum values of V_0 as the pot slider is moved from one end to the other? (*Hint:* With the slider all the way to the right, the circuit acts like a voltage follower.)

8-10. Refer to the graph of Fig. 8-9. What is the open-loop voltage gain of a 741 at a frequency of 100 Hz?

$A_v =$

8-11. If a 741 is used as a voltage follower, what is the highest frequency it can amplify?

$f =$

For the next two problems, refer to Fig. 8-29.

8-29 Circuit for problems 8-12 and 8-13.

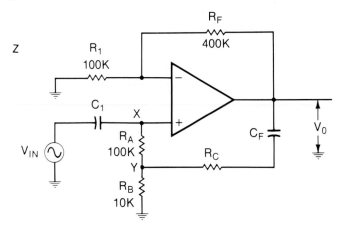

8-12. The voltage follower of Fig. 8-12 uses bootstrapping to increase the input impedance of the circuit. Sometimes a noninverting a-c amplifier is needed to provide more than unity gain. Bootstrapping is again used to increase the input impedance in Fig. 8-29. However, notice that resistors R_C and R_B form a voltage divider from the output to ground. The voltage at point Y is some fraction of the output voltage. Determine a suitable value for R_C. Remember that the signal voltage at point Y should be the same as the voltage at point X.

$R_C =$

8-13. In order for bootstrapping to work, C_F should have a low reactance, approximately $0.1 \times (R_C + R_B)$ at the lowest frequency to be amplified. Suppose you are replacing a defective feedback capacitor C_F whose printed value has been worn off. What value of C would you use if the amplifier is to handle frequencies from 50 Hz to 30 KHz?

$C =$

8-14. Suppose in the circuit of Fig. 8-13, $V_1 = V_2 = 2$ V, $R_F = R_B = 50$ KΩ, and $R_1 = R_A = 2$ KΩ. If the op amp has a CMRR of 30,000, what is the value of the common-mode output voltage?

$V_{cmo} =$

8-15. If the op amp in the circuit of problem 8-14 is replaced with one having a CMR of 75 dB, what will be the value of the common-mode output voltage?

CMRR =

For the next two problems, refer to Fig. 8-30.

8-30 Circuit for problems 8-16 and 8-17.

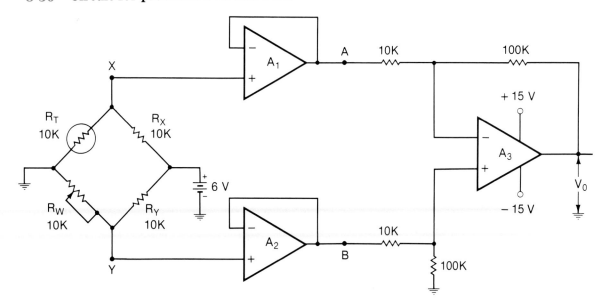

8-16. This circuit shows a variation of the circuit of Fig. 8-14. Notice that the branches of the bridge circuit use fairly high values (10 KΩ) of resistance. What are amplifiers A_1 and A_2 used for in this circuit?

What d-c voltages would you measure at point X and point Y, assuming that $R_W = R_T = 10$ KΩ?

$$V_X =$$

$$V_Y =$$

What d-c voltages would you measure at point A and point B?

$$V_A =$$

$$V_B =$$

What output voltage V_{cmo} would you measure if the CMRR = 25000?

$$V_{cmo} =$$

8-17. Referring to problem 8-16, if the thermistor resistance goes down, because temperature increases, so that the voltage at point X is 0.2 V less positive than at point Y, what would be the value of V_0?

$$V_0 =$$

For the next two problems, refer to Fig. 8-31.

8-31   Circuit for problems 8-18 and 8-19.

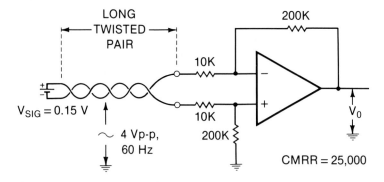

8-18. Differential amps are often used to minimize the effects of a-c common-mode (noise) voltages. Fig. 8-31 shows a differential amplifier being used to amplify a small signal developed by a photovoltaic cell. The cell generates a voltage proportional to the amount of light falling on it. In this particular case, suppose that the cell must be located far away from the amplifier and that the signal leads pass close to electrically noisy machinery. The leads thus pick up a large amount of 60-Hz signal. If the small d-c signal from the cell were fed to a single-ended amplifier, the noise would swamp out the signal. That is, a large noise signal would be present in the output. But since each lead picks up the *same amount* of noise, of the *same*

phase, it can be minimized by the differential amp. For the values shown, what is the amplitude of the d-c output signal V_0 caused by the photocell signal?

$V_0 =$

8-19.   Referring to problem 8-18, how much 60-Hz noise will be present in the output?

$V_{cmo} =$

For the next two problems, refer to Fig. 8-32.

8-32   Circuit and waveform for problems 8-20 and 8-21.

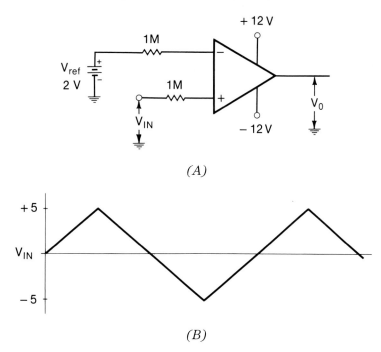

(A)

(B)

8-20.   The waveform is applied to the circuit of Fig. 8-32A. Sketch the output waveform. Assume that the output of the amplifier can swing between $+12$ and -12 V.

8-21. Refer to problem 8-20. Change V_{ref} to -3 V, and sketch the output waveform. V_{in} remains as before.

For the next two problems, refer to Fig. 8-33.

8-33 **Circuit and axis for problems 8-22 and 8-23.**

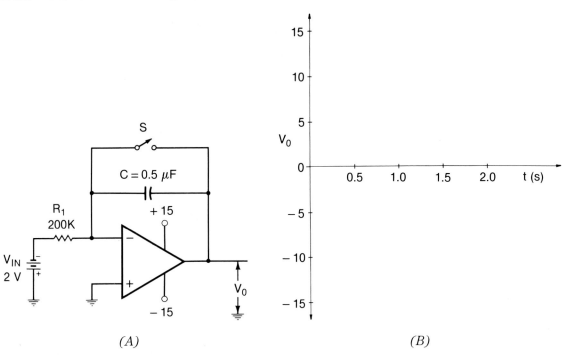

(A)   (B)

8-22. The output voltage V_0 will remain at zero as long as switch S remains closed. Assume that S opens at time t_0. Sketch V_0 on the axis of Fig. 8-33B. Output voltage V_0 can swing between $+15$ and -15 V.

8-23. Referring to problem 8-22, determine a value for R_1 to make V_0 reach $+10$ V exactly 2 s after opening switch S.

$R_1 =$

EXPERIMENT 8-1   SINGLE-ENDED D-C AMPLIFIERS

You will now become familiar with the inverting and noninverting IC op amp configurations, as they are used to amplify d-c signals.

EQUIPMENT

- (2) power supplies, one positive and the other negative polarity (Use either ± 15 VDC or ± 12 VDC or ± 9 VDC)

- voltmeter
- (1) IC operational amplifier, 741 or equivalent
- (1) 100-KΩ pot
- (1) 5-KΩ pot
- (2) 10-KΩ, ½-W ± 5% resistors
- (1) 1-KΩ, ½-W ± 5% resistor

E8-1   Circuits for experiment 8-1.

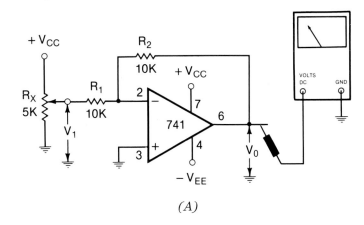

(A)

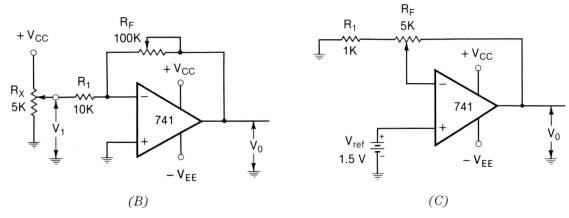

(B)   (C)

PROCEDURE

1.   Build the circuit of Fig. E8-1A. Use either ± 15 V, ± 12 V, or ± 9 V for d-c power supplies V_{CC} and V_{EE}. The actual values are not important here, but use two supplies of the same magnitude, opposite polarity.

2.   Using pot R_X as your input d-c signal source, adjust the pot so that $V_1 = 0$. At this point, the output of the amplifier should measure very near 0 VDC. If it does not, recheck your wiring.

3.   Now adjust pot R_X so that V_1 measures + 1 V. Measure V_0 (observe polarity).

$V_0 =$

According to the values of R_F and R_1, what should be the gain?

$A_v =$

4.   Measure the input differential voltage (voltage between the two amplifier input terminals).

$V_{id} =$

5.   Adjust the pot slider as necessary, and record the output voltage for values shown in Table E8-1, lines 1 through 5. Then remove the pot from V_{CC}, and connect it to $-V_{EE}$. Adjust the slider as necessary and fill in lines 6 through 9.

6.   Next, remove R_F from the circuit, and replace it with a 100-KΩ pot, as shown in Fig. E8-1B.

7.   Adjust the input voltage V_1 to 1 VDC. Then adjust the feedback pot until V_0 reads $+5$ VDC. Without disturbing the pot setting, remove the feedback pot from the circuit, and measure its resistance. Record this value in Table E8-1, above line 10.

Table E8-1

	V_1	V_0
$R_1 = 10\ \text{K}\Omega, R_F = 10\ \text{K}\Omega$		
1	0	
2	$+1$	
3	$+2$	
4	$+5$	
5	$+V_{CC}$	
6	-1	
7	-2	
8	-5	
9	$-V_{EE}$	
$R_1 = 10\ \text{K}\Omega, R_F = \underline{\qquad}$		
10	-1	
11	-2	
12	-5	
13	$-V_{EE}$	
14	$+1$	
15	$+2$	
16	$+5$	
17	$+V_{CC}$	

8. Replace the pot, and adjust R_X to make V_1 equal to each value shown in Table E8-1, lines 10 through 17. Record your measured values for V_0 in each case.

9. Now let's observe the effect of some circuit faults. Adjust V_1 to $+1$ V. Now remove pot R_F, and measure V_0 with the feedback circuit open.

$V_0 =$

10. Replace R_F. Then disconnect the noninverting input, pin 3, from ground. Measure V_0.

$V_0 =$

11. Lastly, connect pin 3 to ground again. Disconnect the $-V_{EE}$ supply from pin. Keep V_{CC} connected, and measure V_0.

$V_0 =$

12. Now let's take a look at a noninverting amp. Build the circuit of Fig. E8-1C. The 1.5 V V_{ref} source can be a separate supply, or it can be obtained by means of a voltage divider from V_{CC}.

13. Adjust the feedback pot R_F so that the resistance between the output and the inverting input is zero. Measure V_0.

$V_0 =$

14. Now readjust R_F so that $R_F = 5$ KΩ between the output and the inverting input. Measure V_0.

$V_0 =$

Also measure the voltage across R_1.

$V_1 =$

QUIZ

1. In step 2, you checked to see if the amplifier output was (a)shorted. (b)nulled.

2. Give the equation for voltage gain that you used in step 3.

$A_v =$

3. For all measured values listed in the Table E8-1, the input differential voltage was negligible. (a)True. (b)False.

4. For lines 1-9 of Table E8-1, did the gain remain constant? (a)Yes. (b)No.

5. According to the measured value of R_F in step 7, what should have been the gain?

$A_v =$

Was it?

(a) Yes. (b) No.

6. Explain why the gain equation did not work in line 12 and 13.

7. In step 9, the gain of the amplifier was—(a)zero, (b)open loop—with R_F removed. Therefore, the amplifier output went to (c)zero. (d)saturation.

8. In step 10, disconnecting pin 3 caused the noninverting input transistor to—(a)cut off, (b)saturate—driving the output to saturation.

9. In step 13, the noninverting amp acted like a/an (a)voltage follower. (b)inverting amp.

10. Step 14 shows that the gain of the noninverting amp is (a)R_F/R_1. (b)(R_F/R_1) + 1. From the voltage measured across R_1, we can conclude that the input differential voltage V_{id} must have been (c)zero. (d)equal to V_{ref}.

EXPERIMENT 8-2   OP AMP A-C SIGNAL AMPLIFIER

The op amp makes an excellent small signal amplifier. In this experiment, you will use a 741 to amplify small audio frequency signals. You will also see how the gain of the circuit affects the frequency response.

The op amp cannot deliver much power to a load. Here you will see how to interface complementary-symmetry transistors to an op amp to get more output power.

EQUIPMENT
- ± 15- or ± 12-VDC dual-power supplies
- oscilloscope
- voltmeter
- audio signal generator
- (1) 741 or LF351 op amp
- (1) TIP 31 or equivalent NPN power transistor
- (1) TIP 32 or equivalent PNP power transistor
- (1) 10-μF, 25-WVDC capacitor
- (1) 10-KΩ pot
- (1) 100-KΩ, ½-W ± 5% resistor
- (2) 10-KΩ, ½-W ± 5% resistors
- (2) 1-KΩ, ½-W ± 5% resistors
- (1) 100-KΩ, ½-W ± 5% resistors
- Optional—8- or 16-Ω speaker

PROCEDURE

1. Build the circuit of Fig. E8-2A. With no signal applied, adjust null pot R_N back and forth a few times while observing V_0 with a voltmeter. You should be able to make V_0 go positive and negative a volt or so with respect to ground as you vary the pot. If you cannot, recheck your wiring.

2. Adjust R_N so that V_0 = 0. Connect the signal generator to the input, and adjust it for V_{in} = 0.1 Vp-p at f = 1 KHz. Measure V_0.

$V_0 =$

Calculate the a-c gain of the circuit.

$$A_v = V_0/V_{in} =$$

Record these values on line 3 of Table E8-2.

E8-2   Circuits for experiment 8-2.

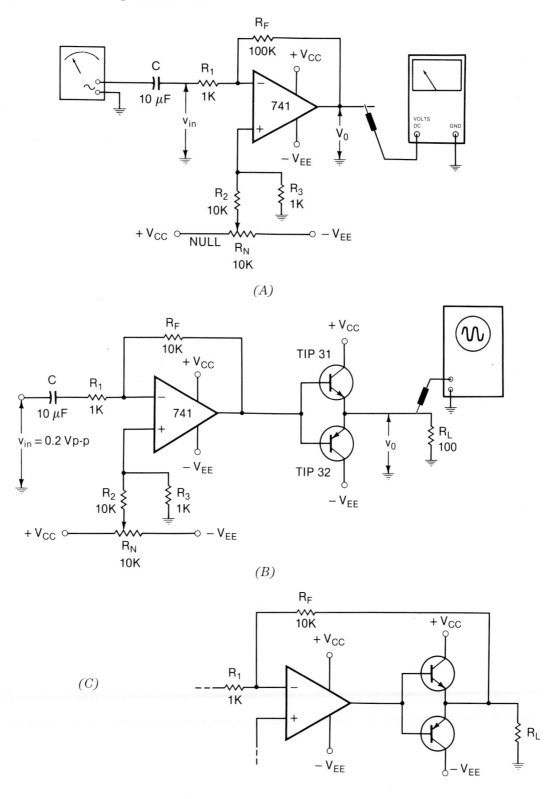

(A)

(B)

(C)

	f (Hz)	V_0	Gain
	$R_F = 100$ KΩ		
1	100		
2	500		
3	1K		
4	5K		
5	10K		
6	20K		
7	50K		
8	100K		
	$R_F = 10$ KΩ		
9	1K		
10	10K		
11	50K		
12	100K		
13	200K		
14	500K		

3.   Now, adjust the generator frequency to each of the values listed in Table E8-2, lines 1 through 8, and record the measured output voltage and the gain.

4.   Next, change R_F so as to make the voltage gain equal to 10 at $f = 1$ KHz.

5.   Repeat your measurements for each frequency listed on lines 9 through 14.

6.   Shut off the power, and add the complementary-symmetry stage to your op amp, as shown in Fig. E8-2B.

7.   Connect a d-c voltmeter to the top of R_L as shown, and apply power. Adjust R_N down while watching the d-c voltage across R_L. It should go positive and negative slightly if your circuit is working properly. Adjust the null control so that the d-c voltage across $R_L = 0$. R_L in this circuit represents the speaker in an audio amplifier. Since no blocking capacitor is used between the amplifier output and the speaker, you can appreciate the need to null the output to zero so as to keep dc out of the speaker.

8.   Apply a 1 KHz signal of 0.2 Vp-p to the input. Now observe the signal across the load with an oscilloscope. Is it distorted?

(a)Yes. (b)No.

Why?

9.   Adjust the NULL control up and down while watching the output. Explain what you observe.

10.   Shut off the power, and reconnect the right end of R_F to the top of R_L, as shown in Fig. E8-2C. Turn on the power again and observe the output. What do you notice?

SUGGESTIONS FOR ADDITIONAL EXPERIMENTATION
1.   Replace R_L with an 8- or 16-Ω speaker, and drive the input with an audio signal from a tape recorder.
2.   Make feedback resistor R_F a 100-KΩ pot, and adjust the gain of the amplifier as needed.

QUIZ

1.   The NULL control varies the (a)gain. (b)a-c signal level. (c)d-c level at the output.

2.   The a-c gain of Fig. E8-1A is equal to $-R_F/R_1$. (a)True. (b)False.

3.   Using a gain of 100, the 741 can amplify frequencies up to approximately _____ Hz.

4.   Using a gain of 10, the 741 can amplify frequencies up to approximately _____ Hz.

5. The complementary–symmetry transistors allow us to deliver more power to a low-resistance load. This is because they (a)increase the voltage gain. (b)act like emitter followers, and thus prevent loading on the op amp.

6. In step 8, severe crossover distortion was evident. This was because (a)both transistors were biased at cutoff. (b)two power supplies were used. (c)no blocking capacitor was used.

7. As the NULL pot was varied in Step 9, the crossover point shifted. This is because (a)the gain changed. (b)as one transistor became more forward biased, the other became more back biased.

8. In Step 10, the crossover distortion (a)increased. (b)decreased. (c)remained the same. This is because (d)a slight forward bias was applied to both transistors simultaneously. (e)the gain of the op amp is very high before either transistor starts conducting, so the op amp makes up for the base-emitter drop by going an additional 0.7 V positive or negative.

9. When you are troubleshooting an amplifier like the one in Fig. E8-2B, you find that the speaker is burned open and that the NPN power transistor is shorted. Before replacing the bad transistor, it is a good idea to (a)try driving the speaker with just the PNP transistor. (b)check the d-c level out of the op amp. (c)put a blocking capacitor between the speaker and the output transistors.

10. The problem in question 9 could have been caused by (a)R_N becoming open or disconnected from V_{EE}. (b)R_F resistance too low. (c)excessive crossover distortion.

EXPERIMENT 8-3   OP AMPS IN THE DIFFERENTIAL MODE

The circuit configuration that uses an IC op amp to its fullest capability is the differential mode amplifier. In this experiment, you will see the op amp's ability to amplify a small desired signal, while rejecting much larger common-mode d-c and a-c voltages. You will also build a control circuit that can be used to maintain a constant preset temperature inside an oven.

EQUIPMENT

Use same equipment as in Experiment 8-2.

PROCEDURE

1. Build the circuit of Fig. E8-3A. Connect a clip lead between points *I* and *II*, shorting them together.
2. Adjust R_X so that $V_1 = V_2 = 1$ VDC. Measure V_0. V_0 should be quite small.

$V_0 =$

3. Change R_X so that $V_1 = V_2 = 3$ VDC. Measure V_0.

$V_0 =$

Do you notice any significant change from Step 2?

(a)Yes. (b)No.

Why?

4. Now remove the clip lead from between points *I* and *II*. Measure V_0.

$V_0 =$

Measure V_1.

$V_1 =$

Measure V_2.

$V_2 =$

> **NOTE** Measure V_1 and V_2 carefully, because there is only a small difference between them. You may want to connect a voltmeter directly across R_Y to make this measurement.

E8-3 Circuits for experiment 8-3.

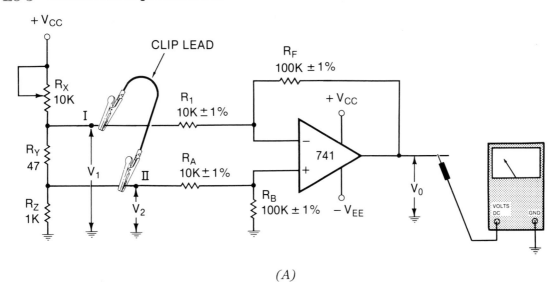

(A)

(B)

E8-3 Continued.

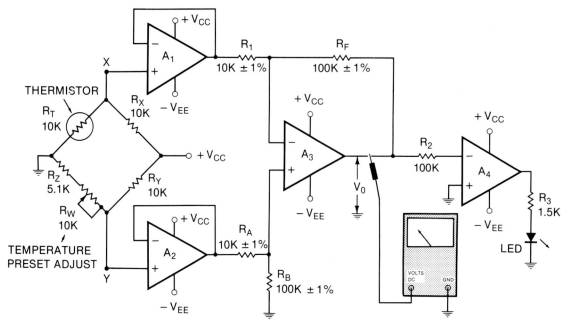

Notes: All op amps 741 or LF 351.
All resistors ½W ± 5% unless otherwise noted.
V_{CC} = 12 to 15 VDC.

(C)

The following material is a continuation of 4 above.
The difference voltage $V_1 - V_2$ is amplified by the circuit to give V_0. Determine the voltage gain.

$A_v = V_0/(V_1 - V_2) =$

Compare this to the theoretical gain.

5.   Connect a signal generator from ground to points *I* and *II* as shown in Fig. E8-3B. Adjust the generator amplitude to 1 Vp-p. Note that both points *I* and *II* receive the full generator voltage, because the two coupling capacitors both connect to the generator output terminal.
    The a-c signal from the generator represents a common-mode noise voltage V_{cm}. Measure the a-c signal at the output of the amplifier using an oscilloscope. (This voltage should be quite small.)

$V_{cmo} =$

6.   What is the common-mode gain of this circuit?

$A_{cm} = V_{cmo}/V_{cm} =$

Now determine the common-mode rejection ratio.

CMRR $= A_v/A_{cm} =$

The circuit of Fig. E8-3C shows a typical application of a difference amplifier to monitor changes in the resistance of a transducer.

7. Build the circuit of Fig. E8-3C. Adjust pot R_W back and forth over its entire range a few times, while watching the difference amp's output with a voltmeter. If your circuit is working properly, the difference amp's output should swing from some positive voltage to some negative voltage as the pot is varied. Explain what happens to the LED as the pot is varied.

Why?

8. Shut off power. Disconnect the end of R_W from point Y. Then using an ohmmeter, adjust R_W so that $R_W + R_Z = 8$ KΩ. Reconnect R_W to point Y and apply power again. Measure V_0.

$V_0 =$

What is the condition of the LED?

(a)On. (b)Off.

9. Now, using a soldering iron, slowly heat R_T for a couple of minutes, while watching the output of A_3. (Do not melt the thermistor!) What happens?

What happens to the LED?

10. Remove the heat source, and watch V_0 and the LED again. What happens to V_0 as the thermistor cools?

What happens to the LED?

SUGGESTIONS FOR TROUBLESHOOTING

Have your lab partner, or lab instructor, open (disconnect) or short, any resistor in the circuit. Then try to locate the faulty component.

> **CAUTION** Do not short either power supply directly to any pin on the chips. This will damage the IC.

When searching for the defective part, try to zero in on its location by making voltage checks with power on. Make as few resistance checks as absolutely necessary. For example, suppose R_1 was open. The LED would be off, regardless of the setting of pot R_W.

So you would backtrack to the output of A_3. Amplifier A_3's output would also always be positive with respect to ground. This is because with R_1 open, A_3 acts somewhat like a voltage follower. And since the noninverting input of A_3 is at the same positive potential as point Y, its output is positive.

Next you would measure the outputs of A_1 and A_2. Here you would notice that A_1's output is different from the voltage at the inverting input of A_3. So you would suspect A_3 and its associated components. A few resistance checks would then pinpoint R_1 as the bad part.

QUIZ

1. In step 2, shorting together points I and II caused the same 2-VDC signal to be applied to both the inverting and noninverting inputs. This signal is called a (a)common-mode signal. (b)differential signal.

2. In step 3, the closed-loop gain of the amplifier was—(a)0, (b)10, (c)infinite— but the d-c output voltage was (d)near zero. (e)ten times the common-mode input signal.

3. Steps 5 and 6 showed that (a)a-c signals cannot be amplified by a difference amp. (b)common-mode a-c signals are rejected by a differential amp.

4. In step 7, when the resistance of $R_W + R_Z$ was less than the resistance of thermistor R_T, point X was—(a)positive, (b)negative—with respect to point Y. The voltage *between* points X and Y is the—(c)common-mode, (d)differential—signal.

5. Referring to question 4, amplifier A_3's output was driven—(a)positive, (b)negative—with respect to ground.

6. Amplifier A_4 is being used as a (a)noninverting amp. (b)zero crossing detector. (c)impedance matching device.

The circuit of Fig. E8-3C represents a temperature control for an oven. Amplifier A_4 drives the heater control. When the LED is lit, the heat in the oven turns on by means of power being applied to a heater coil inside the oven. The thermistor is also mounted inside the oven to sense the oven temperature.

7. Pot R_W in the circuit is used to (a)adjust the voltage applied to the thermistor. (b)preset the desired operating temperature.

8. When the oven is below the preset temperature, the resistance of R_T is— (a)less than, (b)greater than—the resistance of $R_W + R_Z$.

9. Referring to question 8, the LED is—(a)on, (b)off—indicating that power— (c)is, (d)is not—applied to the heater coil.

10. As the temperature inside the oven rises slightly above the preset value, the resistance of R_T becomes—(a)greater than, (b)less than—the resistance of $R_W + R_Z$. This causes A_4's output to switch—(c)positive, (d)negative—thus turning the heat (e)on. (f)off.

Regulated
Power Supplies

9

Practically all of the electronic circuits we have discussed so far needed d-c power supplies. We assumed that if we have a 12-V supply, the terminal voltage of the supply remains at 12 V regardless of any variation in current drawn from the supply. This assumption is usually not true. Due to internal resistance in power supplies, the terminal voltage may vary considerably whenever the current drain changes.

Sometimes a slight change in supply voltage does not cause any problems. But in some circuits, particularly in test instruments or computer circuits, a change in power supply voltage can cause serious problems.

We will now study several circuits, called *voltage regulators*, which are used to maintain a constant voltage in a power supply.

ZENER DIODE REGULATORS

One of the simplest voltage regulators is the zener diode. Fig. 9-1A shows the characteristic curve of a zener diode. Notice that the diode acts like an ordinary silicon diode when it is forward biased. That is, I_F increases rapidly when V_F exceeds about 0.7 V.

9-1 **Zener diode characteristics.**

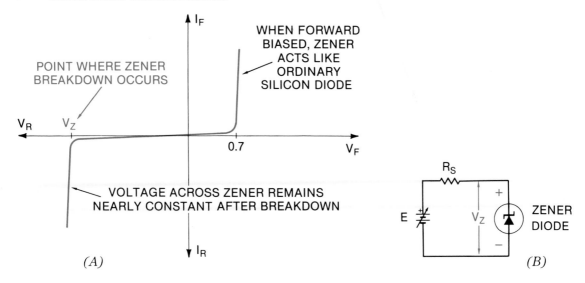

(A)

(B)

As a result of special doping of the crystal, the zener diode has special characteristics when it is reverse biased. As shown in Fig. 9-1B, power supply E reverse biases the diode. As voltage E increases, a point is reached where the diode breaks down. This point is labeled V_Z in Fig. 9-1A. If E is increased beyond V_Z, the voltage across the zener remains essentially constant, so the differences between E and V_Z is dropped across resistor R_S. R_S is necessary to limit the zener current to a safe value. As long as the current through the zener remains below the recommended maximum value, the diode will not be destroyed. In other words, the term "breakdown" does not mean that the zener is damaged. It simply means that the diode has gone into conduction in the reverse direction.

Since the voltage across the zener remains constant as E increases above V_Z, the zener can be used as a voltage regulator. Fig. 9-2 shows a simple zener voltage regulator. Note that load R_L is placed in parallel with the zener. When used in this manner, the diode is called a *shunt* regulator.

9-2 Simple zener regulator.

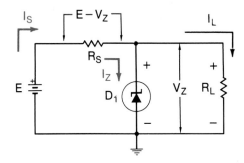

Here's how it works. Resistor R_S, has a voltage across it equal to $E - V_Z$. This voltage difference causes a current I_S to flow through R_S. I_S then splits up to flow through the parallel combination of D_1 and R_L, forming currents I_Z and I_L. That is

$$I_S = I_Z + I_L \qquad\qquad 9\text{-}1$$

If R_L should increase, causing I_L to decrease, the diode will conduct *more* current, keeping the voltage across it essentially constant. Similarly, if I_L should increase, I_Z will decrease accordingly, to keep I_S constant.

> **EXAMPLE 9-1**   In the circuit of Fig. 9-2, $E = 12$ V, $V_Z = 9$ V, $R_S = 60\ \Omega$, and $R_L = 200\ \Omega$. Find I_S, I_L, and I_Z
>
> **SOLUTION**   The voltage across R_S is
>
> $$E - V_Z = 12 - 9 = 3\ \text{V}$$
>
> So
>
> $$I_s = \frac{3\ \text{V}}{60\ \Omega} = 50\ \text{mA}$$
>
> The voltage across R_L is V_Z, or 9 V. So
>
> $$I_L = \frac{9\ \text{V}}{200\ \Omega} = 45\ \text{mA}$$
>
> Then $I_Z = I_S - I_L = 50 - 45 = 5$ mA.

In the circuit of the above example, if R_L increases, say to 300 Ω, the current through R_L drops to $I_L = 9/300 = 30$ mA. The zener diode then turns on harder, carrying the other 20 mA, so as to keep V_Z and, therefore I_S, constant.

On the other hand, if input voltage E increases, say to 14 or 15 V, the zener still holds the voltage across R_L at a constant 9 V. Since the current through R_S increases, the current through the zener also increases. As a matter of fact, a voltage regulator also helps *filter the ripple voltage*. For example, consider the circuit of Fig. 9-3A. Components T_1, D_1, and C_1 form a half-wave rectifier power supply like those in chapter 2. Let's assume that the voltage at point X (across C_1) looks like the waveform shown in Fig. 9-3B. The voltage across C_1 has 2 Vp-p of 60-Hz ripple. This would probably be excessive for use as a good power supply.

9-3 Zener diode reduces ripple.

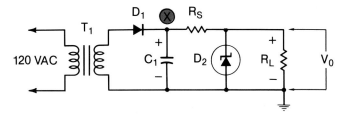

(A) Basic circuit.

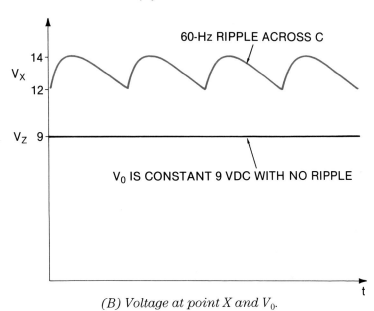

(B) Voltage at point X and V_0.

However the regulator circuit is connected to point X. It consists of R_S and zener diode D_2. Since variations in input voltage to the regulator (left end of R_S) simply cause the zener current to increase or decrease slightly, the voltage across the zener remains essentially constant. That is, the zener effectively filters out most of the ripple.

Zener diodes are available in a wide range of voltages from about three to several hundred volts. The physical packages look like those of ordinary silicon rectifiers. Also, depending on how much current the diode must be capable of carrying, zeners are available in a variety of wattage ratings, such as ¼, ½, 1, 5, 10 W. The power dissipated by a zener diode is the product of zener voltage times zener current, or

$$P_Z = V_Z I_Z \qquad\qquad 9\text{-}2$$

When replacing a defective zener diode, always use one of the same voltage rating as that of the original part. You can use a higher-wattage rating, if the original size is not available.

PROBLEMS WITH ZENER DIODE REGULATORS

Besides the obvious problems of opens and shorts within the diodes themselves, there are two circuit conditions that can cause malfunction of a zener diode regulator.

First of all, *the zener must be conducting in the reverse direction in order to maintain a constant voltage across itself.* At least a few milliamps must be flowing through the zener at all times. If R_S becomes too large, or if E becomes too low, so that I_Z drops below a few milliamps, the zener will not be able to maintain a constant voltage across the load. The load voltage will, therefore, drop below normal.

For example, in the circuit of Fig. 9-3A, the minimum input voltage to the regulator (left end of R_S) is 12 V. Since the zener voltage is 9 V, the minimum drop across R_S is 3 V. If sufficient load current *plus* zener current flows through R_S with 3 V across it, the circuit works fine. On the other hand, if the ripple at point X is excessive, or if R_S is too large, the zener comes out of conduction at the minimum points of the input waveform. The result is that the voltage across R_L is not a smooth d-c voltage, but rather it looks like the waveform shown in Fig. 9-4.

9-4 V_0 **has 60-Hz noise caused by zener coming out of connection.**

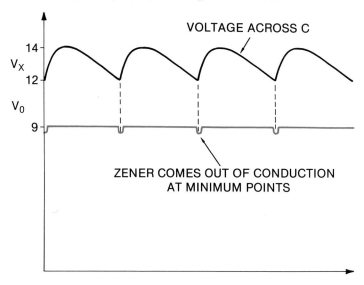

The second condition that can cause problems is excessive zener current. If I_Z becomes too high so that the power dissipated in the zener gets excessive, the zener will be destroyed. This second condition could be caused by input voltage E becoming too high, or by R_S becoming too small.

SERIES REGULATORS

Zener diode shunt regulators are simple and require few parts, but they are inefficient for high-current applications. This is because the zener must carry whatever part of I_S that does not flow through the load. When the load current demand is low, high current flows through the zener, generating a lot of heat.

A more efficient regulator is shown in Fig 9-5. The circuit is basically an emitter follower. The voltage at the emitter is always equal to the voltage at the base, less the diode drop across the base–emitter junction. (As long as $V_{CC} > V_B$.) For example, if $V_B = 12$ V, V_0 will be equal to $12 - 0.7 \cong 11.3$ V. This is true regardless of the value of R_L, or I_L. That is, as R_L changes, causing I_L to vary, the voltage V_0 across R_L remains essentially constant at 11.3 V.

Since in a practical circuit we would not want to use a battery to feed the base of the transistor, a zener diode is used as a reference, as shown in Fig. 9-6A. In this circuit output voltage, V_0 is slightly less than the zener voltage; that is,

$$V_0 = V_Z - V_{BE} \qquad \text{9-3}$$

9-5 Emitter follower used as voltage regulator.

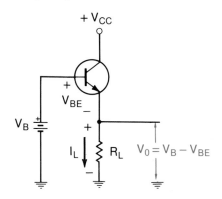

9-6 Simple series regulator.

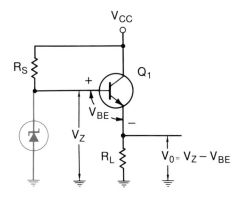

(A) Replacing V_B with a zener diode.

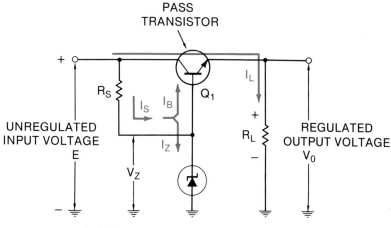

(B) All load current passes through Q_1.

Fig. 9-6B shows the more conventional way of drawing this circuit.

The transistor is in series with the load, so all load current must pass through the transistor. In fact, the transistor is usually called a *pass* transistor. In order to work properly, the pass transistor must *not* be driven into saturation. But as long as the transistor operates in the linear region, the value of V_{CC} does not affect the value of V_0. In other words, *the circuit regulates against changes in input voltage, as well as changes in load current.*

Note that heavy load current does not flow through the zener. The current I_S, flowing through R_S, splits up forming base current I_B and zener current I_Z. As in the regulator of Fig. 9-2, $I_S = (E - V_Z)/R_S$, and it is essentially constant. As the load draws more current, more base current is drawn into the base of the transistor, so I_Z decreases slightly. Conversely, when the load current decreases, I_B decreases. Therefore, more of I_S flows through the zener. In other words, the zener merely has to take up the variation in I_B and not the change in total load current.

The power dissipation in the pass transistor is equal to $I_C \times V_{CE}$. Since $I_C = I_L$ and $V_{CE} = E - V_0$,

$$P_d = (E - V_0) I_L \qquad\qquad 9\text{-}4$$

EXAMPLE 9-2 In the circuit of Fig. 9-6B, $E = 20$ V, $V_Z = 15.7$ V, $R_S = 270$ Ω, $R_L = 50$ Ω, and $\beta = 100$. Find V_0, I_L, I_B, I_S, I_Z, and P_d, the power dissipated in the pass transistor.

SOLUTION

$$V_0 = V_Z - V_{BE} = 15.7 - 0.7 = 15 \text{ V}$$

$$I_L = V_L/R_L = 15/50 = 300 \text{ mA} = I_C$$

$$I_B = I_C/\beta = 300 \text{ mA}/100 = 3 \text{ mA}$$

$$I_S = E - V_Z/R_S = 4.3/270 = 16 \text{ mA}$$

$$I_Z = I_S - I_B = 16 - 3 = 13 \text{ mA}$$

and

$$P_d = (E - V_0) \times I_L = (20 - 15) \times 0.300 = 1.5 \text{ W}$$

Fig. 9-7 shows the power supply section of the Magnavox tape recorder that you studied in chapter 5. A series regulator like the one in Fig. 9-6B is highlighted. The zener diode in this circuit is labeled 25 V, but the voltage at the base of X_2 is measured at 22.8 V. The measured values were recorded by a technician who performed measurements on a specific piece of equipment. The point is that the voltage measured across a zener diode may differ slightly from the nominal value. Zener diodes are available with various tolerances, such as $\pm 10\%$ or $\pm 5\%$, just like resistors.

If the regulator of Fig. 9-6B were perfect, there would be absolutely no change in load voltage as R_L changed, causing I_L to vary. But, of course, there is no such thing as a perfect regulator. As I_L changes, there is some small change in V_0.

Here's why the V_0 changes. The input curves of a typical silicon transistor are shown in Fig. 9-8. Suppose the I_L is such that 20 mA of base current flows, as shown at point A. The corresponding base–emitter voltage, shown at point A', is 0.7 V for this transistor. Now let's suppose the load resistance decreases, causing an increase in I_L. Also assume that the base current must increase to point B (30 mA) in order to deliver the new value of I_L. As can be seen from Fig. 9-8, I_B must increase to point B', or to 0.8 V. Even if the zener diode could maintain a perfectly constant voltage at the base of the pass transistor, the emitter voltage V_0 decreases by 0.1 V because I_L increases.

The amount of load voltage variation caused by load current change is referred to as *percent regulation*. The equation for finding percent regulation is

$$\% \text{ regulation} = \frac{V_{\text{no load}} - V_{\text{full load}}}{V_{\text{full load}}} \times 100 \qquad\qquad 9\text{-}5$$

9-7 **Power supply of Magnavox Model TE3410WA11. From** *Tape Recorder Series No. 173*, **Howard W. Sams Company, Inc.**

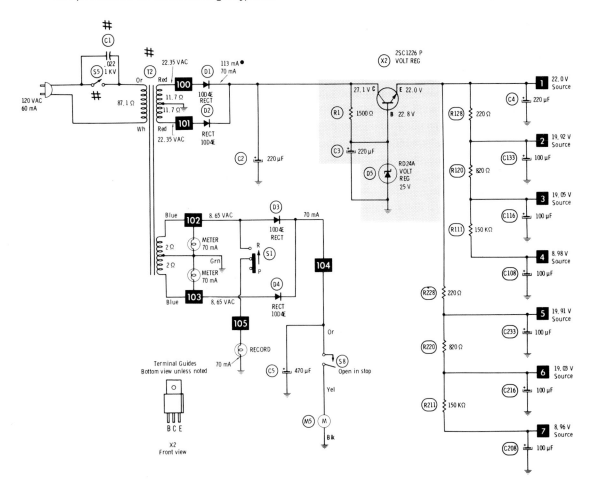

9-8 **Voltage V_{BE} must change in order for I_B to change.**

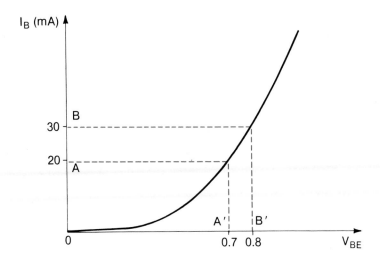

EXAMPLE 9-3 Suppose a regulator, like the one in Fig. 9-6B, has an output voltage of 9.3 V when no load is connected to the output. Then when a full-load current of 100 mA is drawn, the output voltage drops to 9.1 V. Find the percent regulation.

SOLUTION

$$\% \text{ regulation} = \frac{9.3 - 9.1}{9.1} \times 100 \approx 2.2\%$$

For many applications, 2 or 3% is fine, but some test instruments need much better regulation, on the order of 0.1% or so. When testing regulated supplies, you determine percent regulation by measuring the power supply output voltage with no load connected. Then connect a dummy load (resistive load) to the supply so as to draw normal full-load current. Measure V_0 again, and plug the values into equation 9-4. If the measured percent regulation is equal to or less than the maximum permissible value, you are in good shape.

ADJUSTABLE VOLTAGE REGULATORS

Fig. 9-9 shows a voltage regulator capable of better regulation than previous circuits. Note that a pass transistor and reference are used as before. Two new sections, the *sample* circuit and the *comparator* have been added.

9-9 **Improved series voltage regulator.**

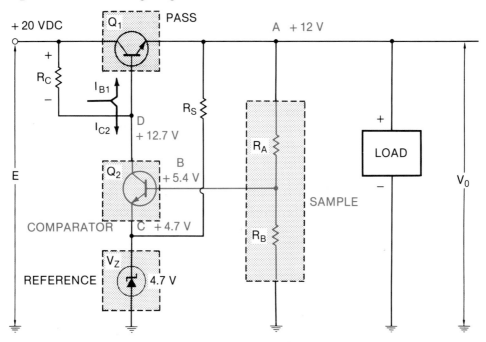

To see how the regulator works, let's assume that the output voltage at point A is 12 V. Resistors R_A and R_B (the sample circuit) form a voltage divider from output to ground. Let's assume that the resistors are chosen so that the voltage at the base of Q_2 (point B) is about 5.4 V. Since the emitter of Q_2 is connected to a constant 4.7 V reference (point C), Q_2 is forward biased. I_{C2} and I_{B1} flow through resistor R_C as shown. Due to the drop across R_C, the base of Q_1 is held at 12.7 V. These voltages represent our starting conditions, with perhaps a small load current flowing.

Now let's suppose the load resistance decreases, drawing more load current. The voltage at point A tends to decrease. Because of the voltage divider, the voltage at point B decreases also. And since the voltage at point C is constant, the forward bias on Q_2 decreases, causing Q_2 to conduct less. Because less I_{C2} is flowing through R_C, the voltage at point D goes up. And since Q_1 acts like an emitter follower, driving its base in the positive direction pulls its emitter (point A) more positive also. The net result is that the voltage at point A does not decrease as much as it would have if the regulator were not used.

Transistor Q_2 is used as a *comparator*, since it compares the sampled voltage to the reference voltage. It is also sometimes called an *error amplifier*, because it amplifies any change, or error, in regulator output.

As an exercise, try to determine what happens to the voltage at points A, B, C, and D, if the load current *decreases*.

Incidentally, notice that zener resistor R_S is connected to the regulated output (point A) rather than to the unregulated input voltage E. Since the voltage at point A is more constant, the zener current is held constant, thus keeping the reference voltage very constant.

Fig. 9-10A shows a modification of the circuit of Fig. 9-9. Pot R_X allows adjustment of the ratio of the voltage at point B with respect to point A. In other words, by moving the pot slider up or

9-10 Adjustable series regulators.

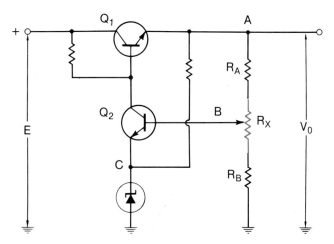

(A) Pot R_X allows adjustment of voltage at point A.

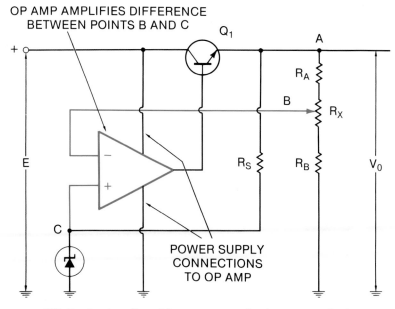

(B) Replacing Q_2 with an op amp for better regulation.

down, the output voltage at point A can be adjusted. Once manually set to some desired value, the output voltage at point A remains constant, even if the load current varies or input voltage E changes. Fig. 9-10B is a variation of the circuit of Fig. 9-10A. Notice that an op amp has higher gain than the single stage Q_2, so it amplifies any slight error voltage more than Q_2 does. The net result is that the op amp regulator has a lower percent of regulation than the discrete component circuit.

CURRENT LIMITERS

In the circuit of Fig. 9-10, if the load resistance drops too low, or accidentally becomes shorted, excessive current will flow through Q_1. This current might burn it out. A refinement of the circuit is shown in Fig. 9-11. Transistor Q_3 acts as a *current limiter*, or *short circuit protection* device.

9-11 **Series regulator with current limiter.**

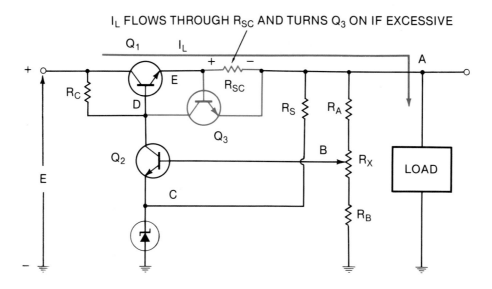

Here's how it works. As I_L flows through R_{SC}, a small voltage drop is developed across the resistor. The value of the resistor is chosen so that the drop across it is too small to turn on Q_3, with normal load current flowing. That is, Q_3 is normally off. But if the I_L becomes excessive, the drop across R_{SC} rises to 0.7 V or more and turns on Q_3. When Q_3 turns on, it acts like a closed switch, effectively shorting point D to point A. The forward bias on Q_1 is reduced accordingly. The net effect is that Q_1 shuts down and does not carry excessive current. The maximum short circuit current that can be drawn by the load is a value that will drop 0.7 V across R_{SC}. That is,

$$I_{SC} = \frac{0.7 \text{ V}}{R_{SC}}$$

9-6

where

I_{SC} = short circuit current

EXAMPLE 9-4 In the circuit of Fig. 9-10, if R_{SC} = 3.5 Ω, what is the maximum current that can be drawn through Q_1?

SOLUTION

$$I_{SC} = \frac{0.7 \text{ V}}{3.5 \text{ }\Omega} = 200 \text{ mA}$$

TROUBLESHOOTING THE SERIES REGULATOR

A circuit like the regulator of Fig. 9-11 is sometimes difficult to troubleshoot because so many components interact with each other. For example, the feedback loop consists of the sample circuit that controls the bias on Q_2, which controls the bias on Q_1, which controls the voltage applied to the sample circuit. This arrangement makes all voltages interdependent. So when a problem arises, the best way to handle it is to break the feedback loop in some way.

Here is one way to handle the problem. By substituting a known good d-c voltage of appropriate value at the top end of R_S, the zener, if good, will go into conduction. See Fig. 9-12. Similarly, by disconnecting the top end of sample resistor R_A from the output and connecting it to the known good supply, comparator Q_2 can be tested. The substitute power supply can be a bench-type regulated supply, or even a battery, since very little current will be drawn from it. The voltage of the substitute supply does not need to be exactly equal to the normal output voltage. It just needs to be sufficiently high to drive the reference zener into conduction, and no higher than the normal output voltage of the regulator being tested.

9-12   Breaking feedback loop to test series regulator.

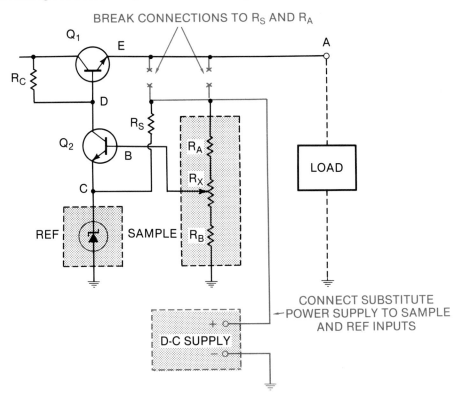

The flowchart of Fig. 9-13 shows a possible test procedure. Let's assume that the output voltage at point A of Fig. 9-11 is abnormal and that adjustment of R_X does not bring V_0 back to normal. We enter the flowchart from the top and come to a block that asks, "Is input voltage E OK?" Here we measure E to make sure that it is normal. Input E should be several volts higher than V_0 to ensure that Q_1 does not go into saturation. E should also not have excessive ripple, otherwise Q_1 will go into saturation on the negative peaks of the ripple voltage. If E has excessive ripple, the output voltage from the regulator will look similar to V_0 in Fig. 9-4.

Getting back to the flowchart, if E measures normal, we remove the load and substitute a dummy (resistive) load. Then we disconnect the top ends of resistors R_S and R_A from the output point and connect them to our substitute supply, as shown in Fig. 9-12.

At this point, we come to a block labeled "Ref OK?" We measure the voltage across the zener. If it is not normal, we follow the NO route. But if it is normal, we adjust pot R_X up and down, while

9-13 Flowchart for troubleshooting series regulator like Fig. 9-10.

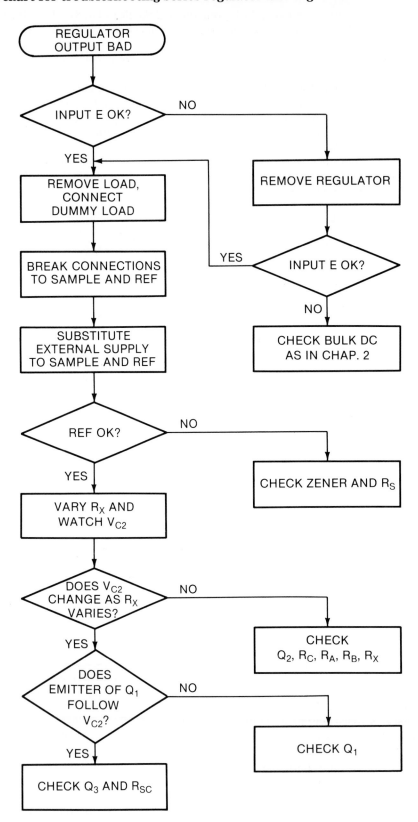

watching the collector of Q_2 with a voltmeter. This step will check the sample circuit as well as Q_2 and its collector load R_C.

The remainder of the flowchart is self-explanatory. The key point here is to try to make the test procedure like those you have used before. That is, apply a normal input to a stage, and check for normal output. In the case of the regulator, you must break the feedback path to apply a test input to any stage.

SINGLE-CHIP REGULATORS

Recently, manufacturers have been putting entire series voltage regulators on single chips. These monolithic regulators contain all of the circuitry of Fig. 9-11, or even more, in one package. Typical packages for some three-terminal, single-chip regulators are shown in Fig. 9-14.

9-14   Typical packages for three-terminal regulators.

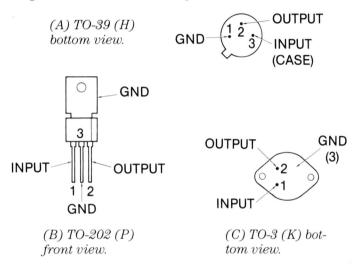

(A) TO-39 (H) bottom view.

(B) TO-202 (P) front view.

(C) TO-3 (K) bottom view.

One version is the 78XX series. Two digits replacing the XX in the number signify the output voltage. That is, the 7805 is a $+5$ VDC regulator, the 7812 is a $+12$ VDC regulator, and so on. Similarly, a 340XX number specifies another series of regulators where the last two digits indicate the output voltage.

Fig. 9-15 shows how a three-terminal regulator can be used in the simplest manner. Input voltage E is the unregulated, but filtered, bulk d-c input, just as in the discrete component circuit of Fig. 9-11. The single-chip regulator contains the reference, sample, pass, comparator, and short-circuit protection components. The output of the regulator is well-regulated d-c voltage. Capacitors C_1 and C_2 are small disc or mylar capacitors of about 0.1 or 0.2 μF. These are needed to prevent the circuit from oscillating.

9-15   Using a three-terminal regulator.

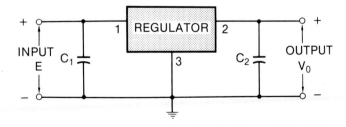

The 78XX and 340XX regulators typically can deliver up to 1 A of direct current to the load.

The manufacturers specify that the maximum d-c input voltage to the regulator is about 30–35 V for 340XX regulators with outputs up to 18 VDC. As with any device, if you are going to modify a circuit for a different application, you should study the complete specs given by the manufacturer. Of course, there is a minimum value for E, otherwise the series regulator will become saturated and drop out of regulation. Typically the drop out voltage for E is about 2 V higher than V_0.

Sometimes an adjustable voltage is needed from a regulated supply. The single-chip regulator can then be used, as shown in the circuit of Fig. 9-16. Here's how it works. The regulator holds the voltage across R_1 constant; therefore, I_1 is constant. Normally I_1 is much larger than the small bias current I_Q that flows through terminal 3. Therefore I_2 is approximately equal to I_1.

9-16 **Adjustable output regulator made with single chip.**

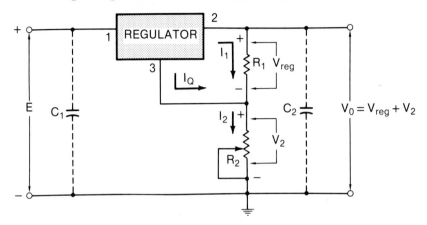

Depending on the value of R_2, some voltage is developed across it. The total value for the output voltage V_0 is then

$$V_0 = V_{reg} + \frac{V_{reg}}{R_1} R_2 \qquad\qquad 9\text{-}7$$

EXAMPLE 9-5 In the circuit of Fig. 9-16, if the regulator is an LM7805, R_1 = 1 KΩ and R_2 = 1.5 KΩ, find V_0.

SOLUTION

$$V_0 = 5 \text{ V} + \frac{5 \text{ V}}{1 \text{ K}\Omega} \times 1.5 \text{ K}\Omega = 12.5 \text{ V}$$

When more current is needed than can be supplied by a single regulator, many equipment manufacturers use separate regulators on each printed circuit card. As shown in Fig. 9-17, the two PC cards have their grounds tied together. They also receive unregulated d-c voltage from a common line. But each card has a separate regulator that supplies regulated d-c voltage V_0 to all circuitry on that card. Note that *outputs of separate voltage regulators should never be tied together!*

There is a way of connecting external components to a single chip regulator to increase its output current capability. This is shown in Fig. 9-18. When power is supplied, the regulator starts delivering current to the load. All regulator current flows through R_2. The current through R_2 causes a voltage to be developed across R_2, thus forward biasing Q_1. Since D_1 has the same amount of voltage across it (0.7 V) as V_{BE} of Q_1, the voltage across R_1 is always equal to the voltage across R_2. And since the current through Q_1 is controlled by the voltage across R_1, we can see that

$$I_1 = \frac{R_2}{R_1} I_{reg} \qquad\qquad 9\text{-}8$$

Of course, the total current through the load is the *sum* of currents $I_1 + I_{reg}$.

9-17   Using separate regulator chips on different cards.

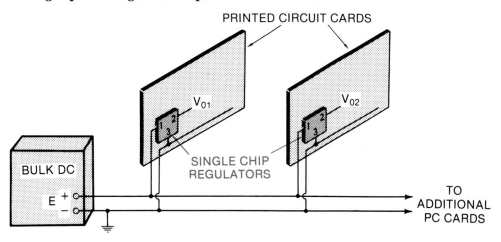

PRINTED CIRCUIT CARDS

9-18   Increasing current capability of regulator.

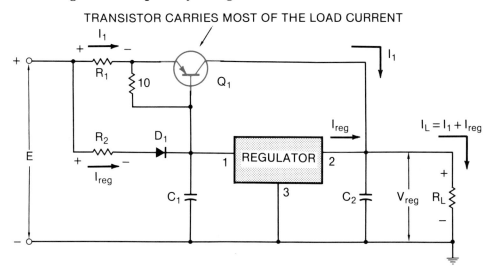

TRANSISTOR CARRIES MOST OF THE LOAD CURRENT

EXAMPLE 9-6   In the circuit of Fig. 9-18, R_1 = 2 Ω, R_2 = 10 Ω, and I_{reg} = 300 mA. What is the total load current?

SOLUTION

$$I_1 = \frac{10\,\Omega}{2\,\Omega} \times 0.3\text{ A} = 1.5\text{ A}$$

So

$$I_{LOAD} = I_1 + I_{reg} = 1.5\text{A} + 0.3\text{A} = 1.8\text{ A}$$

SWITCHING REGULATIONS

In the series regulators, all load current flows through the pass transistor. And if input voltage E is many volts higher than output voltage V_0, a considerable amount of power is dissipated in the pass transistor. This is especially true for high values of load current. The series regulator, in this case, is not very efficient.

EXAMPLE 9-7   Suppose in a circuit like Fig. 9-11, $E = 20$ V, $V_0 = 6$ V, and $I_L = 10$ A. Find the power delivered to the load, and the power dissipated by the pass transistor.

SOLUTION

$$P_L = V_L \times I_L = 6 \times 10 = 60 \text{ W}$$

$$P_d = (E - V_L) \times I_L = (20 - 6) \times 10 = 140 \text{ W}$$

Notice that much more power is dissipated in the pass transistor than is delivered to the load. The circuit is very inefficient.

One technique of improving efficiency, which has become very popular in recent years, is the use of a *switching* regulator. In the switching regulator, the pass transistor does not conduct continuously. Instead, it is switched on and off at a high rate, so as to keep the load voltage at some desired value.

Fig. 9-19 shows very basically how one type of switching regulator works. Suppose a d-c voltmeter is connected across R_L as shown in Fig. 9-19A. Next, switch S is made to close and open alternately at some high rate of speed, say 1000 times per second. Fig. 9-19B shows the square waveform you would expect to see at point A if you observed that point with an oscilloscope. But the d-c voltmeter cannot respond to such rapid variations. The voltmeter tries to go up scale to $+20$ VDC half the time, then down to 0 VDC the other half. The net result is that the voltmeter reads the *average* voltage across R_L, namely $+10$ VDC. The average value is $+10$ V because the voltage at point A is $+20$ V for exactly *half* of the input cycle. The waveform A is said to have a *50% duty cycle.*

Now let's suppose that we change the percentage of time that switch S is closed to a 75% duty cycle. This is shown in Fig. 9-19C. Since point A is a $+20$ V for 75% of the time, the average voltage as read on the d-c voltmeter is 75% of $+20$ V, or $+15$ V. Similarly, Fig. 9-19D shows the condition for a 25% duty cycle.

In each of these cases, the *frequency* of the input waveform at point A was held constant, but the percentage of time the switch remained closed was changed. We can say that *the width of the input pulses at point A controls the average voltage* at point A. This technique is known as *pulse width modulation* (PWM). Although varying the width of the pulse allows us to vary the average output voltage, the total load voltage is far from being smooth, well-filtered direct current.

The circuit of Fig. 9-20 shows a typical filter circuit used with switching regulators. In Fig. 9-20A, switch S is closed. Current flows through switch S, coil L, and load R_L, as shown. A counter electromotive force (cemf) is developed across coil L which prevents the load from increasing too rapidly. Diode D is back biased when switch S is closed. Then, when switch S opens, the collapsing magnetic field around choke coil L induces a cemf of *opposite polarity* across L. This cemf forward biases diode D and keeps current flowing through R_L in the same direction as before. Diode D is needed to provide a path for current flow through L and R_L when S is open.

The action of choke coil L keeping a relatively constant current flowing through R_L, keeps the voltage across R_L more constant. Capacitor C further smooths out variations in voltage (ripple) across R_L. Typically, the switching frequency is high, about 20 KHz to perhaps 100 KHz. The high frequency allows the use of smaller, less expensive filter components than if lower switching frequencies were used.

The choke is often a toroid (donut-shaped) coil, but it is sometimes made by wrapping several turns of wire on a bobbin. The bobbin is usually completely enclosed in a ferrite case to prevent magnetic coupling to nearby circuits.

The simplified circuit of Fig. 9-20 uses a mechanical switch to chop the input voltage into pulses. Obviously, this is not a good final solution, so the mechanical switch is replaced by a switching transistor. Secondly, we need circuitry to sample the output voltage, compare it to some reference, and finally control the switching of the pass transistor. Fig. 9-21 shows a circuit containing all of the required sections.

Of course, all of the compare, reference, and control circuitry can be built with discrete components, but a simpler solution is shown in Fig. 9-22. This circuit uses an IC, an SG1524 pulse width modulator (PWM), to do much of the job. Inside the SG1524 is a sawtooth oscillator, a 5-V reference, an error amplifier, a comparator, and a few other components.

9-19 **Controlling average voltage across R_L by varying the pulse width.**

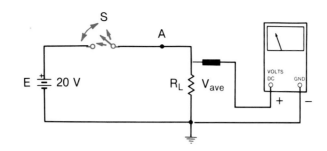

(A) Basic circuit.

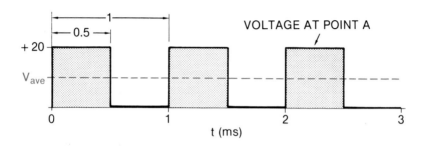

(B) 50% duty cycle, $V_{ave} = +10$ V.

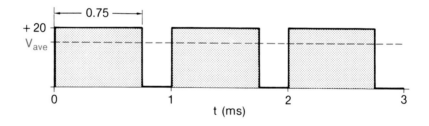

(C) 75% duty cycle, $V_{ave} = +15$ V.

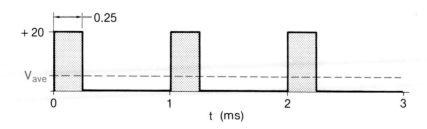

(D) 25% duty cycle, $V_{ave} = +5$ V.

9-20   Filter circuit for switcher.

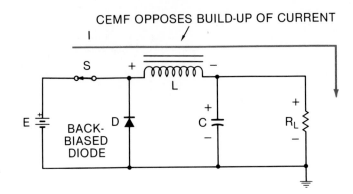

(A) Current flow when switch is closed.

CEMF OPPOSES BUILD-UP OF CURRENT

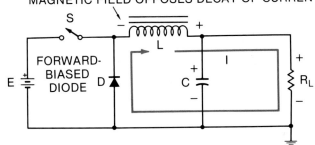

(B) Current flow when switch is open.

KICKBACK VOLTAGE CAUSED BY COLLAPSING
MAGNETIC FIELD OPPOSES DECAY OF CURRENT

9-21   Switching regulator with sample, reference, and control added.

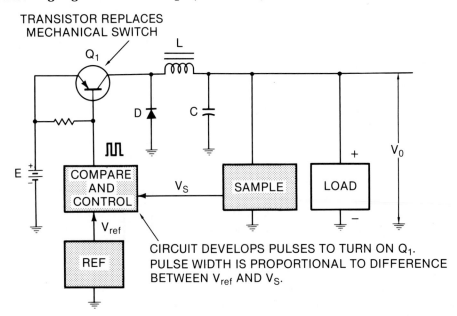

TRANSISTOR REPLACES
MECHANICAL SWITCH

CIRCUIT DEVELOPS PULSES TO TURN ON Q_1.
PULSE WIDTH IS PROPORTIONAL TO DIFFERENCE
BETWEEN V_{ref} AND V_S.

9-22 Pulse width modulator switching regulator.

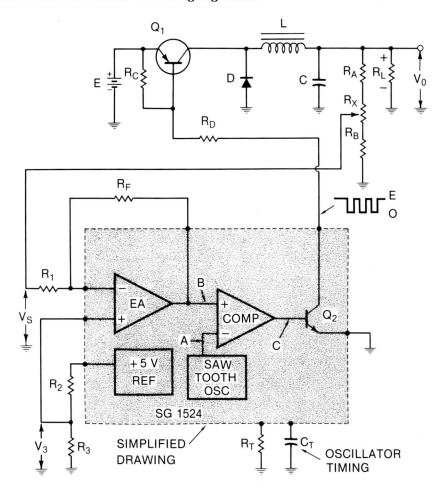

Notice that a sample voltage V_s is picked off from pot R_X, as was done in the series regulators. Voltage V_S is compared to a reference voltage V_3, which is obtained from a voltage divider, R_2–R_3, across an internal 5-V reference supply.

Error amplifier EA amplifies the difference between V_S and V_3. Note that resistors R_F and R_1 control the gain of the error amp, just as they did in the op amp circuit in chapter 8. The voltage at point B inside the SG1524 is the amplified difference between V_S and V_3. This voltage is fed to the noninverting input of the comparator. The more positive V_3 is with respect to V_S, the more positive point B will be. The other input to the comparator comes from the internal sawtooth generator. The frequency of the sawtooth waveform is controlled by external components R_T and C_T.

Fig. 9-23 shows typical waveforms. Whenever the voltage at the noninverting input of the comparator (point B) is more positive than the ramp voltage at the inverting input (point A), the output of the comparator (point C) is positive. Suppose initially that the voltage at point B is $+2$ V. From the time that the ramp passes through $+1$ V until it passes through $+2$ V, the voltage at point C is positive. Incidentally, some circuitry inside the SG1524 (not shown) keeps point C low when the ramp is lower than $+1$ V. This period of the ramp is referred to as "dead" time. The voltage at point C is fed to the base of transistor Q_2, turning it on from time $t = 5$ μs until $t = 15$ μs. When Q_2 turns on, it pulls the base at PNP transistor Q_1 to ground, turning it on.

Now let's suppose that some time later, the load resistance decreases, drawing more load current. Voltage V_S then decreases, causing the voltage at B to increase. As shown in Fig. 9-23B, the net result is that the width of the pulse applied to the base of Q_2 increases. This, of course, increases the duty cycle of Q_1. And, as you know, this will tend to bring V_0 back up to near its original value. Similarly, when the load current decreases, V_S rises, resulting in a shorter pulse width at point C.

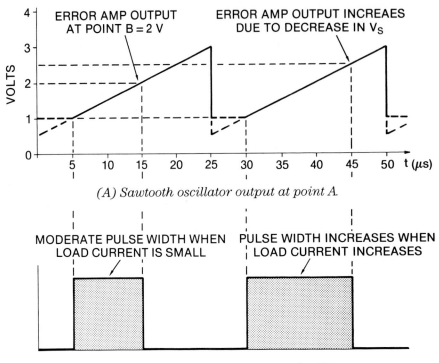

(A) Sawtooth oscillator output at point A.

(B) Output of comparator at point C.

The pass transistor Q_1 is thus switched on and off with a longer or shorter duty cycle, as is needed to maintain a constant output voltage. When the pass transistor is on, it is driven into saturation, acting like a closed switch. During this conduction time, the voltage across Q_1 is very small. So even though a large current flows through Q_1, very little power is dissipated in it.

During the time when the transistor is cut off, no current flows through it, so no power is dissipated in it. All in all, using a power transistor in the switching mode is much more efficient than having it conduct continuously with several volts across it.

Another type of switching regulator is shown in Fig. 9-24. It is part of a μA78S40 switching regulator chip. This circuit is not a pulse width modulator, but rather it pulses pass transistor Q_2 on periodically to recharge a filter capacitor as needed.

There is an internal oscillator, whose frequency is controlled by external capacitor C_T. The oscillator produces a train of rectangular pulses and feeds them to an and gate, and also to the reset input of a flip-flop. The other input to the AND gate comes from the comparator. The action of the comparator is obvious. It compares the sample voltage V_S to an internal $+1.3$-V reference. Whenever V_{ref} is more positive than V_S, the comparator is high (positive). Otherwise it is low (at ground potential).

The action of the and gate follows: Whenever *both* of its inputs are high, its output is high. If either input is low, its output is low.

Whenever the AND gate output goes high, it *sets* the flip-flop. This means that the Q output of the flip-flop goes high. Of course, this causes Q_1 to turn on, which then turns on Q_2. Later, when the oscillator output goes low, it *resets* the flip-flop. This means that the Q output of the flip flop goes low, turning Q_1 and Q_2 off. The action is shown in Fig. 9-25. Note that whenever the comparator output is high, oscillator pulses get through the AND gate. Therefore, current pulses flow through R_L.

A complete step-down switching regulator circuit is shown in Fig. 9-26. Note the L-C filter circuit and sample circuit at the output.

Here's how it works. Fig. 9-27A shows that the oscillator is continuously running. At the start of Fig. 9-27B, we see that a current pulse through Q_2 charges up capacitor C_0 to some voltage slightly above a preset value V_{set}. This causes sample voltage V_s to be higher than V_{ref}. Therefore, no pulses get through the AND gate (as described earlier).

9-24 Part of μA78S40 IC.

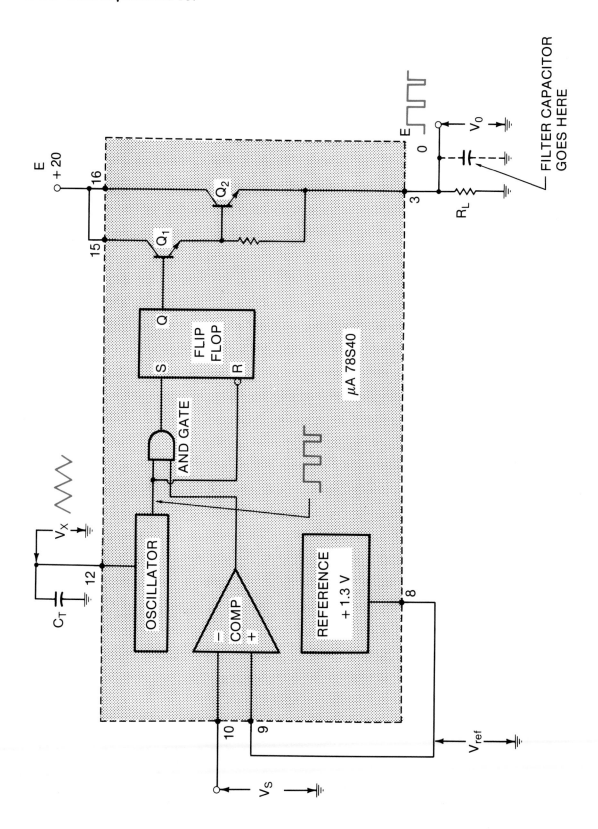

9-25 **Waveforms of μA78S40 of Fig. 9-24.**

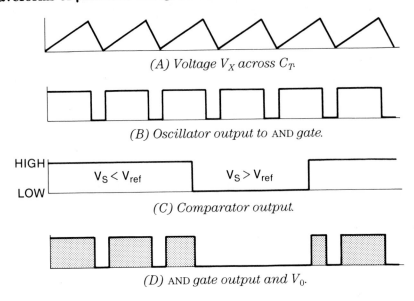

(A) Voltage V_X across C_T.

(B) Oscillator output to AND *gate.*

(C) Comparator output.

(D) AND *gate output and V_0.*

9-26 **Step-down switching regulator using μA78S40 chip.**

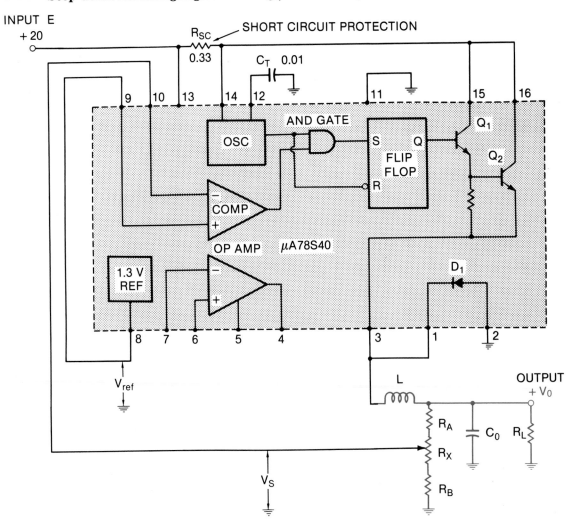

9-27 Waveforms for Fig. 9-26.

(A) Oscillator waveform across C_T.

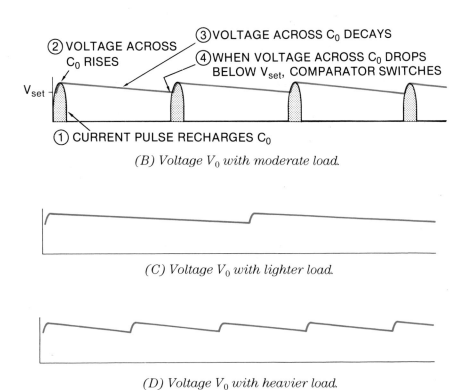

② VOLTAGE ACROSS C_0 RISES

③ VOLTAGE ACROSS C_0 DECAYS

④ WHEN VOLTAGE ACROSS C_0 DROPS BELOW V_{set}, COMPARATOR SWITCHES

V_{set}

① CURRENT PULSE RECHARGES C_0

(B) Voltage V_0 with moderate load.

(C) Voltage V_0 with lighter load.

(D) Voltage V_0 with heavier load.

But because R_L is connected across C_0, the charge of C_0 gradually decays. Eventually, the voltage across C_0 (point 4 in Fig. 9-27B) drops below V_{set}. At this point, V_s becomes less than V_{ref}, so another pulse is allowed to pass through the gate. This pulse turns on Q_2, recharging C_0 again. The process repeats over and over as shown.

Figs. 9-27C and 9-27D show the waveforms across C_0 for lighter and heavier load currents. The overall effect of the μA78S40 is that Q_2 pulses on occasionally as needed to "pump up" the charge on C_0 and thereby maintain a nearly constant voltage across it. Using larger values of C_0 will cause the output to have less ripple. This is true of all filters. Frequencies used for switching regulators are usually quite high, on the order of 5 to 100 KHz. For this reason, filter components can be smaller.

OTHER SWITCHING REGULATOR MODES

So far we have studied only step down switching regulators. The circuit of the last topic gives an output voltage that has the same polarity and is lower in amplitude than input voltage E. This was also the case with the linear type of series regulators. But there are two other modes of operation of switchers that can produce a regulated output voltage that is *higher* in amplitude than the input voltage or is of *opposite polarity* from the input.

The three basic configurations are shown in Fig. 9-28. You are already familiar with the operation of the step-down switcher. Fig. 9-29 shows the operation of the step-up switcher. In Fig. 9-29A, switch S is closed. Current flows from input E through L and S, building up a magnetic field around L.

9-28   Three modes of switching operation.

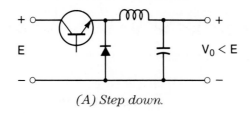

(A) Step down.

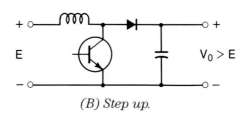

(B) Step up.

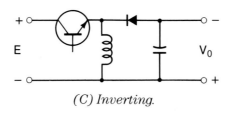

(C) Inverting.

9-29   Operation of step-up switcher.

(A) Current flow with switch closed.

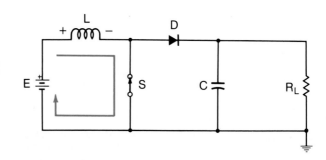

(B) Current flow when switch opens.

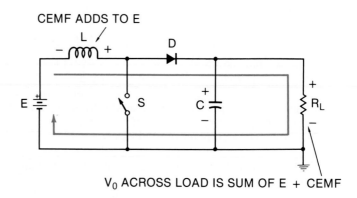

V_0 ACROSS LOAD IS SUM OF E + CEMF

Then on the next part of the cycle, as shown in Fig. 9-29B, S opens, causing the magnetic field around L to start to collapse. The collapsing field develops a cemf, which *adds* to supply voltage E, forward biases D, and develops a voltage across R_L equal to the sum of E *plus* the cemf. Capacitor C also charges to the peak voltage across R_L and maintains voltage across R_L when the switch closes again.

The operation of the inverting switcher is shown in Fig. 9-30. When switch S is closed, current flows through L building up a magnetic field. Diode D is back biased at this time. When the switch opens, the collapsing magnetic field around L develops a cemf, which forward biases D and causes current to flow through R_L as shown. Once again, capacitor C filters out the variations in voltage across R_L, just like in a half-wave rectifier circuit. Each time switch S opens, the charge on C is replenished, or "pumped up" to its peak value again.

9-30 Operation of inverting switcher.

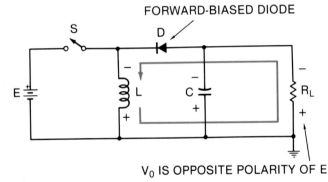

BACK-BIASED DIODE

(A) Current flow with switch closed.

FORWARD-BIASED DIODE

(B) Current flow when switch opens.

V_0 IS OPPOSITE POLARITY OF E

A µA78S40 inverting switcher is shown in Fig. 9-31. Notice that an external pass transistor Q_3 is used, like in the PWM circuit of Fig. 9-22. An external diode is needed in this circuit because a negative voltage appearing on any pin might damage the chip.

The output voltage is fed to the left end of R_1, which drives an internal op amp. The gain of the op amp used in the inverting mode is determined by the ratio of R_F/R_1, where $R_F = R_2 + R_3$. Notice that the gain is less than unity. By adjusting R_3 so that $R_F = 8.125$ KΩ, the output of the op amp, pin 4, will be approximately $+1.3$ V when $V_0 = -12$ V. Therefore, the regulator will hold V_0 at -12 V. Putting it another way, the input signal to the op amp is regulator output voltage $-V_0$. Because of the action of the switcher, the output of the op amp (pin 4) is always held near $+1.3$ V (approximately equal to V_{ref}).

Using the gain equation of the inverting op amp, we see that

$$\frac{V_{(pin\ 4)}}{V_0} = \frac{-R_F}{R_1} = \frac{V_{ref}}{V_0}$$

or

$$V_0 = \frac{-V_{ref}\ R_1}{R_F}$$

9-9

9-31 **Inverting switcher using a μA78S40 chip.**

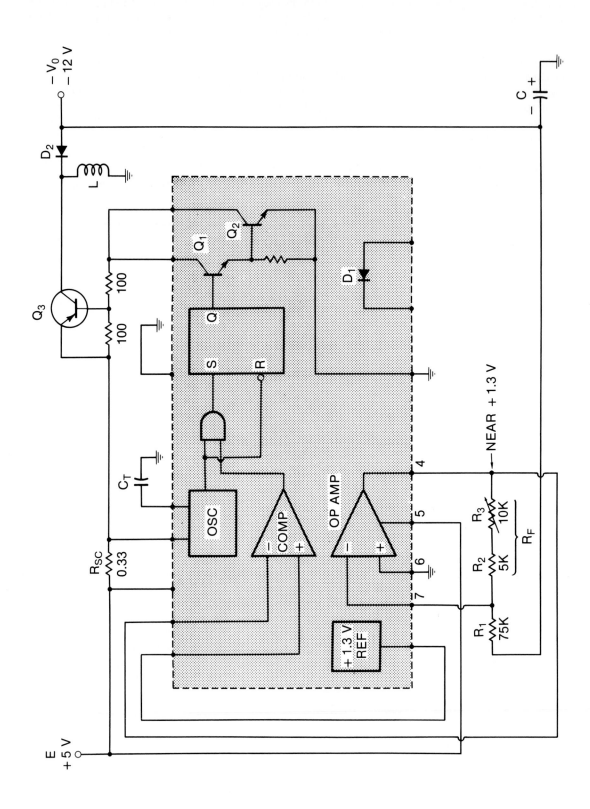

EXAMPLE 9-8 When the pot in Fig. 9-31 is adjusted to 4 KΩ, what is the output voltage of the regulator?

SOLUTION

$$V_0 = \frac{-1.3 \times 75 \text{ K}\Omega}{5 \text{ K}\Omega + 4 \text{ K}\Omega} = -10.8 \text{ V}$$

As an exercise, determine how to wire the µA78S40 as a step-up switcher. (*Hint:* You don't need any more external components than are used in the step-down switcher.)

Troubleshooting a switching regulator is done in much the same manner as troubleshooting the series regulator. You must break the feedback loop and substitute a known value of d-c voltage. For example, if the output of the PWM of Fig. 9-22 is abnormal, you should disconnect the top end of R_A from the output and connect it to a substitute supply. The value of the substitute supply should be near the normal value of V_0.

When this is done, look at the base of Q_1 with an oscilloscope. You should see a high-frequency rectangular wave. Then, as you vary the setting of R_X, the width of the pulses should vary. If they do, you know that the SG1524 circuitry is operating normally. From there on, you can proceed to check the pass transistor and the filter circuit. In the case of the µA78S40 step-down switcher, remove the load from pin 3 and connect a dummy load resistor from pin 3 to ground. Then connect the top of the sample pot to a known good d-c voltage. When V_S is made less than V_{ref}, pulses should appear across R_L, as shown in Fig. 9-25. If they do, the switching circuitry is operating normally.

PROBLEMS

For the first six problems, refer to Fig. 9-32.

9-32   Circuit for problems 9-1 through 9-6.

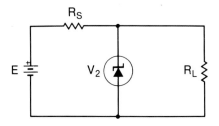

9-1.   If $E = 24$ V, $V_Z = 18$ V, $R_S = 47$ Ω, and $R_L = 200$ Ω, find V_0, I_S, I_L, and I_Z.

$V_0 =$

$I_S =$

$I_L =$

$I_Z =$

9-2.   Referring to problem 9-1, if R_L decreases to 150Ω, find V_0, I_S, I_L, and I_Z.

$V_0 =$

$I_S =$

$I_L =$

$I_Z =$

9-3. If $V_Z = 18$ V, $R_S = 47$ Ω, $R_L = 200$ Ω, and E is increased to 27 V, find V_0, I_S, I_L, and I_Z.

$V_0 =$

$I_S =$

$I_L =$

$I_Z =$

9-4. Assume that $V_Z = 12$ V and $R_S = 30$ Ω. Also assume that the input voltage E has 3 Vp-p of ripple, causing it to vary between 19 and 16 V, and that I_L may vary between 75 and 125 mA. What are the maximum and minimum values of I_S and I_Z?

$I_{S(max)} =$

$I_{S(min)} =$

$I_{Z(max)} =$

$I_{Z(min)} =$

9-5. How much power is dissipated in the zener of problem 9-1?

$P_Z =$

9-6. Suppose the zener diode in the circuit of problem 9-4 became defective. Which of the following power ratings would be sufficient for a replacement? (a)¼ W. (b)1 W. (c)2.5 W. (d)10 W.

For the next six problems, refer to Fig. 9-33.

9-33   Circuit for problems 9-7 through 9-12.

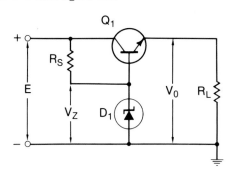

9-7. Assume $E = 18$ V, $V_Z = 12.7$ V, $R_S = 500$ Ω, $R_L = 80$ Ω, β of $Q_1 = 100$, and $V_{BE} = 0.7$ V. Find V_0, I_L, and the power dissipation in Q_1.

$V_0 =$

$I_L =$

$P_d =$

9-8.   In the circuit of problem 9-7, find I_S, I_B, I_Z, and the power dissipation in the zener.

$I_S =$

$I_B =$

$I_Z =$

$P_Z =$

9-9.   In the circuit of problem 9-7, if R_L decreases to 40 Ω, find the new values of I_L and I_Z.

$I_L =$

$I_Z =$

9-10.   Assume $E = 36$ V, $V_Z = 27$ V, $R_S = 910$ Ω, $R_L = 100$ Ω, β of $Q_1 = 100$, and $V_{BE} = 0.7$ V. Find V_0, I_L, and the power dissipation in Q_1.

$V_0 =$

$I_L =$

$P_d =$

9-11.   In the circuit of problem 9-10, find I_S, I_B, I_Z, and the power dissipation in the zener.

$I_S =$

$I_B =$

$I_Z =$

$P_Z =$

9-12.   In the circuit of problem 9-10, if E increases to 40 V, find the new values of the power dissipation in the transistor and the zener.

$P_d =$

$P_D =$

For the next six questions, refer to Fig. 9-7.

9-13.   If R_{128} should burn open, the voltage at TP 1 would (a)go up. (b)go down. (c)remain the same.

9-14.   If C_3 should short, the voltage at TP 1 would (a)go up. (b)go down. (c)remain the same.

9-15.   If zener diode D_5 should open, the voltage at TP 1 would probably (a)go up. (b)go down. (c)remain the same.

9-16. If D_1 should open, the voltage at TP 1 would probably (a)go up. (b)go to zero. (c)resemble the waveform of Fig. 9-4.

9-17. If the voltage at TP 1 is zero, but the voltage at the top of C_3 reads 22.8 V, which component is most likely defective? (a)D_5. (b)X_2. (c)C_3. (d)R_1.

9-18. Suppose the voltage at TP 1 reads much lower than normal, but not zero. Which of the following would be the next logical troubleshooting step? (a)Replace X_2. (b)Disconnect the loads from the emitter of X_2 and connect a dummy load. (c)Replace D_5. (d)Check D_1 and D_2.

For the next three questions, refer to Fig. 9-9. Assume the starting conditions as shown. Fill in the following blanks, using I for increase, D for decrease, and S for remain the same.

9-19. If the load resistance goes up, causing the load current to decrease, the voltage at point A would tend to (a)_____. As a result, the voltage at point B would tend to (b)_____. But since the voltage at point C would (c)_____, the collector current of Q_2 would (d)_____. This would cause the voltage drop across R_C to (e)_____, making the voltage at point D (f)_____. Since Q_1 acts like an emitter follower, the voltage at point A would be forced to (g)_____,back toward the initial value.

9-20. Resistors $R_A R_B$ form a voltage divider from point A to ground, holding point B at about 5.4 V. If R_A decreases in value, the voltage at point B will tend to (a)_____, causing the collector current of Q_2 to (b)_____. The result will be that the voltage at point A will (c)_____.

9-21. By adjusting the ratio of R_A/R_B, the output voltage at point A can be varied. Which of the following is probably the maximum output obtainable? (a)14 V. (b)18 V. (c)22 V. Which of the following is probably the minimum output voltage obtainable? (d)-4.7 V. (e)0 V. (f)$+6$ V.

9-22. Referring to Fig. 9-10B, moving the slider of pot R_X upward makes the voltage at point A (a)go up. (b)go down. (c)remain about the same.

9-23. In the circuit of Fig. 9-11, what value of R_{SC} should be used to limit the short circuit current to 1.5 A?

$R_{SC} =$

9-24. In the circuit of Fig. 9-11, if Q_3 shorts from collector to emitter, the output voltage at point A will (a)increase. (b)decrease. (c)remain the same, but there will be no protection. However, if Q_3 opens, the output voltage at point A will (d)increase. (e)decrease. (f)remain the same, but there will be no protection.

9-25. When using a 7812 single-chip regulator in a circuit like Fig. 9-15, the minimum value for input voltage E should be about (a)7 V. (b)14 V. (c)12 V. (d)78 V.

9-26. You have an LM340-12 regulator chip, and you need a $+20$-V regulated supply. You decide to use a circuit like that of Fig. 9-16. Using R_1 of 1 KΩ, what value should you use for R_2?

$R_2 =$

9-27. You are troubleshooting a circuit like that of Fig. 9-16, which has an output voltage higher than normal. Which of the following might be the cause? (a)R_1 increased. (b)R_1 decreased. (c)R_2 decreased. (d)Input E increased slightly.

9-28. You scope the output of the regulator of Fig. 9-15 and see a high-frequency oscillation of a few volts amplitude. Which of the following might be the cause? (a)This is normal. (b)The internal oscillator in the regulator has too high an amplitude. (c)C_1 or C_2 is open. (d)Input voltage E has too much 60-Hz ripple.

For the next four problems, refer to Fig. 9-22 and answer I for increase, D for decrease, or R for remain the same.

9-29. If the pot slider is moved downward, voltage V_S will (a)_____, but voltage V_3 will (b)_____. As a result, internal voltage B will (c)_____, causing the width of the negative going pulses at the collector of Q_2 to (d)_____. The net effect is that output voltage V_0 will (e)_____.

9-30. If R_3 should open, V_3 would (a)_____, causing V_0 to (b)_____.

9-31. If R_1 should open, the width of the negative-going pulses would (a)_____, causing V_0 to (b)_____.

9-32. If Q_1 should open, the width of the pulses would (a)_____, but V_0 would (b)_____.

9-33. In the circuit of Fig. 9-31, adjusting R_3 to a smaller value causes Q_2 and, therefore, Q_3 to be pulsed on—(a)more, (b)less—frequently. As a result, V_0 becomes—(c)more, (d)less—negative.

9-34. Show how you would wire a μA78S40 switcher to get a regulated output voltage of +15 V from an input of +5 V. You need not show component values, but show all necessary connections to the chip.

EXPERIMENT 9-1 SIMPLE VOLTAGE REGULATORS

You will now become familiar with a simple zener diode shunt regulator. Then you will add a series pass transistor to the circuit to obtain higher output current.

EQUIPMENT

- adjustable, low-voltage, d-c power supply
- VOM
- T1P 31 or equivalent NPN power transistor
- 5.1-V zener diode
- 10-KΩ, ½-W ± 5% resistor
- 5.1-KΩ, ½-W ± 5% resistor
- 1-KΩ, ½-W ± 5% resistor
- 510-Ω, ½-W ± 5% resistor
- 100-Ω, ½-W ± 5% resistor
- 62-Ω, 1-W ± 5% resistor
- 51-Ω, ½-W ± 5% resistor
- 22-Ω, ½-W ± 5% resistor

PROCEDURE

E9-1   Circuits for Experiment 9-1.

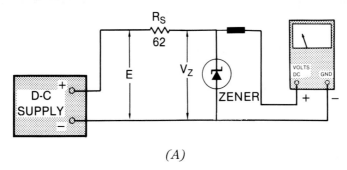

(A)

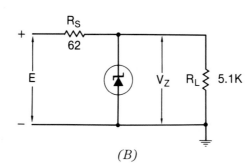

(B)

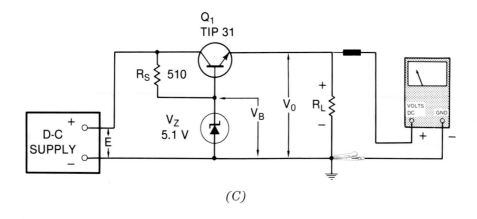

(C)

1. With an ohmmeter, measure the resistance from anode to cathode of the 5.1 V zener diode, connecting the negative lead of the meter to the cathode.

$R_{AK} =$

Then reverse the meter leads, and measure the resistance again.

$R_{KA} =$

If the diode is not shorted or open, it should have a large front-to-back resistance ratio, just like an ordinary silicon diode.

2. In order to measure the breakdown voltage of the zener diode, build the circuit of Fig. E9-1A. Note that the series resistor is necessary to limit zener current to a safe value. Starting with the power supply voltage E at 0 V, record the values of V_Z for each value of E listed in Table E9-1A.

Table E9-1A

E	0	1	2	3	4	5	6	7	8	9
V_Z										

At what voltage does zener breakdown occur?

$V_Z =$

3. Next, shut off the power, and add resistor $R_L = 5.1$ KΩ to the circuit as shown in Fig. E9-1B. Turn on the power supply and adjust E to 9 VDC. Measure V_Z, and record it in the space provided in the first row of Table E9-1B. Then calculate the currents I_S, I_L, and I_Z, and record them in Table E9-1B in the same row.

Table E9-1B

R_L	V_Z	I_S	I_L	I_Z
5.1K				
1K				
510				
100				
51				
22				

4. Change R_L to 1K, 510, 100, 51, and 22 Ω as listed in Table E9-1B, and measure V_Z for each value of R_L. Also calculate currents I_S, I_L, and I_Z, and record them in the appropriate spaces in the table. Be sure to keep E at 9 V for each new set up.

5. Set E to 9 V again, and make R_L = 150Ω. Measure V_Z, and determine values for I_S, I_L, and I_Z, and record these in the first row of Table E9-1C. Then decrease E to 8 V, and fill in the next row. Repeat the measurements for each value of E listed in the table.

Table E9-1C

E	V_Z	I_S	I_L	I_Z
9				
8				
7				
6				
5				
4				

Next we'll study a circuit capable of more output current.

6. Build the circuit of Fig. E9-1C. Starting with R_L = 10 KΩ, record the measured values of V_Z and V_0 and the calculated value of I_L in the first row of Table E9-1D. Then, repeat the measurements for each value of R_L listed in the table. Be sure to keep E = 9 V.

Table E9-1D

R_L	V_Z	V_0	I_L
10K			
1K			
100			
51			
22			

7. Using the measured value of V_0, calculate the percentage of regulation of the circuit of Fig. E9-1D. For $V_{(no\ load)}$, use the voltage measured when $R_L = 10$ KΩ. For $V_{(full\ load)}$, use the voltage measured when $R_L = 22$ Ω. Show your calculations.

% regulation =

Now let's simulate some malfunctions and see what happens to V_0. Using the circuit of Fig. E9-1C, make $R_L = 100$ Ω. V_0 should measure about 4.4 V if the circuit is working correctly.

8. Simulate an open zener by disconnecting one end of it from the circuit. Record V_B and V_0 in Table E9-1E.

Table E9-1E

Trouble	V_Z	V_0
open zener		
shorted zener		
open transistor		
C–E short transistor		
open R_S		

9. Simulate a shorted zener by connecting a clip lead across it, and enter your readings in Table E9-1E.

10. Simulate an open transistor by disconnecting the emitter lead from R_L, and record your readings in Table E9-1E.

11. Fill in the other spaces in Table E9-1E, making appropriate changes wherever necessary.

QUIZ

1. Table E9-1A shows that when a zener is back biased and there is no load across it, the voltage across the zener is always—(a)less than E, (b)equal to E, (c)constant—whenever $E < V_Z$.

2. Referring to question 1, the voltage across the zener is always—(a)less than E, (b)equal to E, (c)constant—whenever $E > V_Z$.

3. Table E9-1B shows that as long as several milliamps flow through the zener, the voltage across it remains constant. (a)True. (b)False. It also shows that $I_S = I_L + I_Z$. (a)True. (b)False.

4. Referring to Table E9-1B, when $R_L = 22$ Ω, V_Z can be increased to 5.1 V by—(a)increasing, (b)decreasing—the value of R_S.

5. Table E9-1C shows that when E drops to a very small value, the zener (a)comes out of conduction. (b)breaks down. (c)regulates at a lower voltage. This is because there is—(d)too much, (e)too little—current flowing through R_S.

6. Table E9-1D shows that more current can be delivered to a load when a series pass transistor is used. The full-load current flows through (a)Q_1. (b)R_S. (c)the zener. What is the power dissipation in Q_1 when $R_L = 22\ \Omega$?

$P_d =$

7. Referring to question 7, how much current flows through R_S? Assuming $\beta = 50$, how much base current flows into Q_1, and how much current flows through the zener?

$I_S =$

$I_B =$

$I_Z =$

For the next three questions, refer to your observations in Table E9-1E.

8. If V_0 reads higher than normal, but V_B reads normal, what is the most likely problem?

9. If V_0 and V_B both read higher than normal, what is the most likely problem?

10. If E reads normal, but V_B and V_0 both read zero, what are two possible malfunctions?

EXPERIMENT 9-2   ADJUSTABLE SERIES REGULATOR

In this experiment you will build and test an adjustable series regulator. You will also observe improved regulation over the simpler regulator of Experiment 9-1 and try some troubleshooting on it.

EQUIPMENT
- 15-VDC supply at 200 mA or more
- VOM
- TIP 31 or equivalent NPN power transistor
- 2N2222 or equivalent NPN transistor
- 5.1-V zener diode
- 1-KΩ pot

- 1-KΩ ½-W ± 5% resistor
- (2) 470-Ω ½-W ± 5% resistors
- (2) 270-Ω ½-W ± 5% resistors
- 47-Ω, 1-W ± 5% resistor
- additional 9-VDC supply or 9-V zener diode and 100-Ω, ½-W ± 5% resistor

E9-2 **Circuit for experiment 9-2.**

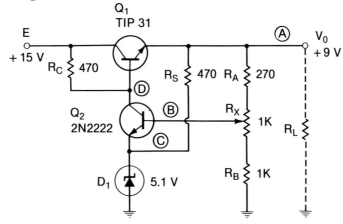

PROCEDURE

1. Build the circuit of Fig. E9-2. Apply an input voltage E of $+15$ VDC, and adjust the pot all the way to one end. Measure V_0.

$V_{0(\text{max})} =$

Then readjust the pot all the way to the other end, and measure V_0 again.

$V_{0(\text{min})} =$

The values recorded are the maximum and minimum we will refer to later. If your circuit can be adjusted so that V_0 varies smoothly from about 6 or 7 V to about 12 V, go on to the next step. Otherwise recheck your wiring.

2. Adjust V_0 to $+9$ V. This will be your no-load voltage.

3. Connect the 47-Ω, 1-W resistor temporarily from ground to point A, measure V_0.

$V_0 =$

Do not leave the load resistor connected too long, unless Q_1 is mounted on a heat sink. The load current will approach 200 mA. Using the measured values of $V_{(\text{no load})}$ and $V_{(\text{full load})}$, calculate the percentage of regulation.

% regulation $=$

4. Before we begin troubleshooting the regulator, measure the normal voltages at points B, C, and D, with point A set at $+9$ V. Use $R_L = 270$ Ω.

$V_B =$

$V_C =$

$V_D =$

5. Now, to simulate an open resistor, disconnect the lower end of R_B from ground. Measure V_0.

$V_0 =$

6. Next disconnect the upper end of R_X and the upper end of R_S from point A, and connect them to an external supply of about 9 V. See Fig. 9-12 for the set up. If you do not have an additional 9-V supply, use a 9-V zener connected to the 15-V supply through a 100-Ω series resistor. In other words, assemble a simple zener regulator like the one in Fig. E9-1A, but use a 9-V zener.

7. Refer to the flowchart of Fig. 9-13. You are now at the block that says "Substitute External Supply. . . ." Measure the reference voltage at point C.

$V_C =$

8. Now vary R_X, and watch the collector voltage of Q_2 with a voltmeter. Is it adjustable up to 9 or 10 V?

(a)Yes. (b)No.

9. At this point, connect the voltmeter to the pot slider. You should notice that the voltage at point B remains at about $+5.8$ V or so, even as R_X is varied. This indicates that Q_2 is probably turned on too hard. A resistance check of R_X and R_B would be the next step to determine the faulty component.

To show that the external supply can be used to feed R_X, connect R_B back into the circuit. Now adjust R_X while watching the collector of Q_2 with a voltmeter. Does it vary as expected?

(a)Yes. (b)No.

Also look at V_0 (point A) while you adjust R_X.

10. Reconnect the circuit as shown in Fig. E9-2, and adjust R_X for $V_0 = 9$ V again. Make each change shown in Table E9-2, and record the voltages at points A, B, C, and D for each change. This will help you understand the function of each component.

Table E9-2

Change	V_A	V_B	V_C	V_D
normal	9			
short R_B				
open D_1				
open Q_2				
open R_A				
open R_C				

> **CAUTION** Do not connect a direct short from point A to ground or Q_1 will be damaged!

QUIZ

Refer to Table E9-2. Answer I (increase), D (decrease), or S (remain the same).

1. Shorting R_B caused the voltage at the base of Q_2 to (a)_____. This caused Q_2's collector current to (b)_____, forcing the voltage at point D to (c)_____.

2. Opening R_B would cause V_B to (a)_____, making the collector current of Q_2 (b)_____ and forcing the voltages at points D and A to (c)_____.

3. Opening D_1 made the collector current of Q_2 (a)_____, which forced the voltages at D and A to (b)_____.

4. If D_1 became shorted, the collector current of Q_2 would (a)_____, forcing the voltages at D and A to (b)_____.

5. Opening Q_2 caused the voltage drop across R_C to (a)_____.

6. Although either an open D_1 or an open Q_2 caused V_A to (a)_____, you can isolate the defective component by measuring V_C. An open Q_2 causes V_C to (b)_____, but an open D_1 causes V_C to (c)_____.

7. An open R_A causes the bias on Q_2 to (a)_____. This causes the base current of Q_1 to (b)_____.

8. An open R_C not only causes the collector current of Q_2 to (a)_____, but it also causes the base current of Q_1 to (b)_____.

EXPERIMENT 9-3 SINGLE-CHIP REGULATORS

This will be a short, simple experiment to familiarize you with the single-chip voltage regulator.

EQUIPMENT

- adjustable, low-voltage, d-c power supply at 200 mA
- VOM
- 7805 or equivalent voltage regulator chip
- (2) 0.1-μF disc capacitors
- 1-KΩ pot
- 10-KΩ, ½-W ± 5% resistor
- 1-KΩ, ½-W ± 5% resistor
- 100-Ω, ½-W ± 5% resistor
- 47-Ω, 1-W ± 5% resistor
- 27-Ω, 1-W ± 5% resistor

E9-3   Circuits for experiment 9-3.

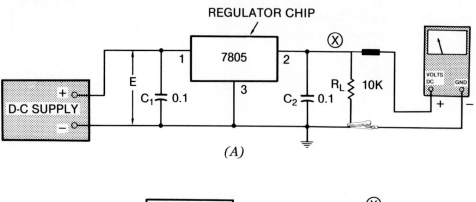

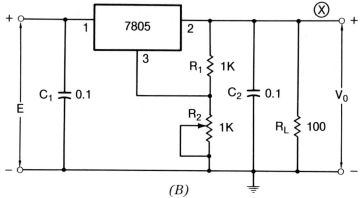

PROCEDURE

1.   Assemble the circuit of Fig. E9-3A. Adjust input voltage E to 9 VDC and apply power. If your circuit is working properly, you should measure 5 V across R_L. If you do, go on to the next step, otherwise recheck your wiring.

The actual voltage you measure at the output may not be *exactly* 5.0 V. The manufacturers state that it may be between 4.75 and 5.25 V, or within 5% of 5 V. That is usually good enough for most applications. Usually it is not important that the power supply voltage be some *exact value*. It is more important that the voltage *remain constant* under changing load and input conditions. Record your measured value in the space provided. Use this as the no-load voltage.

$V_0 =$

2.   Change R_L to 27 Ω, and apply power again. Do not leave the power on more than a few minutes if your regulator is not mounted on a heat sink. Record V_0.

$V_0 =$

Now calculate the load current and the percent of regulation for this measurement.

$I_L =$

% regulation =

3.   Replace R_L with a 47-Ω resistor. This will make I_L a little over 100 mA. Now watch the output voltage with the voltmeter as you decrease input voltage E. Decrease E slowly until you see V_0 begin to drop a few tenths of a volt or more. Measure E at this point.

$E_{(min)} =$

The measured value of E represents the voltage at which the regulator "drops out." Under normal use, E should always be higher than this drop out value.

4.   Crank input E back up to 9 V again. Now with a clip lead, short output point X to ground. Don't worry, it will not hurt the regulator. The chip has built-in current limiting. However, do not leave the short on for too long if the regulator is not mounted on a suitable heat sink. Several seconds will not hurt it, even without a heat sink.

Notice how the output jumps right back up to 5 V when the short circuit is removed.

5.   Now shut off the power and build the adjustable circuit of Fig. E9-3B. Apply power again and adjust input voltage E to 14 or 15 V. With your VOM across R_L, adjust post R_2 from one end to the other. If your circuit is working normally, V_0 should vary several volts. Record the maximum and minimum values.

$V_{0(max)} =$

$V_{0(min)} =$

You should also verify that the circuit still regulates. Connect different values of R_L from point X to ground. That is, you can adjust V_0 to any value between its maximum and minimum values, and it will remain at the preset value under changing load and input conditions.

From your measurement in step 3, what do you suppose the minimum value for E will be if V_0 is set to 9 V?

$E =$

Try it. What does $E_{(min)}$ measure?

$E_{(min)} =$

Digital Circuits 10

Because of advances in economical integrated circuit fabrication, digital electronics has grown overwhelmingly fast in the past several years. Not only have computer and control applications mushroomed, but digital circuits are being used more and more in applications that used to be handled by strictly linear circuits. These include audio amplification, television, tape recording techniques, and test equipment. It is essential, therefore, that all electronics technologists be familiar with digital circuits and their characteristics and with digital testing and troubleshooting techniques.

Since there are many good texts on the theory of logic circuits, we will limit our discussion to the *hardware* of these circuits. We will first look at simple gates and then examine the characteristics of those gates with which testers and troubleshooters must be familiar in order to do an effective job. Then we will look at the commonly used testing tools. Later, we will expand our techniques to more complex digital circuits and systems.

10-1 OR gate.

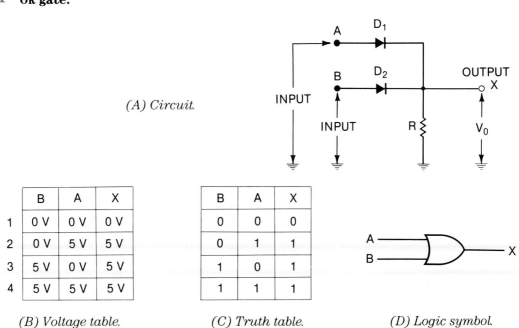

(A) Circuit.

	B	A	X
1	0 V	0 V	0 V
2	0 V	5 V	5 V
3	5 V	0 V	5 V
4	5 V	5 V	5 V

(B) Voltage table.

B	A	X
0	0	0
0	1	1
1	0	1
1	1	1

(C) Truth table.

(D) Logic symbol.

LOGIC GATES

A gate is a circuit with two or more inputs and one output. It belongs to a class of logic circuits called *combinational logic*. This means that the output of the circuit depends on the combination of inputs. One such circuit, called an OR gate, is shown in Fig. 10-1A.

The inputs to the gate are not continuously variable (analog) voltages. They are instead either of two levels, high or low. Let's assume that the high level is equal to $+5$ V and the low level is equal to 0 V. Since there are two inputs, A and B, there are four possible combinations of these inputs, as shown in Fig. 10-1B.

When both A and B are low (0 V), as in row 1, the output at point X is low. But, as shown in row 2, if the input to point A is high ($+5$ V), the output at X is also high. This is because D_1 conducts when its anode is made positive, causing current to flow through R. Similarly, rows 3 and 4 show that if B is high, or if both A and B are high, output X is high.

The voltage shown for point X is $+5$ V when A or B is $+5$ V. Actually, because of the forward drop across the conducting diode, the output at X is closer to $+4.3$ V. However, in logic circuits, we are seldom concerned with the actual voltage, but we do want to show whether the output is high (near $+5$ V) or low (near 0 V). For this reason, a simpler method is used to show the circuit's output.

Fig. 10-1C shows the *truth table* for an OR gate. Notice that we simply replace the high voltage level with a '1,' and low voltage level with a '0.' In this way, we do not get bogged down with slight variations in voltage.

Fig. 10-1D shows the logic symbol for an OR gate. This symbol is always used. It does not matter whether the circuit is built with diodes or with other components, such as transistors.

Another commonly used gate, called the AND gate, is shown in Fig. 10-2A. In this circuit, the output is high, only when *both A and B are high*. As shown in Fig. 10-2B, when A is high and B is low (row 2 in the table), diode D_2 is forward biased. Current flows from V_{CC} through R and D_2 to ground. Since D_2 is forward biased, the voltage across it is 0.7 V, or *near* 0 V. Therefore, the output at Y is at a logic 0. The same would be true if A alone were conducting or if *both* diodes were conducting. The only condition that makes output Y go high is when both A *and* B are high (connected to $+5$-V inputs). The logic symbol for the AND gate is shown in Fig. 10-2E.

10-2 AND gate.

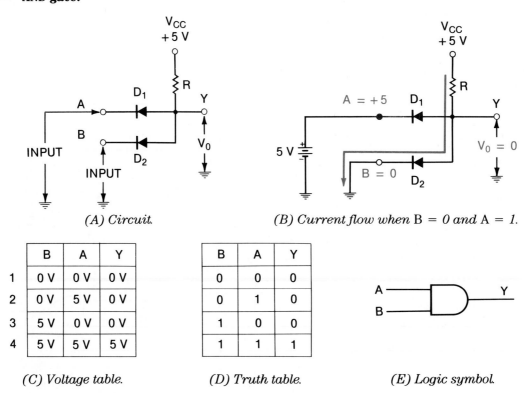

(A) Circuit.

(B) Current flow when $B = 0$ *and* $A = 1$.

	B	A	Y
1	0 V	0 V	0 V
2	0 V	5 V	0 V
3	5 V	0 V	0 V
4	5 V	5 V	5 V

(C) Voltage table.

B	A	Y
0	0	0
0	1	0
1	0	0
1	1	1

(D) Truth table.

(E) Logic symbol.

Fig. 10-3 shows a typical application for an AND gate. Instead of merely labeling the inputs *A*, *B*, etc., they are given signal names, which usually describe what the signals do. Notice that the inputs to the AND gate are signals CLOCK and ENABLE.

10-3 AND gate application.

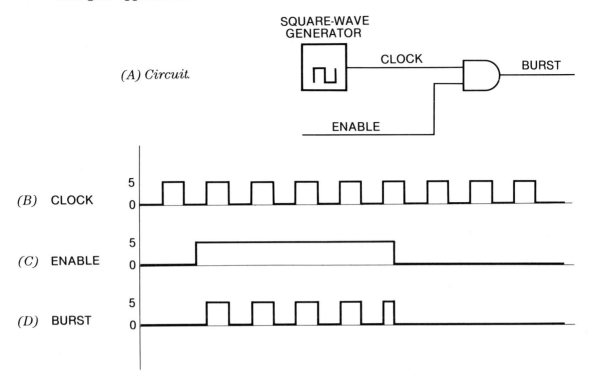

The CLOCK signal comes from a square-wave generator. It is simply a continuous train of rectangular pulses. Signal ENABLE is used to control the gate. As with any AND gate, the output BURST will be high only when both inputs are high. Thus the ENABLE signal allows several of the CLOCK pulses to "pass through" the gate when ENABLE is high. See Figs. 10-3B, 10-3C, and 10-3D. In logic diagrams, ground and V_{CC} connections are omitted for simplicity, but keep in mind that the actual circuits must be connected to the d-c power supply.

INVERTERS

It is often necessary in logic circuits to change a logic high to a low, and vice versa. A circuit that does this job is called an inverter. Fig. 10-4A shows a simple transistor inverter.

When the input (point *X*) is low (switch in position 2), there is no base current flowing into the transistor. Therefore, collector current is cut off, so the collector voltage is near $+V_{CC}$ (high). Then when the switch is thrown to position 1, base current flows, saturating the transistor. The collector voltage then drops to near 0 V (low). In most logic circuits, the transistors are operated either cut off or saturated. That is, they are used simply as *switches,* not as linear amplifiers. Notice that the output of the inverter is labeled $\overline{X}$, and is usually pronounced "not X" or "X bar." The truth table and logic symbol for the inverter are shown in Figs. 10-4C and 10-4D.

If we feed our BURST signal to the input of an inverter, as shown in Fig. 10-5A and 10-5B, the output $\overline{\text{BURST}}$ is the inverse, or *complement* of the input, as shown in Fig. 10-5C.

NAND GATES

If we combine an AND gate followed by an inverter, as shown in Fig. 10-6A, we get another commonly used circuit called a NAND (not and) gate. The NAND gate symbol is shown in Fig. 10-6B, and its truth table, in Fig. 10-6C. Notice that the table for $\overline{C}$ (NAND output) is the opposite of *C* (AND output).

10-4   Inverter.

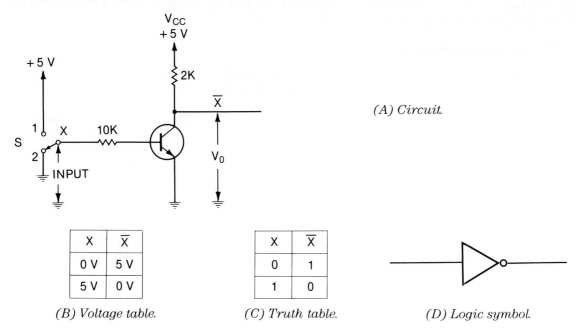

(A) Circuit.

X	$\overline{X}$
0 V	5 V
5 V	0 V

(B) Voltage table.

X	$\overline{X}$
0	1
1	0

(C) Truth table.

(D) Logic symbol.

10-5   Using an inverter.

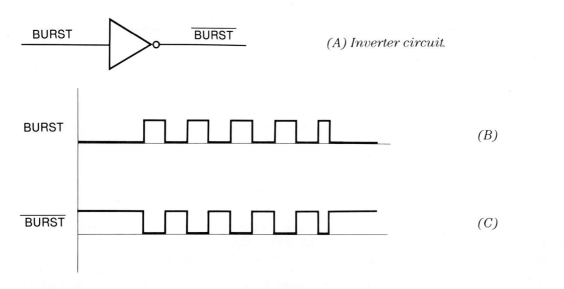

(A) Inverter circuit.

(B)

(C)

We will study the internal construction of the NAND gate, because it is the basic "building block" of a commonly used family of integrated circuits known as TTL (transistor–transistor logic).

First, examine the basic circuit of Fig. 10-7A. Notice the AND gate formed by R_1, D_1, and D_2. Following the AND gate is the transistor inverter. Diode D_3 is added to the circuit to ensure that the transistor is cut off when either input A, or B, is low. That is, if input A is low, D_1 is forward biased. But the forward-biased voltage across D_1 would be enough to begin to turn the transistor on if it were applied directly between base and emitter. By placing diode D_3 in series with the base, point C must reach approximately $+1.4$ V before the transistor turns on. So with D_1 conducting, and point C near 0.7 V, the transistor is definitely cut off.

Now, let's look at Figs. 10-7B and 10-7C. Diodes D_1 and D_3 can actually be replaced with a single transistor. You will remember from earlier work, that a transistor acts like two diodes connected back to back. Since a single transistor actually takes up less space on a chip than two separate diodes

10-6 NAND gate.

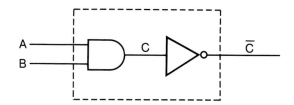

(A) AND *gate followed by an inverter.*

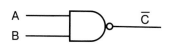

(B) NAND *symbol.*

(C) *Truth table.*

B	A	C AND	$\overline{C}$ NAND
0	0	0	1
0	1	0	1
1	0	0	1
1	1	1	0

10-7 Simplified NAND gate.

(A) *Basic circuit.*

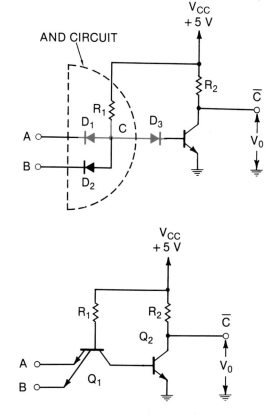

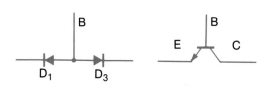

(B) *Part of basic circuit.*

(C) *Transistor.*

(D) *improved circuit.*

and the associated interconnections, more circuitry can be packed into a smaller space by using the transistor instead of the two diodes.

When the artwork for the IC is being made, it is as easy to put *two* emitters on a transistor as it is to put one on. So diodes D_1 and D_2 are both replaced by using a two-emitter transistor, as shown in Fig. 10-7D. It is important to remember when testing these circuits, that Q_1 acts like three diodes and is not being used to amplify any input voltage. Since the circuit of Fig. 10-7D is composed mainly of transistors, it is called a TTL circuit.

The output section of the NAND gate still needs some improvement in order to speed up the rise time of the output signal. Fig. 10-8A shows the output transistor working into load capacitance C_L. All loads have some capacitance of the circuit. (Review the section on high-frequency response of amplifiers, if necessary.)

10-8   Collector resistance increases reset in time of V_0.

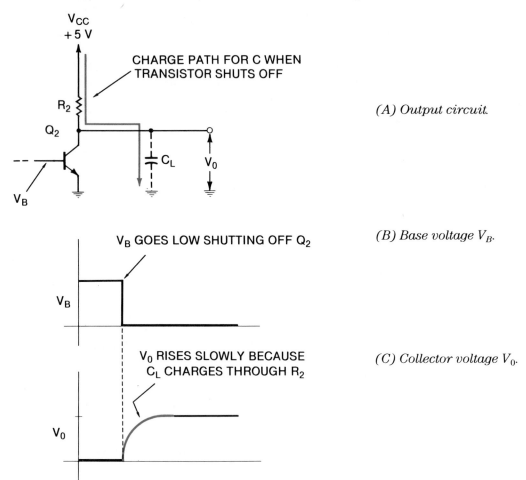

(A) Output circuit.

(B) Base voltage V_B.

(C) Collector voltage V_0.

Now when the base signal goes low, shutting off Q_2 as shown in Fig. 10-8B, load capacitance C_L must charge to V_{CC} through R_2. Therefore, output voltage V_0 has the typical exponential rise shape as in any resistor-capacitor (R-C) circuit. See Fig. 10-8C. The larger the R-C time constant, the longer it takes for V_0 to reach V_{CC}. The rise time could be decreased by making R_2 smaller. But then the current flow through R_2 would be high when Q_2 conducts, resulting in a waste of power.

This problem was solved by replacing Q_2 and R_2 with a *totem pole* circuit, as shown in Fig. 10-9A. You saw totem pole outputs in chapter 6. In the TTL circuit, either Q_3 or Q_4 conducts, but never both at the same time.

Here's how it works. If either input A or B or both are low, Q_2 is cut off. Since no current flows through R_3, there is no forward bias on Q_4, so Q_4 is cut off. At the same time, since Q_2 is cut off, its collector voltage is high, applying bias to the base of Q_3. This turns Q_3 on. The equivalent of the output is shown in Fig. 10-9B.

Then when inputs A and B are both high, Q_2 conducts. Current flow through R_3 produces a positive voltage at the base of Q_4, turning it on. But at the same time, since Q_2 is saturated, there is

10-9 TTL NAND gate.

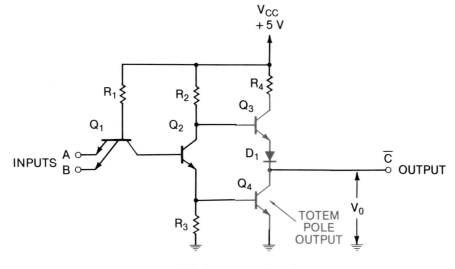

(A) Complete circuit.

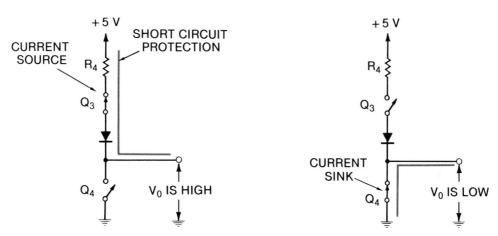

(B) Equivalent when Q_3 conducts (input A or B low).

(C) Equivalent when Q_4 conducts (inputs A and B both high).

practically 0 V across it. Therefore, there is no forward bias on Q_3, so it is cut off. The equivalent circuit for this condition is shown in Fig. 10-9C. Q_3 and Q_4 never both conduct at the same time.

When Q_3 conducts, any load connected to the output terminal can draw current through Q_3 and R_4 in series. Q_3 is called the current *source* for the load. R_4 is small, usually near 100 Ω, and is used simply for short-circuit protection. On the other hand, when Q_4 conducts, it connects any load to ground and is, therefore, called the current *sink*.

Normally TTL circuits are not built to source very much current. But a standard NAND gate output can sink up to 16 mA. If more than 16 mA is allowed to flow through Q_4, the voltage across it gets too large to be considered a good logic low. Q_4 may be damaged if the current through it is excessive.

A whole family of logic circuits is available using the basic construction of the TTL NAND gate. It is called the 7400 series. The number 7400 actually refers to a quad two-input NAND, having four of these gates in a single DIP package. Other members of the 7400 family are numbered 74XXX, where the last two or three numbers identify a specific device. A 54XXX series is also available, having identical pin outs and functions as the 74XXX. But it also has a wider operating temperature range and so is primarily used for military applications. The package and pinouts for the 7400 are shown in Fig. 10-10.

10-10   Package and pinouts for 7400.

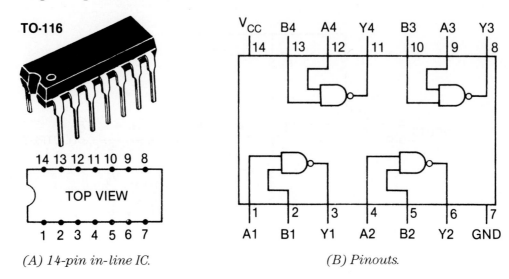

(A) 14-pin in-line IC.   (B) Pinouts.

There are certain characteristics of TTL circuits that you should be familiar with in order to test and troubleshoot them. We are fortunate that the important characteristics of practically all TTL circuits are so similar to those of the 7400. By studying this one circuit, you will be able to test them all. In fact, some circuits that are not TTL are said to be TTL *compatible*. This means that they are capable of driving a TTL input and are capable of being driven by a TTL output.

TTL CHARACTERISTICS

Power Supply

TTL circuits are designed to operate with a V_{CC} supply of 4.5 to 5.5 V, nominally $+5$ V.

Unit Load

When an input is pulled low, current flows through the conducting input diode. See Fig. 10-2B. The current necessary to forward bias the diode is typically 1.6 mA and is referred to as one unit load (UL).

Fan Out

One TTL gate can drive up to ten other TTL gates. That is, as shown in Fig. 10-9C, when Q_4 conducts, it connects all loads tied to its output to ground. If those loads happen to be the inputs of other gates, Q_4 must sink 1.6 mA for each gate. It is designed to carry up to 16 mA, or 10 times the UL. The gate is said to have a *fan out* of 10.

V_{OL} (Low Output Voltage)

When Q_4 conducts, the voltage across it will be less than 0.5 V even with 16 mA flowing through it.

V_{OH} (High Output Voltage)

When Q_3 conducts, it connects the output to the power supply. But actually, the output voltage will not be $+5$ V. Because of the internal components, the logic high output voltage will be at least $+2.4$ V and more typically about $+3$ V or so.

V_{IL} (Low Input Voltage)

To be recognized as a good logic low, the input to any circuit must be less than 0.8 V.

V_{IH} (High Input Voltage)

To be recognized as a good logic high, the input to any circuit must be at least $+2$ V.

Propagation Delay

When an input to any circuit changes, for example when an inverter input goes from low to high, it always takes some small amount of time before the output changes. This time is referred to as *propagation delay*. There is no standard propagation delay, because more time delay occurs in more complex circuits. But for the 7400, it is typically around 20 ns (nanoseconds). This is extremely fast. To get some appreciation for how long one nanosecond is, consider that

1 nanosecond is to 1 second, as 1 second is to 32 *years*!

Table 10-1 summarizes the characteristic of the 7400 IC.

Table 10-1
Summary of Typical 7400 Characteristics*

power supply	+4.5 to +5.5 V
unit load	1.6 mA
fan out	10
V_{OL}	0.5 V
V_{OH}	2.4 V
V_{IL}	0.8 V
V_{IH}	2.0 V
prop delay	20 *ns*

*These characteristics represent typical values. The values can differ slightly with different types of devices and with different manufacturers.

There are also other TTL families related to the standard 7400. These have identifying letters in the part number, such as:
- 74HXXX—High-power TTL
- 74LXXX—Low-power TTL
- 74SXXX—Schottky TTL
- 74LSXXX—Low-power Schottky TTL

While the chips are not 100% interchangeable, the logic function and pinouts of a 7400 and a 74LS00, for example, are identical. The differences in the families occur mainly in the input and output current requirements, as well as in the propagation delay. The Schottky series, for example, is considerably faster than standard TTL, but it does require slightly more power. Generally speaking, it is *not* a good idea to replace a defective chip of one family with a chip of a different family. This usually requires a thorough knowledge of circuit design and is beyond the scope of this text.

SOME DIGITAL TEST EQUIPMENT

Although oscilloscopes and voltmeters can be and are used to test digital circuits, just as with analog circuits, some special tools have been developed that can save time in testing and troubleshooting logic circuits. One of these special tools, as shown in Fig. 10-11, is called a *logic monitor* or a *logic clip*. The unit clips directly over a DIP IC while the IC is still mounted in the circuit. It has separate connections to up to 16 pins and shows the state (high or low) of each pin by lighting or not lighting a corresponding LED. If any input to the clip exceeds 2 V, the LED lights.

10-11 **Logic monitor. Courtesy of Global Specialties Corp.**

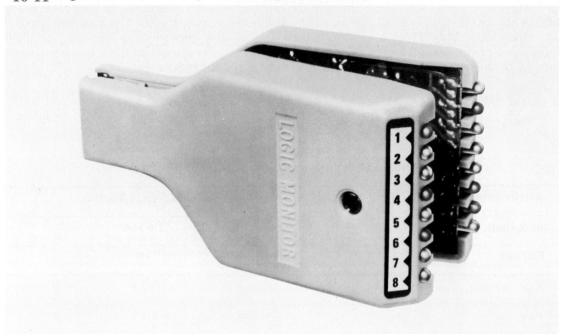

On the monitor shown, each input has an impedance of 100 KΩ, so no loading of the circuit being tested occurs. What's more, you do not have to connect a power supply to the monitor. An internal power-seeking gate network automatically finds power supply leads and feeds them to the internal circuitry. With the monitor, you can at once see the states of all inputs and outputs of every gate in the package.

Another very useful tool is the *logic probe*, shown in Fig. 10-12A. It also indicates the state of any point in a circuit by lighting an LED. It can be used to test single points, not normally accessible to the logic monitor. The logic probe has two clip leads, which must be connected to power supply and ground.

10-12 **Logic probe. Courtesy of Global Specialties Corp.**

The probe has two logic level LED, one to indicate a high, the other to light up when the probe touches a good logic low. There is a reason for two separate indicators. Suppose you touch the probe to an output pin of a gate and the output voltage is, say, 1.5 V. Which lamp will light? *Neither one* will light. The voltage is not a good high or low. If it were a good high, the voltage would be above 2.4 V, and if it were a good low, the voltage would be below 0.5 V. So the fact that *neither* indicator lights shows you that you have a problem at that point in the circuit.

There is one more valuable feature of the probe. It has a blinking pulse detector. You can connect the probe tip to a circuit that occasionally produces a short duration pulse, say less than 1 µs wide. You would never see that pulse with a voltmeter, neither is the pulse active long enough to view with a simple LED monitor. But the probe has a built-in "pulse-stretcher" that is triggered by the short pulse and causes an LED to blink on long enough for you to see it. When viewing continuous low-frequency square waves with the probe, the high and low LED blink on and off. Economy versions of the logic monitor and logic probes can be found in the $25–50 price range. Faster versions are also available at somewhat higher costs.

A third tool, which is used in conjunction with the logic probe, is the *digital pulser*, shown in Fig. 10-13B. It is used to inject a pulse anywhere in a digital circuit. You simply touch the pulser tip to the desired point and press a pushbutton. The pulser then generates a pulse, of proper polarity, to cause the level at the desired point to change (either high or low as needed). Holding the pulse button down generates a continuous pulse train. The DP-1 can source or sink up to 100 mA, which is sufficient to drive almost any input or output to its opposite state without desoldering.

10-13 Digital pulser. Courtesy of Global Specialties Corp.

For example, suppose you suspect that gate II in Fig. 10-14 is defective. (The gates are parts of a 7408 quad 2-input AND chip.) Also suppose that the voltage at pin 5 is high but that the voltage at pin 4 is low. In order to make the output of gate I go high, you would have to make both inputs of gate I high. But maybe the inputs come from some randomly generated signals that are not present yet.

10-14 Testing a gate with a digital pulser and a logic probe.

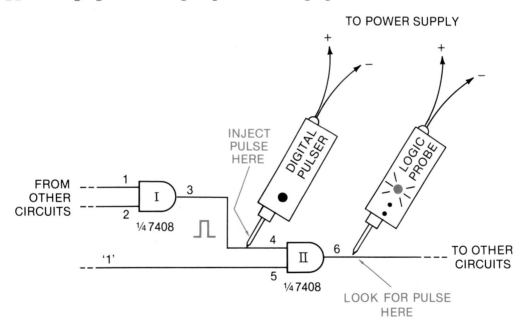

Here's what you do. Connect the logic probe to the output of gate II. Now touch the tip of the pulser to pin 4 and press the pulse button. The pulser will drive pin 4 high momentarily. By observing the logic probe connected to pin 6 of gate II, you will see the pulse indicator blink on if gate II is OK. Otherwise, you know that you have located a defective area. Of course, you still can't jump to conclusions about the gate. It is possible that a short circuit exists, perhaps in the printed circuit foil from pin 6 to ground or from pin 4 to ground. But at least you will have localized the fault. Incidentally, the DP-1 can pulse into a direct short circuit indefinitely without damage.

10-15 CMOS inverter.

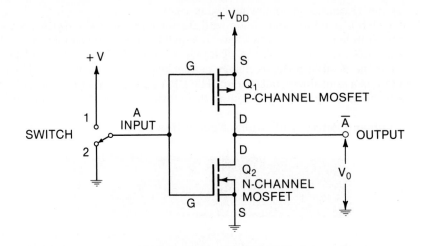

(A) Basic circuit.

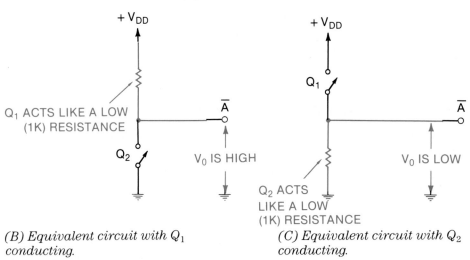

(B) Equivalent circuit with Q_1
conducting.

(C) Equivalent circuit with Q_2
conducting.

MOS DIGITAL CIRCUITS

So far we have looked only at bipolar transistor logic circuits. Field effect transistors are, of course, also used to make digital IC.

In chapter 4, you learned that MOS transistors are essentially voltage-controlled devices and that both P-channel and N-channel MOSFET are used. There is a special family of digital IC called CMOS (complementary MOS). This means that both P-channel and N-channel devices are used in the same circuit, much like complementary-symmetry amplifiers. CMOS digital circuits are the most efficient (low-power consumption) IC of any logic family and are, therefore, valuable in portable or battery-operated equipment.

Fig. 10-15 shows a typical CMOS inverter. When the switch at input A is in position 2 as shown, no forward bias is applied between the gate and source of the N-channel device. The N-channel, therefore, acts like an open switch. But at the same time, since the gate of Q_2 is returned to ground (negative with respect to Q_2's source), Q_2 conducts and acts like a low resistance of approximately 1 KΩ. The equivalent circuit is shown in Fig. 10-15B. Note that the output $\overline{A}$ is pulled high through Q_1.

Then when the input at point A is driven high, by placing the switch in position 1, Q_1 turns off and Q_2 conducts. The equivalent circuit is shown in Fig. 10-15C. Note that this time the output is pulled low through Q_2. Like the totem pole output circuits used in TTL, only one transistor conducts at a time. But unlike the TTL circuits, no input current is necessary to drive the input nor is any intermediate driver stage needed. Therefore, the power consumed by the CMOS inverter is negligible. Since the output stages of CMOS circuits look like approximately 1-KΩ resistances, they cannot directly drive TTL circuits without some buffering or matching networks. Therefore, TTL and CMOS are not directly interchangeable, but they can be interfaced if some provisions are made.

Besides inverters, of course, other logic elements can be built with CMOS components. Fig. 10-16 shows a typical CMOS NOR gate. Note that two P-channel transistors, Q_1 and Q_2, are in series, while two N-channel transistors are in parallel. The gates of Q_1 and Q_3 are tied together, and similarly for Q_2 and Q_4.

10-16 CMOS NOR gate.

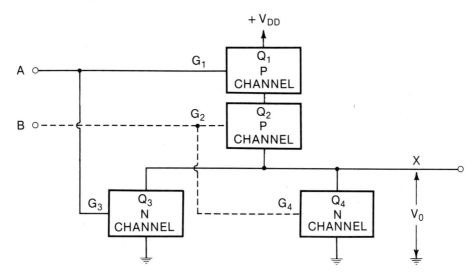

(A) Basic circuit.

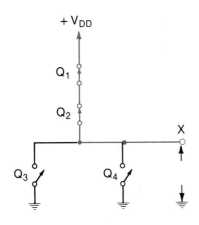

(B) Equivalent when A = 0, B = 0.

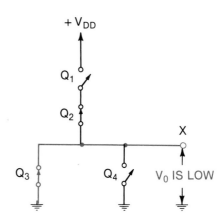

(C) Equivalent when A = 1, B = 0.

	B	A	X
1	0	0	1
2	0	1	0
3	1	0	0
4	1	1	0

(D) Truth table.

(E) Logic symbol.

Here's how it works. When A and B are both low, the two P-channel transistors conduct, and the N-channel transistors are cut off. A simplified equivalent of this situation is shown in Fig. 10-16B. The conducting transistors are shown as closed switches for simplicity, but keep in mind that they each have approximately 1-KΩ resistance.

With both Q_1 and Q_2 conducting, V_0 is pulled high to V_{DD}. This gives us the entry in row 1 of the truth table shown in Fig. 10-16D.

Fig. 10-16C shows the effect when input A is made high, while B is low, giving us the entry for row 2 in the table. Now try to determine which transistors conduct for the other combinations of inputs A and B. You can see by examining the truth table, that output X is exactly the *opposite* of what it would be for an OR gate. Therefore, the circuit is a NOR (not-or) gate. The symbol for the NOR gate is shown in Fig. 10-16E.

CHARACTERISTICS OF CMOS DEVICES

Power Supply
CMOS circuits will operate with power supplies from $+3$ to $+15$ V.

Input Capacitance
The CMOS device does not require any input current to keep its output high or low. It simply needs an input voltage. But the input acts like a small capacitance of approximately 5 pF on each gate.

Fan Out
Since the CMOS device does not have to deliver any current to other CMOS gates, it can drive many of them, perhaps 50 or more. However, each time the input to a gate is switched high or low, the input capacitance must be charged or discharged. If many gates are driven, a considerable increase in rise time or fall time can result.

Input Threshold Voltage
CMOS gates switch either high or low approximately whenever the input voltage passes through 50% of the power supply voltage. However, it is recommended, for efficient operation, that the logic high input voltage be greater than 90% of V_{DD} and that the logic low input be less than 10% of V_{DD}.

Propagation Delay
CMOS gates typically are slower than TTL gates. CMOS gates have a propagation delay of 30–50 ns per gate. Table 10-2 shows the important characteristics of CMOS devices.

Table 10-2
Summary of Typical CMOS Characteristics

power supply	**$+3$ to $+15$ V**
input capacitance	**5 pF per gate**
fan out	**50**
V_{OH}	**$0.9 \times V_{DD}$**
V_{OL}	**$0.1 \times V_{DD}$**
propagation delay	**50 ns**

The logic probes and pulsers discussed earlier can be used equally well to test CMOS devices. RCA has a series of CMOS devices numbered CD4XXX where the X's identify the specific device. For example, a CD4007 is a Hex CMOS inverter chip. National Semiconductor has a line of CMOS chips with numbers like the 7400 series, but identified with a C. For example, a 74C02 is a quad 2-input CMOS NOR gate, in a 14-pin DIP package, having the same pinouts as the 7402 TTL circuit.

HANDLING CMOS DEVICES

Although many small-scale CMOS chips have built-in diodes to protect their inputs from being damaged by static electricity, it is still a good idea to follow some precautions when working with CMOS. These precautions are listed here.

1. CMOS device leads should remain in contact with a conductive foam, except when being used.

2. Metal tools, such as soldering irons, should be grounded before making contact with a CMOS part. In fact, the person handling the CMOS part should be grounded, possibly by wearing a grounding wrist band.

3. Never insert or remove CMOS parts with the power on.

4. Never apply input signals with the power off.

5. All unused inputs must be connected to either ground or to V_{DD}.

In large-scale IC, such as memories and microprocessors, other variations of MOS circuits are used, such as NMOS (N-channel) and PMOS (P-channel) circuits. Although they may not use both N- and P-channel devices simultaneously, the characteristics and handling precautions of NMOS and PMOS are essentially the same as for CMOS.

FLIP-FLOPS AND COUNTERS

As mentioned earlier, gates and inverters are used to perform combinational logic; that is, the output of the circuit depends on the combination of inputs. We will now discuss some circuits whose outputs depend not only on the combination of inputs but also on some previous conditions. In other words, the operation of these circuits, which are used in what is called *sequential* logic, depends on the sequence of events. They essentially "remember" what happened previously.

Fig. 10-17A shows how to connect two NOR gates to build a set-reset (R-S) flip-flop. Note that the circuit (within the dashed lines) has two outputs, Q and $\overline{Q}$, and two inputs, R and S. Resistors R_1 and R_2 are called *pull-down* resistors because they pull input pins 2 and 6 low.

10-17   R-S flip-flop built with NOR gates.

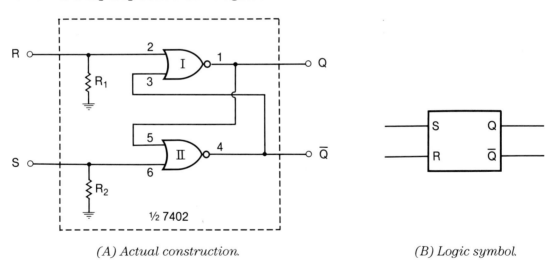

(A) Actual construction.   (B) Logic symbol.

Let's assume initially that pin 1 is low. With pin 1 low, both inputs to gate II are low, so its output, pin 4, is high. With pin 4 high, pin 3 of gate I is held high, which holds pin 1 low as a result. In other words, the circuit is stable with the Q output low and $\overline{Q}$ high. When the Q output is low, the flip-flop is said to be *reset*, or *cleared*.

If we drive the reset *(R)* input high, Q remains low. This is because pin 3 is already high, and any high into a NOR gate makes it output low.

But when we drive the set *(S)* input high, pin 4 goes low. Since at this time both pins 2 and 3 are low, the Q output goes high. The flip-flop is said to be *set* when the Q output is high. Then if we remove the high signal from the set input, the flip-flop *remains* set because the high on pin 1 holds pin 5 high, thus keeping pin 4 low.

We can reset the flip-flop by driving the R input high, and we can set the flip-flop by driving the S input high. Once the circuit is either set or reset, it will remain in that state until we force it into the other state. The logic symbol for an R-S flip-flop is shown in Fig. 10-17B. R-S flip-flops are often used as switch debouncers or as single-bit memories.

R-S flip-flops are also built with cross-coupled NAND gates, as shown in Fig. 10-18. But notice that the inputs are overbarred. The overbars mean that when $\bar{S}$ is driven low, the flip-flop will set (Q will go high), and when $\bar{R}$ is driven low, the flip-flop will reset. The logic symbol shows small circles or "bubbles" on the $\bar{S}$ and $\bar{R}$ inputs. The bubbles on logic circuit inputs indicate that those inputs must be driven low to activate the circuit. They are called *active low* inputs. If bubbles do not exist on the inputs, as in the circuit of Fig. 10-17, the inputs are said to be *active high*.

10-18   R-S flip-flop built with NAND gates.

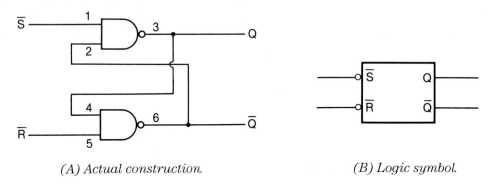

(A) Actual construction.   *(B) Logic symbol.*

In fact, the symbols used for the NAND gates in the R-S flip-flop are usually drawn in a different form to better show the active low inputs. This is a significant point, as we will see later, so let's expand on it here. In Fig. 10-19A we see the familiar AND gate followed by an inverter, which we recognize as a NAND gate. Fig. 10-19C shows the truth table for the circuit in columns 1 through 4.

10-19   NAND gate symbols.

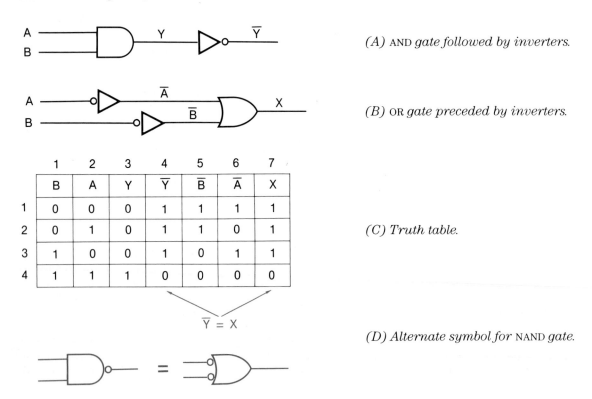

(A) AND gate followed by inverters.

(B) OR gate preceded by inverters.

	1	2	3	4	5	6	7
	B	A	Y	$\bar{Y}$	$\bar{B}$	$\bar{A}$	X
1	0	0	0	1	1	1	1
2	0	1	0	1	1	0	1
3	1	0	0	1	0	1	1
4	1	1	1	0	0	0	0

$\bar{Y} = X$

(C) Truth table.

(D) Alternate symbol for NAND gate.

In Fig. 10-19B, inputs A and B are fed through inverters to obtain $\bar{A}$ and $\bar{B}$. (The bubble can be drawn before or after the triangular buffer symbol to show inversion.) Columns 5 and 6 of the truth table show $\bar{A}$ and $\bar{B}$. We then feed $\bar{A}$ and $\bar{B}$ into the OR gate and obtain output X, as shown in column 7. Notice that columns 4 and 7 are identical for every input combination. Identical output columns in truth tables mean that the two circuits are logically equivalent. That is, the circuit of Fig. 10-19A can be replaced by the circuit of Fig. 10-19B.

Fig. 10-19D shows the compressed logic symbols for Figs. 10-19A and 10-19B. The AND-form symbol on the left tells us that the output of the gate will be low only when *both* inputs are high. The OR-form symbol on the right tells us that the output of the gate will be high if *either* input is low. We will see much more of this in chapter 11.

Fig. 10-20 shows an alternative diagram for the R-S flip-flop using NAND gates and emphasizing active low inputs. Note that *pull-up* resistors R_1 and R_2 hold the $\bar{S}$ and $\bar{R}$ inputs inactive high until an active low signal is applied.

10-20 NAND **gate R-S flip-flop showing active low inputs and pull-up resistors.**

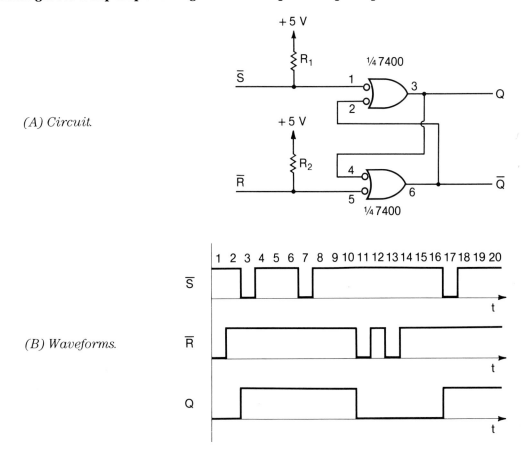

(A) Circuit.

(B) Waveforms.

Fig. 10-20B shows typical waveforms for the NAND gate. At time 1, the $\bar{R}$ input is low, so the Q output is low. Even though the $\bar{R}$ input goes high at time 2, the Q output remains reset until time 3 when the $\bar{S}$ input goes low. At time 7, the Q output does not change, since it is already set. Study the remainder of the waveforms until you are sure you understand what is happening.

Another commonly used circuit is the D flip-flop shown in Fig. 10-21A. The small arrowheads on the connecting lines indicate which lines are inputs and which are outputs. The two inputs marked PR (preset) and CLR (clear) operate exactly the same as the $\bar{S}$ and $\bar{R}$ inputs of the NAND gate R-S flip-flop. Also Q and $\bar{Q}$ are always in opposite states. Input CK (clock) causes the Q output to go either high or low depending on the level of the D input during the *rising edge* of the CK input. If D is high when CK goes high, Q will set. If D is low when CK goes high, Q will reset. See the waveforms of Fig. 10-21B. Fig. 10-21C shows the pinouts of the 7474 dual-D flip-flop.

10-21 *D flip-flop.*

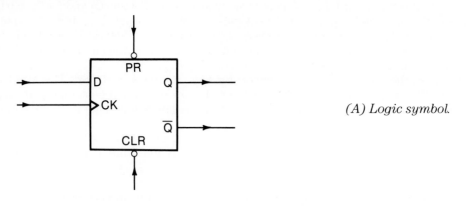

(A) Logic symbol.

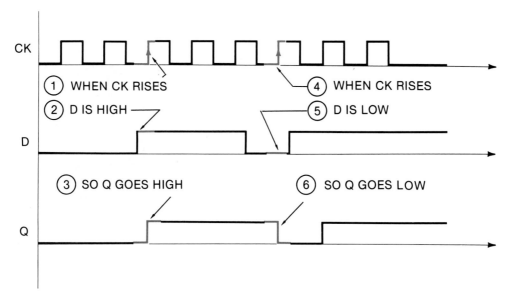

1 WHEN CK RISES

2 D IS HIGH

3 SO Q GOES HIGH

4 WHEN CK RISES

5 D IS LOW

6 SO Q GOES LOW

(B) Waveform.

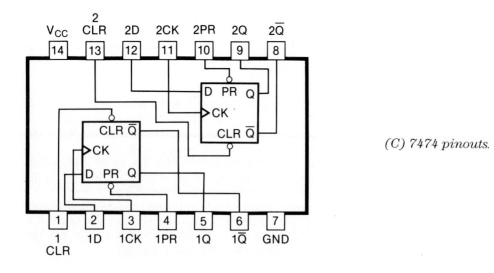

(C) 7474 pinouts.

If the *CLR* and/or *PR* inputs are not used in a particular application, they must be tied high, usually through pull-up resistors of about 1 KΩ or so. *Never leave inputs on logic circuits unconnected, because they might pick up noise and cause false operation.*

Another commonly used circuit is the *J-K* flip-flop, shown in Fig. 10-22. It is somewhat more complex than the *D* flip-flop and is more flexible. Like the *D* flip-flop, it has *CLR* and *PR* inputs. But the *J-K* circuit also has two inputs, labeled *J* and *K*, which set up the operating mode. In other words, these inputs determine what the circuit will do *when the next clock pulse is applied.*

10-22 *J-K* flip-flop.

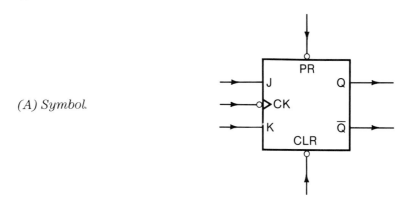

(A) Symbol.

J	K	MODE	DESCRIPTION
0	0	INHIBIT	Q OUTPUT WILL NOT CHANGE WHEN CLOCKED
0	1	RESET	Q OUTPUT WILL GO LOW WHEN CLOCKED
1	0	SET	Q OUTPUT WILL GO HIGH WHEN CLOCKED
1	1	TOGGLE	Q OUTPUT WILL CHANGE TO OPPOSITE STATE WHEN CLOCKED

(B) Table of operation.

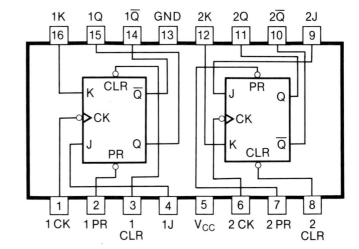

(C) Pinouts of 7476 dual J-K flip-flops.

Notice the bubble on the clock input. The bubble shows that the circuit is activated by a low clock input. In addition the arrow > at the CK input shows that the falling *edge* of the *CK* is what triggers the circuit. The table in Fig. 10-22B describes the operation of the *J-K* flip-flop for all combinations of *J* and *K*. The pinouts for the 7476 dual-*J-K* flip-flop are shown in Fig. 10-22C.

One interesting application of the *J-K* flip-flop is its use in a *binary counter,* as shown in Fig. 10-23. To simplify the diagram, no connections are shown to the *J, K, PR,* or *CLR* inputs. The unused inputs, however, would be tied high in the actual circuit. With *J* and *K* both high, all flip-flops are in the *toggle* mode.

10-23 Binary counter.

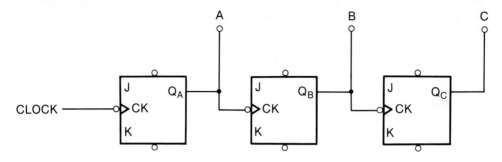

(A) Three-bit binary counter circuit.

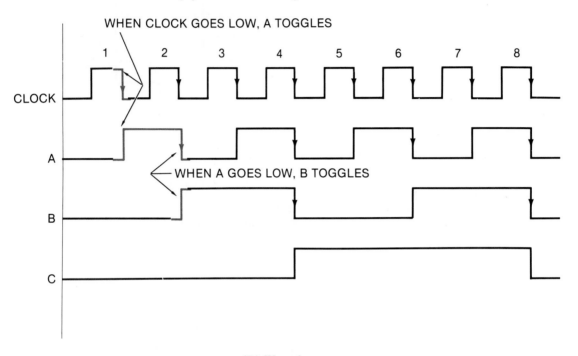

(B) Waveforms.

C	B	A	CLOCK
0	0	0	0
0	0	1	1
0	1	0	2
0	1	1	3
1	0	0	4
1	0	1	5
1	1	0	6
1	1	1	7
0	0	0	8

(C) Truth table.

In this circuit, the Q output of flip-flop A acts as the clock input to flip-flop B, and the output of B acts as the input to C. As can be seen in Fig. 10-23B, each time the clock input goes from high to low, flip-flop A toggles. It changes states. Similarly, each time A goes from high to low, B gets toggled, and when B goes low, C gets toggled. The truth table of Fig. 10-23C shows that the counter generates a binary number sequence.

When testing the binary counter, a CLOCK signal is fed to the *CK* input of flip-flop *A*. Then a logic probe or oscilloscope is connected to output *A*. The output at *A* should be a square wave at *half* the frequency of the input *CK*. The output at *B* should be a square wave at half the frequency of output *A*, and so on. In other words, testing a series binary counter is much like testing a cascaded amplifier. Inject an input signal, then use an oscilloscope or probe to look for appropriate outputs. Once you find a stage where you have the correct input and an incorrect output, you have localized the faulty stage.

Since binary counters are used frequently, manufacturers make single-chip counters, like the 7493 shown in Fig. 10-24. Fig. 10-24 shows that four *J-K* flip-flops are inside the chip. They are used in the toggle mode. Note that flip-flops *B*, *C*, and *D* are connected in series but that *A*'s output is not internally connected to *B*'s input. This makes the circuit more flexible. If you need a counter to develop eight counts (called a MOD-8 counter), simply use the *B* input. But if you need 16 counts (MOD-16), then connect the Q_A output to input *B*, forming a 4-bit counter.

10-24 7493 four-bit binary counter.

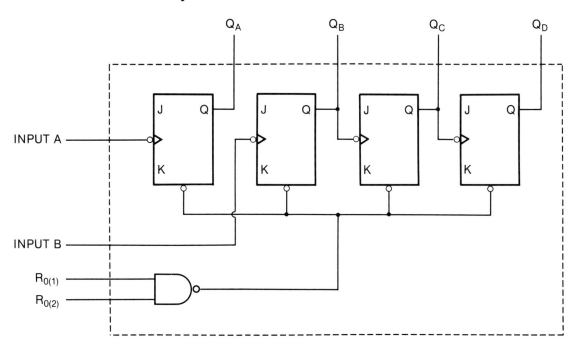

(A) Functional block diagram.

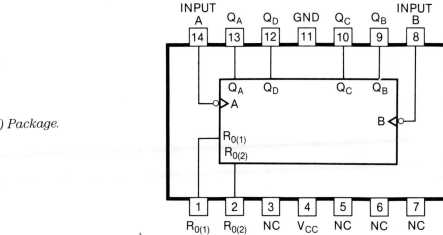

(B) Package.

Also included inside the package is a NAND gate tied to all *CLR* inputs. When both reset inputs $(R_{0(1)})$ and $(R_{0(2)})$ are driven active high, all flip-flops are reset. The package of the 7493 is shown in Fig. 10-24B. Whenever it is convenient to count in multiples of ten, the 7490 *decade counter* is used. This counter is internally wired to reset to zero after every tenth input pulse.

The package for the 7490 is shown in FIg. 10-25. It is similar to the 7493, but also has two "reset nine" (R_9) inputs. When R_9 inputs are driven high, flip-flops *A* and *D* are set, while *B* and *C* are cleared. Since these inputs are active high (no bubbles), you must ground them if they are not to be used.

10-25 **Package of 7490 decade counter.**

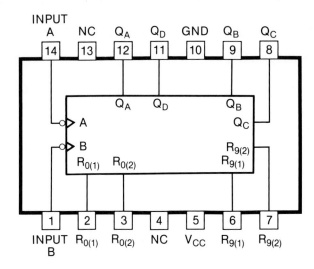

To be able to read out the numerical equivalent of the BCD (Binary Coded Decimal) value in the 7490, we must add two more chips. These are the decoder/driver and the seven-segment display. A typical common-anode seven-segment display chip is shown in Fig. 10-26. Notice that each segment acts like two LED in series. In order to light any segment, we simply forward bias it. For example, pulling the cathodes of segments *b* and *c* low causes them to light up, thus displaying the numeral 1.

10-26 **Seven-segment display.**

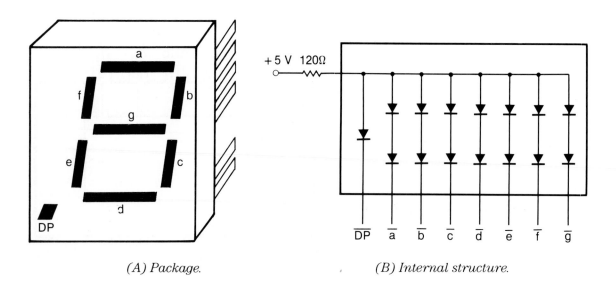

(A) Package.   (B) Internal structure.

In order to convert the BCD code into a seven-segment code, another chip is needed, such as the 7447 decoder/driver. Fig. 10-27A shows how the 7447 chip is connected between the 7490 decade counter and the seven-segment display. Fig. 10-27B shows the table of operation for the 7447.

Pins 3, 4, and 5 of the decoder/driver are shown tied inactive high. Pin 3 is the lamp test (LT). When pulled low, all outputs a–f go low, causing the number 8 to light, regardless of the inputs on A–D. The other two pins are ripple blanking input (RBI) and ripple blanking output (RBO). They are used to suppress (blank out) leading zeros in a multidigit display. To use blanking, tie the RBI of the most significant digit (MSD) to ground. Then "daisy chain" the remaining 7447s by connecting the RBO of the MSD to the RBI of the next significant digit, and so on down the line. If the BCD for the MSD is all zeros, that digit will blank out. Also, the RBO of the 7447 for the MSD will go low,

Connecting a seven-segment display to a counter with a 7447 decoder/driver.

10-27

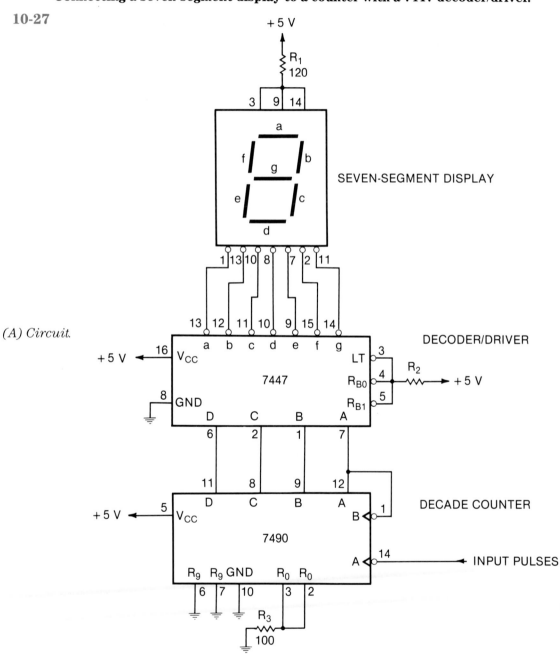

(A) Circuit.

Count	BCD Code				Decoder Outputs							Display
	D	C	B	A	a	b	c	d	e	f	g	
0	0	0	0	0	0	0	0	0	0	0	1	0
1	0	0	0	1	1	0	0	1	1	1	1	1
2	0	0	1	0	0	0	1	0	0	1	0	2
3	0	0	1	1	0	0	0	0	1	1	0	3
4	0	1	0	0	1	0	0	1	1	0	0	4
5	0	1	0	1	0	1	0	0	1	0	0	5
6	0	1	1	0	1	1	0	0	0	0	0	6
7	0	1	1	1	0	0	0	1	1	1	1	7
8	1	0	0	0	0	0	0	0	0	0	0	8
9	1	0	0	1	0	0	0	0	1	0	0	9

(B) Table.

thereby informing the next significant digit that the MSD is blanked. If the code for the next significant digit is all zeros, it will also be blanked. But any digit other than zero will be displayed. The sequence is continued all the way down the line.

Of course, there is an enormous variety of digital IC available. We will not attempt to cover them all here. The only reason for discussing the circuits that we have looked at so far is to point out the significance of the symbols, as well as some of the internal hardware characteristics. You can better test and troubleshoot these circuits when you realize the significance of the presence or absence of bubbles on inputs and outputs. You will also know whether an input is affected by an *edge* (high-to-low or low-to-high transition), as in the case of the *CK* input ($>$), or by a *level*, as in the case of the *CLR* input.

PROBLEMS

10-1.   Assume that you are testing various points in a TTL circuit using a voltmeter. Tell whether each of the following readings represent a logic high, logic low, or undefined, by writing *H*, *L*, or *U*, respectively.

Reading	Logic level
(a) 4.5 V	
(b) 0.2 V	
(c) 1.8 V	
(d) 1.0 V	
(e) 3.2 V	
(f) 0.4 V	

For the next three problems, refer to Fig. 10-28.

10-28 **Circuit for problems 10-2 through 10-4.**

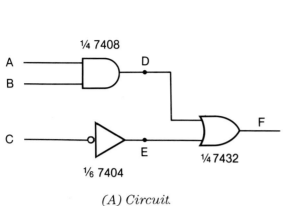

C	B	A	D	E	F
0	0	0			
0	0	1			
0	1	0			
0	1	1			
1	0	0			
1	0	1			
1	1	0			
1	1	1			

(A) Circuit.

(B) Table.

10-2. Fill in the truth table for all indicated points in the circuit. Use a 1 to represent a high and 0 to represent a low.

10-3. Again referring to Fig. 10-28, suppose you measure the following voltages: $A = 0.2, B = 3.0, C = 0.5, D = 0.4, E = 0.2, F = 0.3$. Which component would you suspect as being defective? (a)AND gate. (b)Inverter. (c)OR gate.

10-4. You could verify the defective circuit of problem 10-3 by using a pulser and a logic probe. Which of the following is the best test? (a)Inject a positive pulse at E and observe F. (b)Inject a negative pulse at D and observe F. (c)Inject a positive pulse at A and observe D. (b)Inject a negative pulse at B and observe D.

For the next two problems, refer to Fig. 10-29.

10-5. Fill in the truth table for the circuit.

10-29 **Circuit for problems 10-5 and 10-6.**

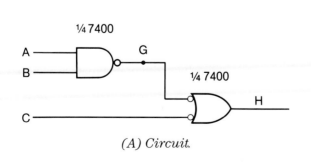

C	B	A	G	H
0	0	0		
0	0	1		
0	1	0		
0	1	1		
1	0	0		
1	0	1		
1	1	0		
1	1	1		

(A) Circuit.

(B) Table.

10-6.   The circuit of Fig. 10-29 is built with two NAND gates, whereas the circuit of Fig. 10-28 uses three different elements. Could you use the circuit of Fig. 10-29 to replace that of Fig. 10-28?

(a)Yes. (b)No.

Explain.

For the next six problems, refer to Fig. 10-30. Choose circuit A, B, C, D, E, or F. Remember that a bubble on an input indicates that the input must be low to activate that input, whereas the absence of a bubble indicates that a high is required. Similarly, a bubble on the output of a circuit means that the output goes low when the inputs to the circuit are satisfied, whereas the absence of a bubble at the output indicates an active high output.

10-30   Circuit for problems 10-7 through 10-12.

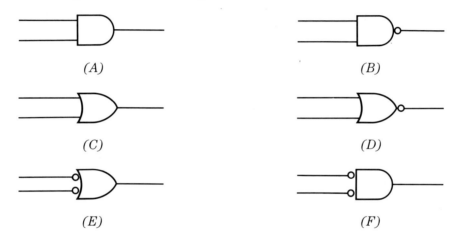

(A)   (B)

(C)   (D)

(E)   (F)

10-7.   Which circuit will have a high output if *either* of its inputs is high?

10-8.   Which circuit will have a low output if *either* of its inputs is high?

10-9.   Which circuit will have a low output only when *both* of its inputs are high?

10-10.   Which circuit will have a high output only when *both* of its inputs are low?

10-11. Which circuit will have a high output only when *both* of its inputs are high?

10-12. Which circuit will have a high output only if *either* of its inputs is low?

For the next two problems, refer to Fig. 10-31.

10-31 **Circuit for problems 10-13 and 10-14.**

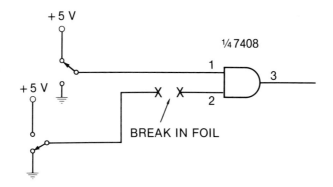

10-13. The two switches are in the positions shown, but there is a break in the printed circuit foil connected to pin 2 of the gate. What logic level would you read at pin 3? (*Hint:* This question is tricky, but it is very important in troubleshooting. Refer to the diode gate of Fig. 10-2 for help.)

10-14. Referring to problem 10-13, a clue to the problem could be found by connecting a voltmeter, or logic probe, to pin 2. What level would you expect to read? (a)High. (b)Low. Pulsing pin 2 at a—(c)high, (d)low—would cause a pulse to appear at pin 3.

For the next four problems, refer to Fig. 10-32.

10-15. When both inputs to gate A are high as shown, the output of gate A is low. The dashed lines indicate the current paths from the inputs of each of the other circuits. How much total current does gate A have to sink?

$I_A =$

10-16. List the output states, high or low, for each of the other circuits.

$B =$

$C =$

$D =$

$E =$

$F =$

10-32   **Circuit for problems 10-15 through 10-18.**

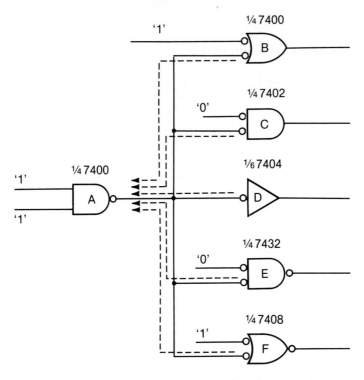

10-17.   If one of the inputs of gate *A* goes low, how much current will flow through the output of gate *A*? (a)1.6 mA. (b)8 mA. (c)More than 10 mA. (d)Much less than 1 mA.

10-18.   If all of the circuits were replaced with 74C series chips, that is, 74C00, 74C02, etc., how much current would circuit *A* have to sink?

10-19.   The input signals shown in Fig. 10-33 are applied to a 7474 *D* flip-flop. Assume that the *PR* input is tied high. Sketch the *Q* output waveform.

10-33 **Waveforms for problem 10-19.**

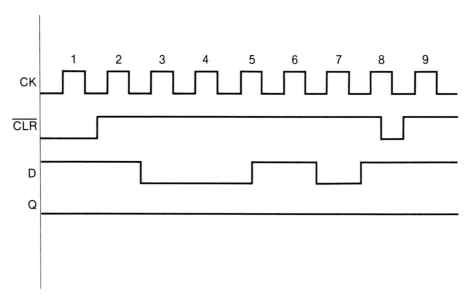

10-34 **D flip-flop for problem 10-20.**

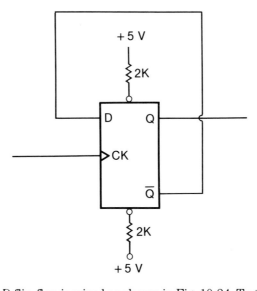

10-20.   A D flip-flop is wired as shown in Fig. 10-34. To test the circuit using a logic probe and a digital pulser, connect the probe to the Q output and the pulser to the CK input. Explain what you would expect to see on the probe as you slowly pulse the input several times. Assume that Q is initially low.

For the next seven problems, refer to Fig. 10-27.

10-21.   The easiest way to test the seven-segment display is to (a)short pin 3 of the 7447 to ground. (b)short all inputs of the seven-segment display to ground. (c)tie all inputs of the display high.

10-22.   If you short pin 3 of the 7447 to ground and the number 0 lights up, which of the following should you do next? (a)Nothing, this is normal. (b)Connect a logic probe to pin 11 of the display to see if it is low. (c)Check the outputs of the 7490.

10-23.   What numeral should be displayed if you drive pins 2 and 3 of the 7490 high?

10-24.   Suppose that you connect a logic monitor (clip) over the 7490 and then reset the chip. Next, you connect a pulser to pin 14 and depress the pushbutton ten times. What should you see on the seven-segment display?

10-25.   Referring to question 10-24, suppose the seven-segment display alternated between numerals 0 and 1 each time you pressed the pushbutton. Looking at the logic monitor, you notice that pin 12 alternated between high and low, each time pin 14 blinks. But no other pins change states. Which of the following could be the problem? (More than one may be correct.) (a)Defective 7490. (b)Defective 7447. (c)Defective seven-segment display. (d)Open circuit between pins 1 and 12 of the 7490.

10-26.   The quickest way to find the fault described in problem 10-25 is to (a)Replace the 7490. (b)Replace the 7447. (c)Replace the seven-segment display. (d)Apply pulses to pin 1 of the 7490.

10-27.   Suppose that the numeral 0 remains lit, even though you pulse pin 14 of the 7490. The logic monitor on the 7490 shows that pins 2, 3, and 5 are high, pin 14 blinks, and all the remaining pins are low. Which of the following is the most likely cause? (a)Defective 7490. (b)Defective 7447. (c)Open R_3. (d)Open R_1.

EXPERIMENT 10-1   LOGIC GATES

This experiment will familiarize you with the typical voltage values that represent logic highs and lows in TTL circuits. You will also build a simple logic probe and a digital pulser to get familiar with their use. The test instruments you will build are rather primitive and, of course, do not have all of the features of commercially available equipment. But they will give you an idea of what you can learn from the tools. Leave the logic probe and pulser assembled when you finish this experiment because they will also be used in Experiment 10-2, as well as in chapter 11.

All circuits discussed in these experiments can be easily assembled on a breadboarding socket strip, such as an Assembly Products ACE-1.

EQUIPMENT

- 5-VDC power supply
- oscilloscope

- VOM
- (1) 7400 quad 2-input NAND
- (1) 555 timer
- (3) NPN transistors, 2N2222 or equivalent
- (4) SPDT toggle switches
- (1) SPST momentary pushbutton switch
- (2) LED
- (1) 220-KΩ, ¼-W ± 5% resistor
- (1) 100-KΩ, ¼-W ± 5% resistor
- (1) 10-KΩ, ¼-W ± 5% resistor
- (1) 4.7-KΩ, ¼-W ± 5% resistor
- (2) 2-KΩ, ¼-W ± 5% resistors
- (3) 1-KΩ, ¼-W ± 5% resistors
- (2) 220-Ω, ¼-W ± 5% resistors
- (1) 1.0-µF capacitor at 15 WVDC
- (1) 0.1-µF capacitor

E10-1 **Circuits for experiment 10-1.**

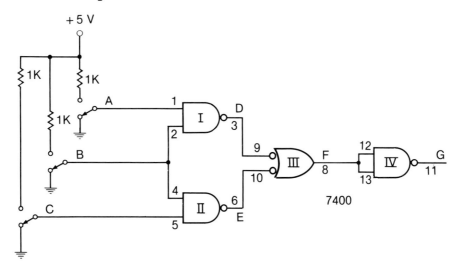

(A) Circuit.

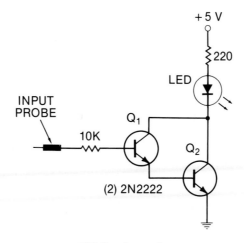

(B) Logic probe.

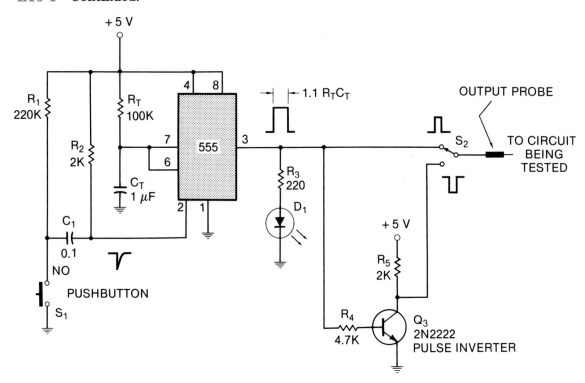

(C) Digital pulser.

PROCEDURE

1.   Build the circuit of Fig. E10-1A. Be sure to connect pin 7 of the chip to ground and connect pin 14 to V_{CC}. Flip the input switches A, B, and C high or low, as indicated in Table E10-1, and fill in the blanks in the table with your measured *voltage* for each input combination. Record the voltage to the nearest tenth of a volt, do not simply write 1 or 0.

2.   To see the effect of an open input lead, disconnect the lead from pin 1 to switch A. Place switch B in the low (0) position. Measure pin 3.

$D =$

Now flip switch B high (1), and measure pin 3.

$D =$

With a voltmeter, measure pin 1.

$V_1 =$

Reconnect pin 1 to switch A for later tests.

3.   It is usually more time consuming, and often unnecessarily confusing, to measure the output levels of gates with a voltmeter. We will use a logic probe now. Build the logic probe circuit of Fig. E10-1B. The circuit can be assembled on a breadboard socket strip for simplicity, but you may eventually want to package it in a more convenient container, such as an empty ball point pen housing.

Test the logic probe by applying power to it. Then alternately touch the input probe to ground and to V_{CC}.

The LED lights when the probe is touched to a—(a)high, (b)low—level.

C	B	A	D	E	F	G
0	0	0				
0	0	5				
0	5	0				
0	5	5				
5	0	0				
5	0	5				
5	5	0				
5	5	5				

4. Flip switches A, B, and C high or low as needed, to try every input combination in Table E10-2. Using your logic probe, measure the indicated points, and record a 1 or 0 as observed. Compare this table with that of Table E10-1.

Does every point agree? (a)Yes. (b)No.

5. Next, build the pulser of Fig. E10-1C. The heart of this pulser is a 555 timer chip, wired as a one-shot multivibrator. Each time pushbutton switch S_1 is pressed, the output, pin 3 of the chip, goes high for a short time. The duration of the pulse is controlled by the time constant of the two components R_T–C_T. When pin 3 goes high, LED D_1 should blink. Apply power, and test the circuit to see that D_1 blinks.

Transistor Q_3 is used as an inverter. When pin 3 goes high, Q_3 gets turned on, causing its collector to go low. Switch S_2 allows you to select either a positive-going or a negative-going output pulse.

Observe the output probe with an oscilloscope, while you depress S_1 several times. Have S_2 set for a positive pulse. Draw the output below, showing the high and low voltage levels, as well as the time duration of the pulse. Repeat with S_2 set for a negative-going output pulse.

C	B	A	D	E	F	G
0	0	0				
0	0	1				
0	1	0				
0	1	1				
1	0	0				
1	0	1				
1	1	0				
1	1	1				

6.   Now we will use the pulser on the circuit of Fig. E10-1A. Before connecting the pulser to the circuit, flip switch A low, B high, and C low. Test points D, E, F, and G with your logic probe, and record the measured levels (1 or 0).

$D =$

$E =$

$F =$

$G =$

If your circuit is working correctly, D, E, and G should read high, and F should read low.
    Let's see how to test gate III of the circuit. The symbol for gate III indicates that output F should be high when either input D or E is low. So flip S_2 on the pulser to give a low-going pulse.
7.   Connect the input of the logic probe to point F and the pulser tip to point D. Now depress pushbutton S_1 a few times. What do you see on the logic probe?

Repeat the test, applying the pulser to point E. What do you see on the logic probe?

8. Without changing switches A, B, and C, test gate IV using the logic probe and pulser.

If you performed these tests satisfactorily, you should realize that using a logic probe and pulser, you can easily check through combinational logic circuits to see if they are operating normally, without removing any components from the circuit board.

Change input switches A, B, and C to various combinations, and use your pulser and probe to test the gates. Before you connect the pulser to any pin, be sure to test the normal level on that pin to determine whether you need a positive or negative pulse.

QUIZ

1. According to manufacturing specs, a good logic low in TTL circuits should measure less than 0.8 V. Do Tables E10-1 and E10-2 agree on the logic low states? (a)Yes. (b)No.

2. According to specs, a good logic high should be at least +2.4 V. Do the tables agree? (a)Yes. (b)No.

3. Step 2 of the experiment shows that unconnected inputs of TTL circuits act like—(a)high, (b)low—input levels.

4. The logic probe uses two transistors in a darlington arrangement. What is the minimum input voltage that would cause Q_1 to begin to conduct?

$V_{min} =$

5. Referring to question 4, will the LED be definitely off whenever the input probe is connected to a logic low, say 0.8 V? (a)Yes. (b)No.

6. When testing gate IV with a pulser and logic probe, the resting (normal) state of the input pins was (a)high. (b)low. Therefore, the pulser was set to give a— (c)high-going, (d)low-going—pulse.

7. Gate IV is being used as an (a)OR circuit. (b)inverter. (c)AND circuit.

8. If you needed a shorter duration pulse from the pulser, you could—(a)increase, (b)decrease—the value of R_T or C_T or both.

EXPERIMENT 10-2 FLIP-FLOP AND COUNTERS

You will now work with J-K flip-flops to see how they are set, reset, and toggled. Then you will build a three-bit binary counter and test it with the logic probe and pulser.

EQUIPMENT
- 5-VDC power supply
- logic probe and pulser (from previous experiment)
- (2) 7476 J-K flip-flop
- (6) 2-KΩ, ¼-W ± 5% resistors
- (1) 1-KΩ, ¼-W ± 5% resistor
- (1) SPDT toggle switch

PROCEDURE

1. Build the three-bit counter of Fig. E10-2A. Use (2) 7476 J-K flip-flop chips. One of the flip-flops in one package is not used.

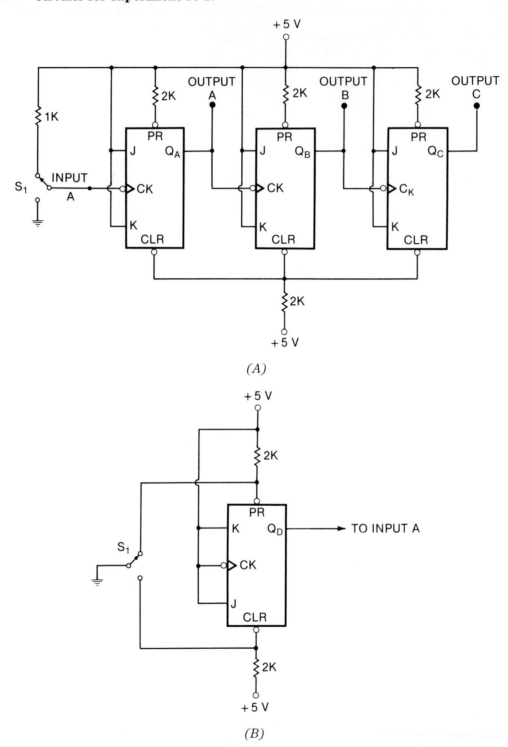

(A)

(B)

2. With input switch S_1 in the up (high) position, reset all flip-flops by applying a pulse to the *CLR* inputs using your pulser. Determine whether you need a high-going or low-going pulse. Check each Q output with the logic probe to see if clearing has occurred. If so, go on to the next step, otherwise recheck your wiring.

3.   Next, see if each flip-flop can be *set* independently, by applying a pulse to each *PR* input and checking each output with the logic probe. If each can be set independently, go on to the next step. Otherwise recheck your wiring.

4.   Reset all flip-flops again in preparation for counting.

5.   All flip-flops are wired in the toggle mode (J and K both high). Therefore, each time switch S_1 goes from high to low, flip-flop A should change states. Flip switch S_1 down, then back up again. Record the observed levels (1 or 0) for each flip-flop in row 1 of Table E10-3.

Table E10-3

Pulse	C	B	A
0	0	0	0
1			
2			
3			
4			
5			
6			
7			

6.   Throw switch S_1 low, then high again, and record the observed levels in row 2 of Table E10-3. Repeat for the remainder of the table. Don't be concerned if the sequence of outputs does not make sense. Just record the observed levels. What can you determine from the sequence of entries in the table?

7.   Next, disconnect S_1 from input A, and connect the output of flip-flop D to input A, as shown in Fig. E10-2B. Also wire S_1 to the *PR* and *CLR* inputs of flip-flop D, as shown. Reset all flip-flops using your pulser.

8.   Now throw switch S_1 down and up, and record the levels of flip-flops A, B, and C in row 1 of Table E10-4. Repeat the operation for the remainder of the table. What do you see different in this table from the previous one? Why?

Pulse	C	B	A
0	0	0	0
1			
2			
3			
4			
5			
6			
7			

9.   You have already used the pulser to set and reset the flip-flops. Now experiment on your own, feeding the pulser into the CK input of each flip-flop, and observe the output of the clocked flip-flop with the logic probe. The purpose of this step is to get more familiar with *in-circuit* testing of logic circuits.

QUIZ

1.   In order to reset the *J-K* flip-flops, the *CLR* input must be driven (a)high. (b)low.

2.   In order to set the *J-K* flip-flops, the *PR* inputs must be driven (a)high. (b)low.

3.   With the *J* and *K* inputs both high, the flip-flops are in the—(a)inhibit, (b)reset, (c)set, (d)toggle—mode.

4.   Referring to question 3, with the flip-flops in this mode, each time the *CK* input goes—(a)low, (b)high—the output of that flip-flop (c)goes low. (d)goes high. (e)changes states.

5.   Table E10-3 shows that the outputs—(a)did, (b)did not—change in a binary progression when input *A* was fed with a toggle switch. This was because (c)the voltage levels at *A* were not TTL compatible. (d)the switch bounced, causing input *A* to change many times each time S_1 was thrown.

6.   In step 7, flip-flop D was used as a (a)switch debouncer. (b)free running clock. (c)toggle-mode flip-flop.

7.   Whenever S_1 was flipped down, making the CLR input of flip-flop D low, the Q_D output went (a)high. (b)low. If S_1 bounced, causing the CLR input to go alternately high and low many times, the Q_D output (c)went high and low many times also. (d)went low only once, and stayed low until the PR input was driven low.

8.   If an extra flip-flop were not available, the circuit of Fig. E10-2B could be replaced by a couple of cross-connected NAND gates, as shown in Fig. 10-19. (a)True. (b)False.

9.   If the Q_A output was low and you wanted to test flip-flop B with a pulser, you would apply a short duration—(a)positive, (b)negative—pulse to the CK input of flip-flop B.

10.   If you performed the test of question 9, a logic probe connected to output B would show (a)a short duration pulse. (b)that B changed states when the pulse was applied.

Digital Systems

We will now study how the digital electronic circuits of chapter 10 are combined to build complete digital systems, systems that can measure time or frequency or that can control some manufacturing process. We will discuss how to analyze large systems based on the information given on the diagrams. Finally, we will examine some common system problems and how to find them.

UNDERSTANDING DIGITAL SYSTEMS

The key to successful troubleshooting of digital systems is understanding the symbols, signal names, and other useful information found on system diagrams. These will help you recognize the *inputs, outputs,* and *control* sections of various parts of the system.

Fig. 11-1 is a block diagram of a typical digital system. It can also represent the diagram of a computer system. All digital systems can be shown to have system block diagrams similar to this one. A block labeled *"inputs"* feeds a *control* section. The control block may have both input and

11-1 Block diagram of a digital system.

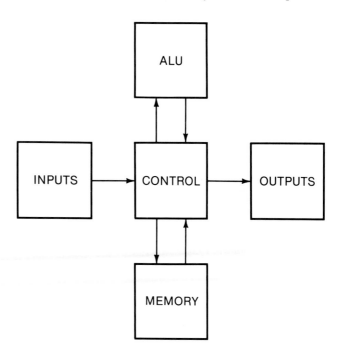

output connections, to a *memory* and to an ALU (arithmetic logic unit). The memory is sometimes omitted. Finally, all digital systems have some sort of *output* section. Here they perform the task for which they were built.

As an example of a small digital system, let's examine the time-interval counter of Fig. 11-2. This system, which is used as an "electronic stopwatch," measures the time from the instant the start switch is pressed, until the stop switch is pressed.

Here's how it works. Suppose the operator wants to time some event, say a runner doing a 50-meter sprint or a dragster doing a quarter-mile run. The operator first presses the reset button (S_1) to clear all displays. Then, at the start of the event, when the gun goes off, the operator presses the start button (S_2). The control flip-flop is thus set, causing ENABLE to go high. As long as ENABLE is high, the CLOCK pulses "pass through" the NAND gate and show up as $\overline{\text{COUNT}}$ pulses, which are fed to the 7490 decade counters. When the tenth $\overline{\text{COUNT}}$ pulse comes along, the least significant counter overflows. This causes the next higher-order counter to be incremented, and so on. When the operator presses the stop button (S_3), the control flip-flop is cleared. This causes ENABLE to go inactive low and, thus, stops the counting process.

The number appearing on the seven-segment display is the total elapsed time in seconds, to the nearest $1/100$ s, because the CLOCK frequency is 100 Hz. Usually, a decimal point is lit to the left of the middle digit, thus displaying a maximum time of 9.99 s.

Comparing Figs. 11-1 and 11-2, we see that the reset (S_1), start (S_2), and stop (S_3) pushbuttons are all external *inputs*. The control flip-flop, the clock, and the NAND gate make up the *control* section. The decade counters are the *arithmetic unit* here. Finally, the decoder/drivers and seven-segment displays form the *output* of this system.

Although the system of Fig. 11-2 is very simple, the ideas of isolating various blocks can be clearly seen. And by separating the system into these blocks, testing and troubleshooting are simplified. For example, to test the control section, simply observe the $\overline{\text{COUNT}}$ signal with either an oscilloscope or logic probe, while you alternately make $\overline{\text{START}}$, then $\overline{\text{STOP}}$, go low. If nothing happens when you depress the start switch S_2, check to see if the $\overline{\text{START}}$ signal is low when S_2 is pressed. If it is not, your trouble is in the input switch. If $\overline{\text{START}}$ is low when S_2 is pressed, the problem is inside the *control* block, either the flip-flop, the clock, or the AND gate. The point is to try to isolate the major block first, then zero in on the defective circuits.

Similarly, if when you press S_2 you see $\overline{\text{COUNT}}$ pulses at the output of the NAND gate, but the displays either do not charge or are all blanked, find out whether the output section or the arithmetic section is at fault. Remember, you can easily test seven-segment displays by applying a low signal to the lamp test pins on the 7447 drivers. Similarly, by watching the outputs of the 7490 with an oscilloscope, logic probe, or logic monitor, you can quickly verify their operation.

Each major block of the system can be tested independently by applying the proper inputs and looking for appropriate outputs. The *input* block, for example, has no inputs to it, but it has outputs RESET, START, and $\overline{\text{STOP}}$. The *control* block has inputs $\overline{\text{START}}$ and $\overline{\text{STOP}}$ and its output is $\overline{\text{COUNT}}$. The arithmetic block has inputs RESET and $\overline{\text{COUNT}}$, and its outputs are three sets of BCD signals to the decoder/drivers. Lastly, the *output* section has three sets of BCD inputs, and its output is numerals lit up on the seven-segment displays.

As mentioned in chapter 10, using signal names greatly simplifies testing and troubleshooting. For example, suppose that the time interval counter of Fig. 11-2 is modified to be used to time a downhill skier. When the skier trips a wand while leaving the starting gate, timing should begin. Then when the skier crosses the finish line, a beam of light is broken. The beam shines into a photo transistor that generates a $\overline{\text{STOP}}$ pulse.

Of course, the arithmetic and display sections must be expanded to measure longer times, but the general arrangement remains pretty much the same. Should the wand generate a high or low signal as the person leaves the starting gate? How about the photo transistor?

You can see that if the major blocks of the system were built on separate printed circuit cards, or modules, you would be able to find a defective module quickly simply by applying appropriate inputs while looking for the proper outputs. And you can do this *even if you do not know what is inside the module, or how it works!*

BUS-ORIENTED SYSTEMS

In Fig. 11-2, the connections from the 7447 decoder/drivers to the seven-segment displays are shown as large arrows, rather than by seven individual lines. The slash line through the arrow,

11-2 Simple time interval counter.

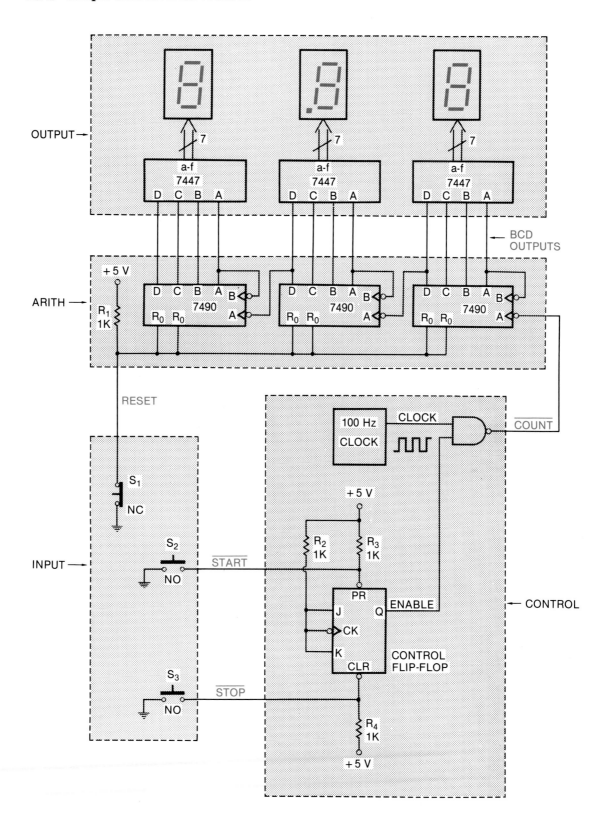

accompanied by the number 7 nearby, indicates that there are seven lines routed together as a group. Drawing bundles of wires, or groups of printed circuit foils, in this manner make the diagram less cluttered and easier to read.

Often, the same group of foils or wires are used to transfer information from any one of several sources to any one of several destinations. Such a group of common lines is called a *bus*. Busses are used to transfer several bits of digital information simultaneously, say from a microprocessor unit (MPU) to memory, from memory to MPU, from the MPU to an output port, or from an input port to the MPU.

Whenever two or more outputs are tied to the same point, or line, problems arise. For example, Fig. 11-3 shows the outputs of gates 1 and 2 both tied to point X, which connects to an input of gate 3. What would be the logic level at point X with the input conditions shown, assuming that the two gates were standard TTL circuits with totem pole outputs?

11-3 Outputs of two gates connected to a common point.

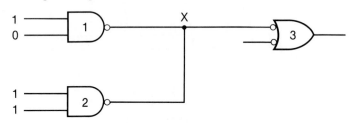

Gate 1 tries to pull point X high, while at the same time, gate 2 tries to pull it low. If these gates have totem pole outputs, like the 7400, they will work against each other. So point X may be neither a good high nor a good low, but rather some indeterminate level. Gate 3 will not know how to interpret that level. The point is that we *cannot* tie outputs of standard TTL gates together when their inputs may differ. However, there are special circuits designed to solve this problem.

One gate of a 7401 quad two-input NAND gate is shown in Fig. 11-4. The 7401 has *open-collector* outputs. Notice that the upper transistor of the totem pole is omitted, so there is no active pull-up circuitry. An external pull-up resistor is needed to pull the output high when Q_3 is cut off.

If we replace gates 1 and 2 of Fig. 11-3 with open-collector NAND gates and tie an external pull-up resistor to V_{CC}, only the gate whose output transistor is conducting will pull point X low. If neither output transistor is conducting, point X is pulled high by the external pull-up resistor. Any number of open-collector outputs can thus be tied to point X, and only one pull-up is needed.

Fig. 11-5 shows how open-collector NAND gates are used to place either of two possible sets of inputs onto a 4-bit data bus. For example, if inputs W_0–W_3 are to be placed onto the bus, a high LOAD W signal must be applied along with data inputs W_0–W_3. At the same time, load X should be kept inactive low.

11-4 Open-collector NAND gate as in 7401.

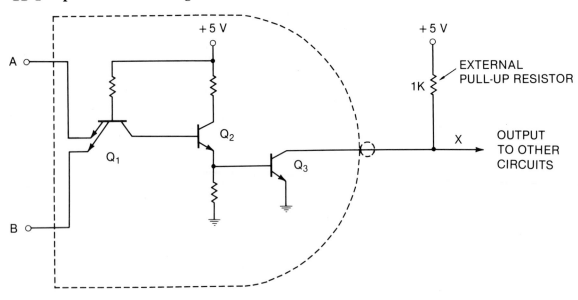

Actually, since NAND gates are used, the *complement* of data bits W_0–W_3 appear on the bus. But when the data is read from the bus by one of the output ports, Y or Z, the data again gets inverted by means of the NOR gates.

If LOAD W is made high, while $\overline{\text{STORE } Z}$ is made low, input data bits W_0–W_3 will appear at outputs Z_0–Z_3, respectively. Control signals LOAD X and $\overline{\text{STORE } Y}$ would both be kept inactive during this time. By using the common 4-bit data bus, either set of inputs can be fed to either output port, by simply applying the proper control signals. The number of input and output ports hanging on the bus can be increased, up to the fan-out limits of the input gates. In this manner, much hardware and wiring is eliminated.

An even better technique to tie outputs of several circuits together is the use of tristate outputs. Fig. 11-6 shows a simplified TTL inverter with a tristate output. Notice that the circuit is essentially the same as a totem pole inverter circuit. The only additional component is diode D_1.

When the C (control) input (cathode of D_1) is made positive (high), D_1 is back biased, and the circuit behaves exactly like any other TTL inverter. That is, D_1 has no effect. But if the C input is made low, D_1 becomes forward biased, pulling the collector of Q_2 to ground.

With the collector of Q_2 low, there is no forward bias on the base of Q_3, so Q_3 is cut off. Likewise, since the collector of Q_2 is at ground potential, no collector current flows through Q_2. Since no emitter current flows through R_3, there is no forward bias at the base of Q_4. So Q_4 is also cut off. In other words, *both* transistors Q_3 and Q_4 are cut off at the same time. The equivalent of the output is shown in Fig. 11-6B. Point $\overline{A}$ is effectively disconnected from both ground and V_{CC}. The output point $\overline{A}$ is then said to be in its *high-impedance* state. Fig. 11-6C shows the symbol for a tristate inverter.

Sometimes the control (enable) inputs of several circuits are tied together on a single chip. Fig. 11-7A shows the pinouts of a 74240 chip. When pin 1 is made low, the inverters whose inputs

11-5   Transfer bus using open collector NAND.

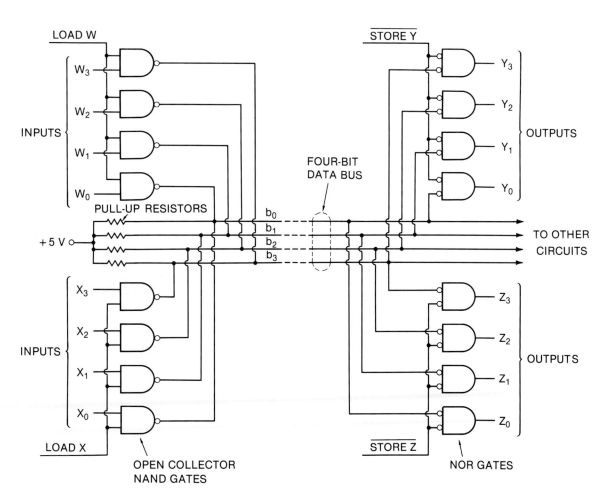

11-6   TTL tristate inverter.

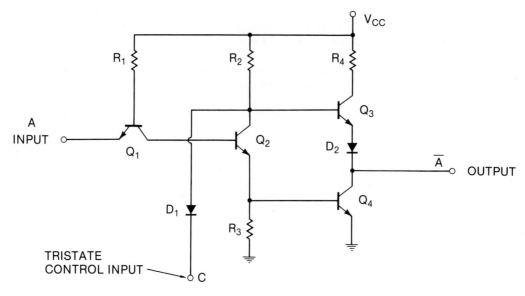

(A) Simplified tristate inverter.

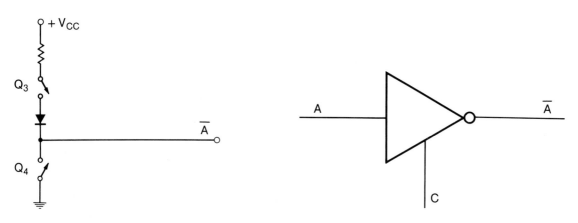

(B) Equivalent circuit of output when control input C is low.

(C) Logic symbol for tristate inverter.

connect to pins 2, 4, 6, and 8 are all enabled. But when pin 1 is made high, those inverters all go to their high-impedance output states. The other four inverters function similarly. The inverters in the 74240 are called line drivers/line receivers, since they are often used to drive lines on a data bus or receive data from the bus. The 74244, shown in Fig. 11-7B, has noninverted outputs. Otherwise the pinouts are the same. Circuits with noninverted outputs are often used as *buffers* to increase drive capability.

Fig. 11-8 shows how a single 74240 chip can be used to replace all eight input gates of Fig. 11-5. Study the circuit until you are sure you understand how it works.

Besides using a data bus, most large digital systems also use an *address* bus. The address bus carries the address, or device number, of the source or destination of the data. The address, which is simply a binary number, is usually fed to some sort of address decoder circuit. Such a decoder is shown in Fig. 11-9.

The function table for the 74155, when wired as shown, shows that depending on the inputs A, B, and C, one and only one output is low at any time. For example, when C and A are high, while B is low, output 5 goes active low. All other outputs remain inactive high. In addition the $\overline{\text{STROBE}}$ (G) inputs must be both low in order for *any* output to go active low.

11-7   Pin assignments of 74240 and 74244.

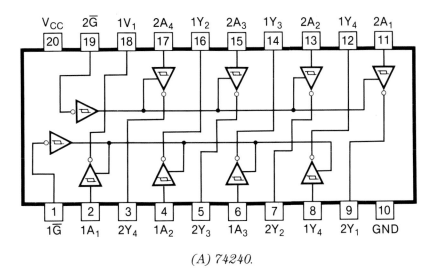

(A) 74240.

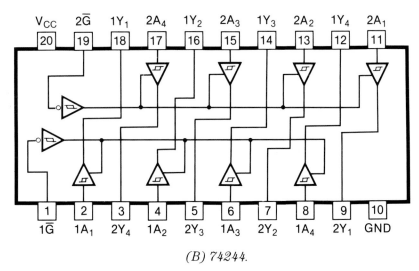

(B) 74244.

Fig. 11-10 shows a simple application of the address decoder being used to select data from port W or port X. If address lines A_2, A_1, and A_0 are 010, respectively, and if $\overline{\text{STROBE}}$ is low, output 2 of the decoder goes low. This low signal is fed to the tristate control input ($\text{LOAD }W$) of port W, thus enabling the buffers of port W. Whatever data appears on $W_0 = W_3$ at this time is placed on the data bus. What logic levels should appear on A_2, A_1, and A_0 in order to place port X data on the data bus?

Although pull-up resistors are not necessary at the outputs of tristate circuits, they are often used so that bus lines, such as those in Fig. 11-10, will be in some definite state when all tristate outputs are in their high-impedance state.

PROBLEMS IN DIGITAL SYSTEMS

RINGING AND REFLECTIONS

We have already discussed some possible problems that occur at the chip level, such as a defective chip, open inputs, and excessive loading. In addition to these, there are certain problems that begin to show up at the system level, that is, when several chips are used simultaneously. One such problem is *ringing* and/or *reflections* caused by long interconnecting lines.

11-8 Replacing open collector NAND gates with 74240 tristate line driver.

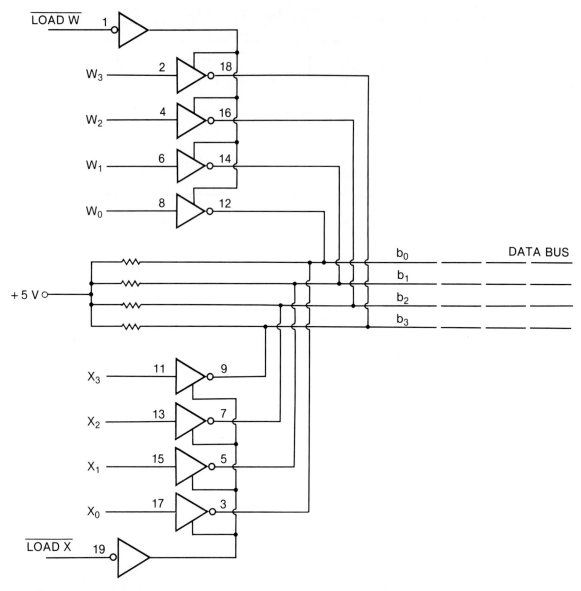

From communications theory, we know that when a transmission line becomes some appreciable fraction of a wavelength long, the load impedance should be matched to the source impedance. If it is not, reflections of the transmitted signal will occur. This means that some of the signal sent down the transmission line will bounce back when it reaches the load.

The problem is that the reflected signal interferes with new signals, or pulses, being sent down the line. The reflected signal might be in phase or out of phase with the signals being transmitted. The net effect is that when the two signals meet on the transmission line, they add or subtract to form a new signal, which is different from the one being transmitted. Thus, the signal seen at the receiving end may be considerably different from the originally transmitted signal.

This problem occurs in logic circuits whenever the source and destination are placed fairly far apart, say a few feet or so. How can we relate digital circuits to transmission line theory? First of all, the rectangular pulses generated by logic gates can be shown to be made up of many very high-frequency harmonics. The high-frequency components must all be passed on the transmitting line, or foil, if the pulse shape is to remain unchanged. Secondly, in regard to matching impedances, a very messy problem occurs. When the input level to a gate is high, the input diode is back biased and looks like a very high impedance. But when the input level is low, the input diode is forward

11-9 74155 decoder.

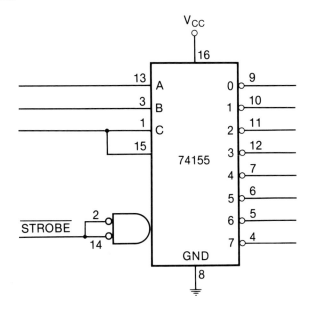

(A) 1 of 8 decoder.

G	C	B	A		0	1	2	3	4	5	6	7
1	X	X	X		1	1	1	1	1	1	1	1
0	0	0	0		0	1	1	1	1	1	1	1
0	0	0	1		1	0	1	1	1	1	1	1
0	0	1	0		1	1	0	1	1	1	1	1
0	0	1	1		1	1	1	0	1	1	1	1
0	1	0	0		1	1	1	1	0	1	1	1
0	1	0	1		1	1	1	1	1	0	1	1
0	1	1	0		1	1	1	1	1	1	0	1
0	1	1	1		1	1	1	1	1	1	1	0

(B) Function table.

biased and looks like a low impedance. In other words, the transmission problem is aggravated by the fact that the input impedance to the gate (the load end of the transmission line) *changes* drastically. Therefore, matching source and destination impedances is virtually impossible.

Fig. 11-11 illustrates the reflection problem. The input signals to the OR gate at the left are as shown. So the output of the OR gate should look like the waveform shown in dashed lines at its output. The OR gate acts as the transmitter on this line.

The connecting line between the source (OR gate) and destination (NAND gate) might be a long printed circuit foil, or a lead in a cable, etc. When the signal reaches the NAND gate, some of it is reflected, and possibly shifted in phase (time). The dotted lines show some signal being reflected back. When the transmitted and reflected signals collide, the net result might look like the resultant signal shown at the input to the NAND gate. It is this resultant signal, which can be observed on an oscilloscope, that actives the NAND gate.

Notice that the resultant signal differs drastically from what it would have looked like if the two gates were close together. Obviously, this distorted signal can activate the NAND gate differently from the way that an undistorted signal would, thereby causing erroneous outputs.

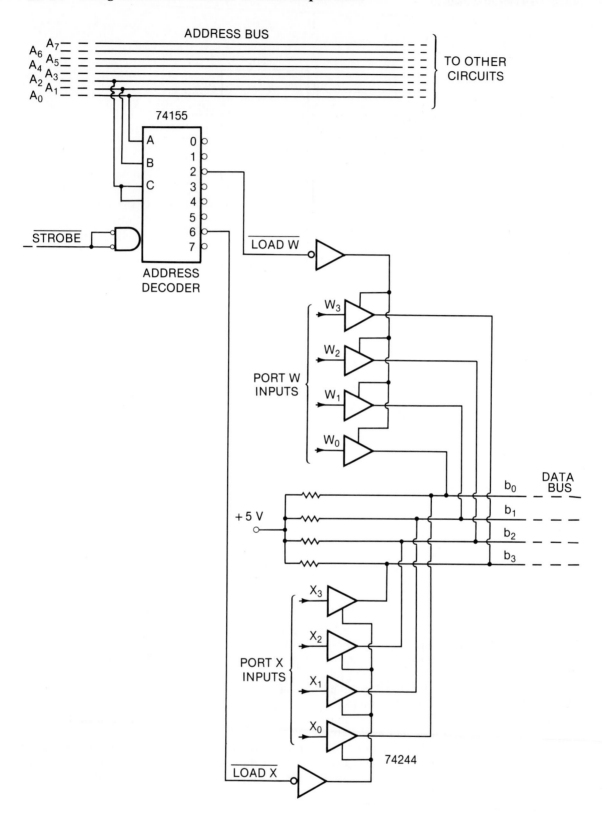

11-11 **Long transmission line causes ringing and distortion.**

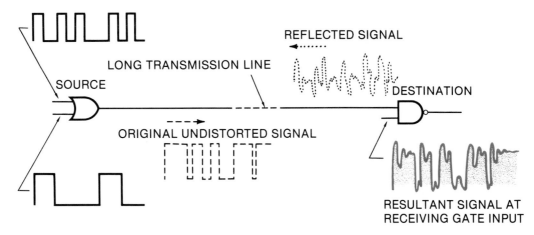

One solution to the problem is to keep connecting lines short, whenever possible. For example, if you are testing two or more modules that must be linked together, keep the connecting lines very short. Otherwise you may introduce problems that will cause malfunctions. It is also a good idea to connect the ground leads of test instruments as closely as possible to the circuit under test. This should prevent ground loops.

Designers sometimes purposely slow down the rise and fall times of signals, which must travel over long cables, to avoid the reflection problems. They do this by connecting a small resistor in series with the output of the driving source, followed by a small capacitor to ground. The procedure is similar to the technique used to roll off high frequencies in linear amplifiers.

Long connecting lines are also susceptible to noise pickup and crosstalk. If a connecting wire passes near an electrically noisy area, such as a drill motor or arc welder, noise pulses may be induced into the wire. These noise pulses interfere with the transmitted signal, causing erroneous data to be received. This is especially true when using high-impedance circuits. The best solution is to shield the system both electrically and magnetically from the source of noise by enclosing the system in a grounded metal cabinet. If two systems must communicate over long distances, use shielded cables.

POWER SUPPLY GLITCHES

Remember that the power supply for TTL logic circuits must be at 5 V ± 10%. If the supply voltage changes, unpredictable results can occur. So normally a power supply regulator, such as 7805 or a similar device, is used. These hold the d-c supply voltage constant satisfactorily. However, when several chips are used in a system, problems can occur as a result of rapid *changes* in current drawn through common power supply leads.

Fig. 11-12 shows several chips being fed from a single-chip regulator. All of the power supply input pins of the chips are tied to the common + rail. Similarly, all ground pins are tied to the common ground foil.

Suppose, because of some combination of signals, IC-5 suddenly switches on and draws an additional 50 mA or so of current. The increased d-c drop across the printed circuit foil resulting from the additional flow is negligible. But because of the *rate of change* of the current, considerable voltage is developed along the foil strip due largely to an *inductive* effect. The induced voltage can be estimated by using equation 11-1.

$$v_L = L \, di/dt \qquad\qquad 11\text{-}1$$

where

v_L = voltage developed across an inductance
L = inductance in henries
di/dt = rate of change of current in amperes per second

11-12 Power supply glitches.

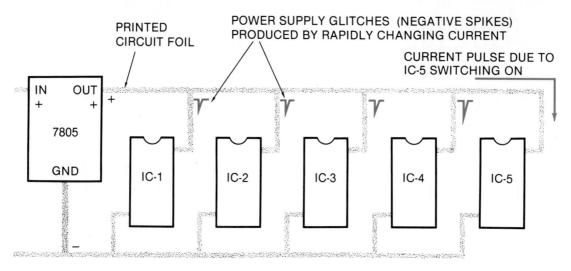

EXAMPLE 11-1 Suppose the inductance of the positive power supply foil is 1 μH (microhenry). What will be the voltage drop if current through it changes 50 mA in 20 ns?

SOLUTION

$$V_L = L\, di/dt = 1 \times 10^{-6} \times \frac{50 \times 10^{-3}}{20 \times 10^{-9}} = 2.5 \text{ V}$$

Is the 2.5 V drop across the foil constant for a long time? No, the drop only exists as a very short spike, called a *glitch*. It only exists while the current is *changing*. However, this glitch does appear at the power supply inputs of the other ICs. (Actually IC-1 sees a smaller glitch than IC-4.) The glitch in supply voltage can cause the outputs of IC-1 through IC-4 to change unpredictably. Obviously, this causes problems elsewhere.

The best way to avoid glitches on power supply leads is to use *despiking* (decoupling) capacitors, as shown in Fig. 11-13. Very close to the power supply input pins of *each chip*, there is a small disc or mylar capacitor. Usually 0.01-μF capacitors are sufficient for small-scale chips, and 0.1 μF, for large chips. Since the voltage across a capacitor cannot change instantly, the caps hold the voltage constant until the glitch-producing current pulse is gone.

Another possible cause of power supply glitches is temporary overloads (excessive current pulses) on the supply. Whenever circuits begin acting erratically, be on the lookout for glitches on the power supply pins.

Common ground foils, such as the one shown in Fig. 11-12, can also cause problems, particularly in high-current circuits. The current returning to the regulator through the ground foil from IC-5 interacts with the ground currents of the other chips. If the ground foil, or bus, is fairly large and the

11-13 Using despiking caps to filter out power supply glitches.

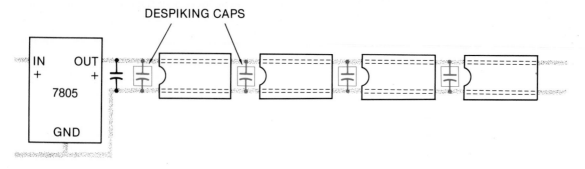

ground currents are small, there is usually no problem. But high-amplitude pulses through a narrow foil spell trouble. Proper circuit board layout is usually the best way to avoid such problems. Fig. 11-14 shows the general idea. The ground leads of each circuit are connected to *one common ground point*, thereby eliminating the problem before it starts.

11-14   Minimizing ground loop problems by using one common ground point.

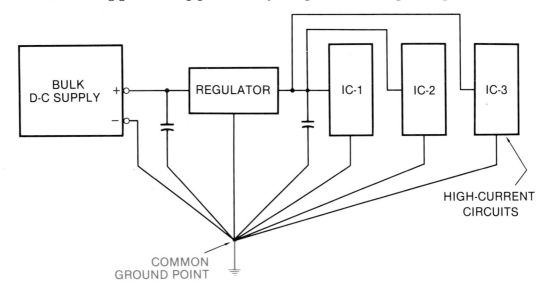

"Glitch" refers to any unwanted or unexpected or transient occurrence anywhere in a circuit. Let's look at another source of glitches. Fig. 11-15A shows a two-stage binary ripple counter with a couple of outputs feeding AND gate G. Fig. 11-15B shows the ideal waveforms we would expect to see at the various points shown. The waveforms are perfectly rectangular, with no rise time, fall time, or propagation delays shown.

The ideal waveforms show that at exactly the instant CLOCK pulse #4 goes low, Q_A and Q_B go low also. However, real flip-flops and gates *do* have propagation delays, which can cause unexpected results. Fig. 11-16 shows the actual situation occurring at the falling edge of CLOCK pulse #4. From the time CLOCK pulse #4 reaches a low level until the Q_A output reaches a low level, there is a propagation delay of about 20 to 40 ns (shown as 30 ns).

Since the C_P input of flip-flop B is fed by the Q_A output, flip-flop B cannot begin to toggle until Q_A reaches a low level. So we see that the Q_B output really does not reach a low level for at least another 30 ns *after* Q_A goes low. The point is that for about 30 ns or so, both $\overline{Q_A}$ and Q_B are high, causing the output of AND gate G to go high for a short duration.

Our idealized waveforms show that gate G's output is high for one clock period following pulse #2 and again following pulse #6. But as can be seen in Fig. 11-16, gate G puts out a short-duration spike (glitch) following pulse #4. Fig. 11-17 shows the position of the glitches in relation to all other waveforms. These glitches are very narrow and easy to miss when viewing the waveforms with an oscilloscope. *Watch out for glitches whenever all inputs to a gate change at almost the same time.* This is sometimes called a *racing* problem.

This brings up another interesting point. Depending on the relative propagation delays of the flip-flops and the gate, the glitch may or may not be seen. For example, if the AND gate has a long propagation delay, say 40 or 50 ns, but the flip-flops being used have a short delay of say 20 ns, no glitch will appear. But if later someone replaces the AND gate with a high-speed version, or if the flip-flops are replaced with slower versions, *voila!* We have bugs!

Temperature changes can also cause glitches to appear or disappear mysteriously. Propagation delays change when temperature changes. Even rerouting wiring can introduce sufficient delays. Here, too, glitches begin to appear or disappear without reason.

Running down glitches can be tricky. It is almost impossible to give a step-by-step procedure that will work in every case. However, if you are troubleshooting a system that worked at one time but that suddenly becomes erratic, here are some points to consider which might lead you to suspect glitches as being the culprits.

11-15 **Two-stage ripple counter with outputs gated.**

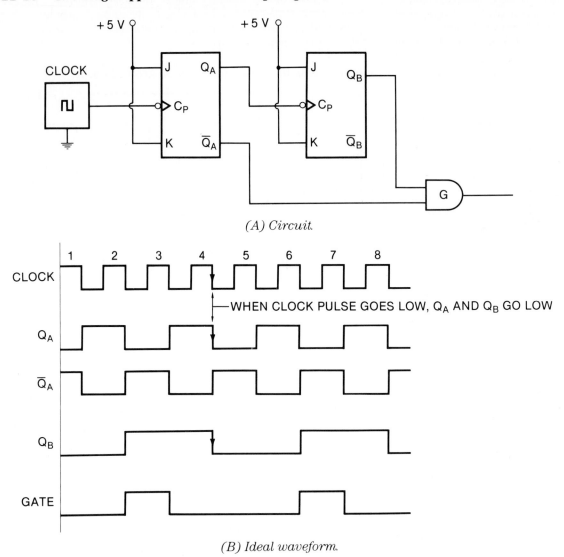

(A) Circuit.

(B) Ideal waveform.

11-16 **Timing diagram during fourth clock pulse showing propagation delays.**

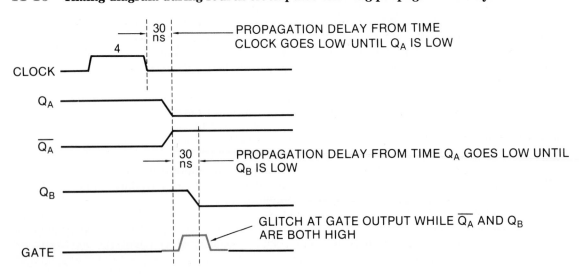

11-17   Waveforms of the circuit in Fig. 11-15 showing possible glitches.

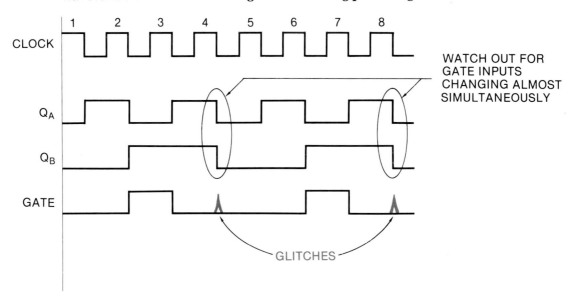

CHANGES IN LAYOUT

If a printed circuit board or cabling for the system were changed, check whether rerouting might have changed time delays.

CHANGES IN COMPONENTS

Did anyone change components, substituting those of a different family, say 74SXX for a 74LXX, etc.? This can cause glitches. Sometimes using chips that have exactly the same part number but are made by a different manufacturer can change delays.

CHANGES IN TEMPERATURE

This type of problem should be caught at the systems test level of quality control. The entire system should be cycled from the lowest to the highest anticipated operating temperature a few times to see the overall effect. As mentioned, changes in temperature can cause time delay changes.

SPOTTING GLITCHES

How do you spot glitches? The most common way is with an oscilloscope. When you suspect that a certain point in a system may be developing glitches, watch that point with an oscilloscope while simulating the conditions that seem to cause the glitches.

Glitches are extremely narrow. You will have to crank up the beam intensity and look for thin vertical lines, which are the glitches. Sometimes it is possible to rig up a test flip-flop or one shot to watch for an occasional glitch on a normally quiet line. Connect the C_P or trigger input to the point suspected of generating the glitch. The flip-flop or one shot can effectively "stretch out" the glitch, so that it can be seen more easily.

High-speed logic probes, like those discussed in chapter 10, can also be used to "stretch out" a glitch on a line that normally rests at one state. For example, suppose a particular gate should normally have a steady-state (constant high or low) output when a particular random event occurs elsewhere in the system. But you suspect that this gate is producing a glitch when that event occurs. Touch the logic probe to the gate output and watch the pulse LED. If a glitch occurs, the pulse LED will blink on long enough (a few tenths of a second) for you to see it.

TROUBLESHOOTING DIGITAL SYSTEMS

The basic techniques of troubleshooting digital systems are much the same as those of trouble-shooting analog systems. Applying typical inputs, successively split the system into smaller and smaller sections. Look for circuits that have normal inputs but abnormal outputs. Start by looking at the signals somewhere near halfway between the input and output. Then each successive split should cut the remaining circuitry approximately in half.

We shall now examine a simple frequency counter to illustrate our troubleshooting technique. Fig. 11-18 shows a simplified block diagram of the system. Notice that the major blocks include the *input, control, arithmetic, memory,* and *output* sections, as discussed earlier.

Briefly, here's how the counter operates. The input signal, whose frequency is to be measured, is fed to the squaring block. The squaring circuit essentially amplifies the input signal and forms it into a rectangular wave, which is TTL compatible. Before the start of a measuring period, the timing and control circuitry generates a RESET pulse, then it permits the squared input signal to pass through a gate for exactly 1 s. The output signal from the gate is a series of $\overline{\text{COUNT}}$ pulses, at the same frequency as the original input signal.

The $\overline{\text{COUNT}}$ pulses are fed to decade counters, which total up the number of cycles of the input wave that occur in 1 s. At the end of the 1-s period, the count gate is disabled, preventing any more pulses from getting through. Then a STORE pulse is generated by the control circuitry. It causes the outputs of the counters to be stored in the latches. Latches are essentially D flip-flops, which are used as temporary memory locations. They hold the previous count while the counters are being updated.

The latch outputs feed decoder/driver chips, which are connected to seven-segment displays. The numbers read on the displays represent the number of cycles per second (hertz) of the incoming signal.

Fig. 11-19 shows a detailed chip-level diagram of the frequency counter. The timing diagram is shown in Fig. 11-20. The clock (IC-3) is a 555 timer chip used as an astable multivibrator. The output at pin 3 is a rectangular wave whose positive interval is calibrated to exactly 1 s by adjusting pot R_1.

11-18 Block diagram of frequency counter.

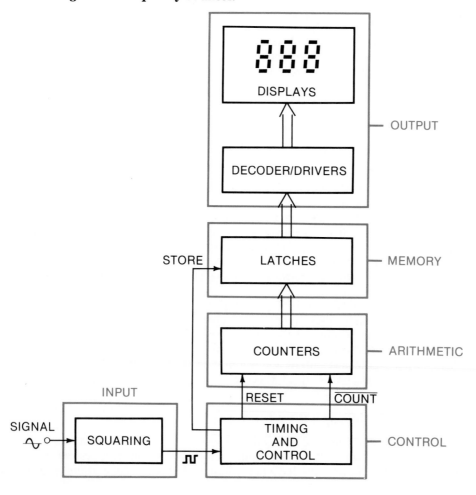

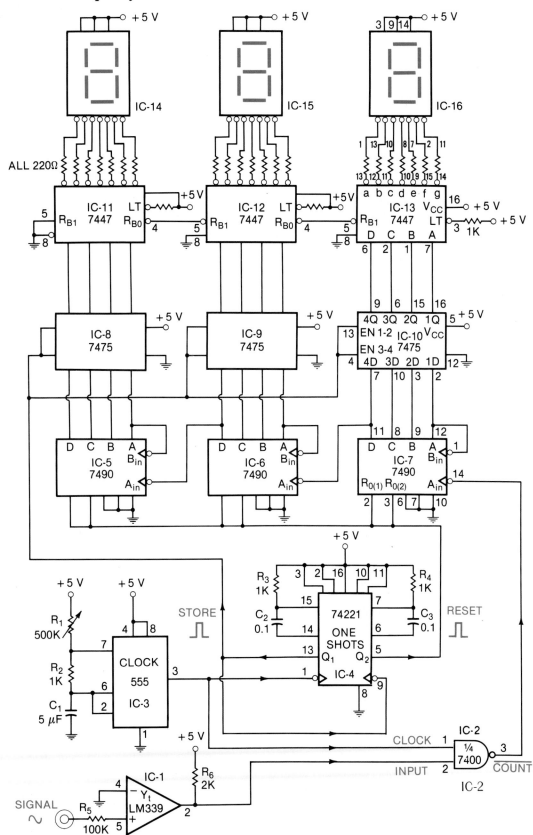

11-19 MSI frequency counter.

The falling edge of the CLOCK signal triggers a one shot in IC-4, causing pin 13 to go high for a short-duration pulse. The pulse width, which is controlled by R_3–C_2, is approximately 100 μs long. This pulse, called STORE, is fed to the enable inputs of the 7475 latches and causes the data on the D inputs of the latches to be stored at the Q outputs.

The falling edge of the STORE pulse triggers another oneshot in the 74221 package. The second monostable generates the RESET pulse, which is fed to all counters. Thus, the counters are cleared prior to the next counting period.

Notice that the $\overline{\text{COUNT}}$ waveform in Fig. 11-20 consists of a series of pulses at the frequency of the input signal. Since the gate (IC-2) is enabled for exactly 1 s, the number of pulses fed to the 7490 decade counters is equal to the frequency of the input signal. The input frequency must be less than 1000 Hz for this system.

Now let's discuss how to troubleshoot the frequency counter of Fig. 11-19. You start by feeding a signal into the input jack. Let's say the signal is at a frequency of a few hundred hertz. By observing the seven-segment display, you can tell quite a bit about any possible problems. For example, if the least significant digit appears to work properly but the second and third digits do not light or do not change when the input signal frequency is changed, you know that all of the input and timing and control circuitry must be working. Also IC-7, IC-10, and IC-13 must be working. So, the place to look for the problem is from IC-6 on.

Now suppose that none of the displays are working normally. Where do you begin to test? One good method is to split the system in half, say by looking at the outputs of the arithmetic circuits. Do this by clipping a logic monitor over IC-7. Then, if the outputs are *not* normal, split the first half of the system again, possibly by looking at the inputs to the counters, then at the clock, and so on. If the outputs of the counters are normal, split the second half of the systems by looking at the outputs of the latches. Then look at the outputs of the decoder/drivers, etc.

Fig. 11-21 is a flowchart indicating possible tests to make on the frequency counter of Fig. 11-19. The procedure assumes that an input signal is applied, and either no outputs or abnormal outputs are observed in the displays. Study the flowchart to see how it works. Try to imagine a specific fault and see if the flowchart leads you to it. Obviously, no detailed test procedure is 100% foolproof. There will always be some problems that cannot be solved in any general procedure. The flowchart should stimulate you to think about how to zero in on a defective area with the least amount of unnecessary effort.

As mentioned, you can clip a logic monitor, like the LM-1, over each chip that you wish to test and see if its outputs are normal. However, if the outputs are changing periodically, or if you wish to look at the outputs and inputs of several chips simultaneously, a more useful test instrument is the LM-3 triggerable logic monitor, shown in Fig. 11-22.

The LM-3 can simultaneously monitor up to 40 channels, or test points, which are connected to the instrument through a ribbon cable. Tiny clips at the end of each input wire allow you to clip on to any point in the system. A front panel toggle switch allows you to select the logic high input level as TTL/DTL, CMOS, or variable, which is controlled by a threshold control pot.

One powerful feature of the LM-3 is that the data displayed on the 40 LED are *latched* into the display by the transition if a trigger signal from high to low or from low to high. For example, if you connected input clips to all 7490 outputs and then used the STORE signal (pin 13 of IC-4) as the trigger, you would latch and display all of the counter outputs at the time of the positive transition of STORE. This would allow you to compare the counter outputs to the numerals displayed on the seven-segment readouts.

In addition, you could also latch in the outputs of the 7475 chips, 7447 chips, clock, reset, or any combination of these. It is also possible to use the LM-3 in the run mode, which means that input data are not latched. This allows you to observe up to 40 points in the system simultaneously and see what each point is doing in real time. This is particularly useful when testing systems in which major events are occurring slowly enough to be observed by a human but where many points must be monitored simultaneously. The LM-3 collects all test point data and displays them on one panel.

As usual, testing specific single points or signals is probably best done with a logic probe or oscilloscope. For example, the short-duration STORE and RESET pulses can be easily seen after being "stretch out" by the logic probe. But $\overline{\text{COUNT}}$ and INPUT are best observed with an oscilloscope.

The digital pulser again becomes useful to apply a STORE, RESET, or single $\overline{\text{COUNT}}$ pulse at various points in the system.

The modern trend in digital design is to use large-scale integrated (LSI) circuits to perform multiple functions. Fig. 11-23 is a diagram of an inexpensive frequency counter using a 7217 four-

11-20 **Timing diagram for frequency counter of Fig. 11-14.**

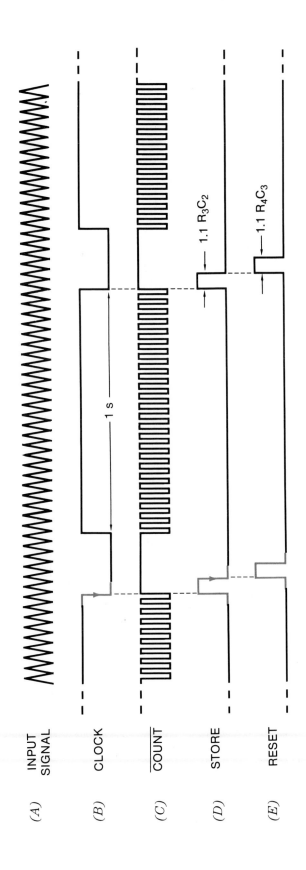

11-21 **Flowchart for troubleshooting the frequency counter of Fig. 11-19.**

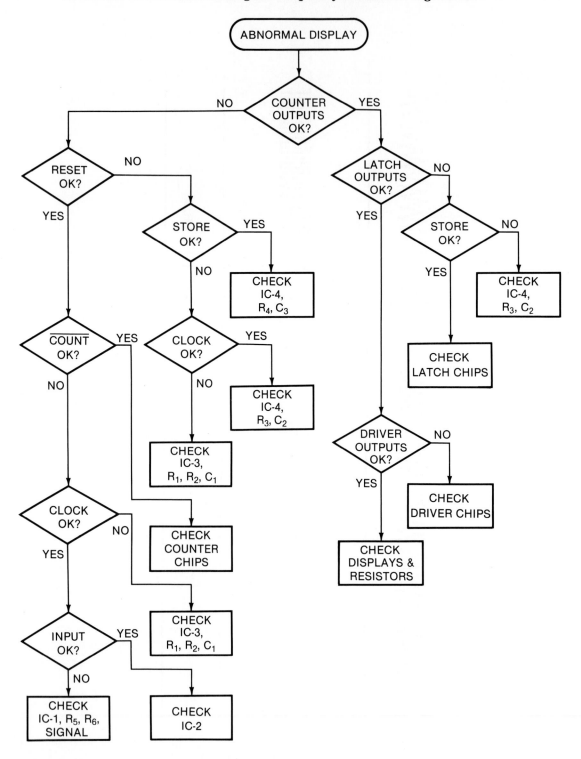

digit CMOS up/down counter/display driver chip. On a single 28-pin chip are four decade counters, latches, and multiplexed display drivers. Although LSI circuits are quite complex internally, they need not be excessively difficult to troubleshoot. However, you must be able to observe inputs and outputs.

11-22 LM-3 logic monitor. Courtesy of Global Specialties Corp.

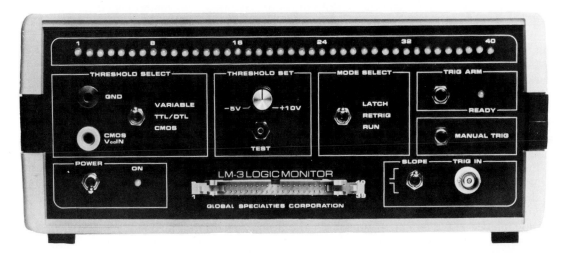

Notice that the input signals are COUNT, $\overline{\text{RESET}}$, and $\overline{\text{STORE}}$. These do essentially the same job as in the circuit of Fig. 11-19. Timing for the control signals is accomplished by the 556 dual timer chip. The 556 acts like two 555 chips. One of the internal timers, whose output is pin 5, is set up as a clock. Components $R_1, R_2,$ and C_1 determine the clock frequency. Pot R_1 is adjusted to give a positive clock period of exactly 1 s. The other half of the 556 (output pin 9) is used as a one shot. The pulse duration of STORE is controlled by R_4–C_4. The timing diagram for the circuit of Fig. 11-23 is shown in Fig. 11-24.

As with the circuit of Fig. 11-19, the counters are reset prior to a counting period. Then the CLOCK signal enables input gate 1 for exactly 1 s. The number of input cycles occurring within 1 s are counted by internal counters. The falling edge of the CLOCK signal triggers the one shot, generating the STORE pulse. Gate 4 inverts the pulse and applies an active low $\overline{\text{STORE}}$ pulse to the chip. The count is thus latched into the outputs, leaving the counters free to be reset and to count again.

The outputs of the 7217 are multiplexed to minimize the number of output pins. Here's how the multiplexing works. After a count has been latched in by a $\overline{\text{STORE}}$ pulse, the segment code for the most significant digit is output on pins a–g. At the same time, a high level is emitted on the D_4 output pin, applying the anode voltage to the most significant display chip. D_1 through D_3 are all low at this time. A short time later, the segment code for the third digit is emitted, and D_3 goes high, while D_1, D_2, and D_4 are made low.

This procedure is repeated for digit 2, then digit 1. Then digit 4 is displayed again. The process is repeated over and over, displaying digits 4-3-2-1-4-3-2-1-4-etc. at a rate so high that no flicker is seen. An internal scan oscillator, which operates at about 10 KHz, generates the required scan signals.

All a–g segment inputs to each display chip are connected in parallel. This multiplexing scheme allows all four display chips to be driven with just 11 pins. Without multiplexing, 28 output pins would be needed.

Even though the 7217 is quite complex internally, you test it by applying normal inputs, such as $\overline{\text{RESET}}$, $\overline{\text{STORE}}$, and COUNT, while observing the outputs. You could, for example, connect an instrument like the LM-3 to outputs a–g. Then use the positive transition of the D_1 output to latch the signals into the display. Then repeat using $D_2, D_3,$ and D_4 as the trigger.

TESTING AND TROUBLESHOOTING MICROCOMPUTER SYSTEMS

Microprocessors and programmable logic are rapidly taking over as the most effective way of designing complex digital systems. By using programmable logic, the hardware design is more standardized, and variations in system operation are accomplished by modifying the programs. Although analyzing microprocessor systems is beyond the scope of this text, there are a few things that can be tested, even by persons not familiar with microprocessors or programming.

Fig. 11-25 shows a small microcomputer system using Motorola's MC6800 microprocessor. Notice that it has all of the major blocks referred to at the beginning of this chapter. The MPU contains all of the control and arithmetic sections. The 74LS244 is a tristate octal input buffer. It connects the external inputs (e.g., switches) to the data bus when its enable pins 1G and 2G are made active low. Output from the system is accomplished via the 74LS373 octal latch. The 2716 EPROM is an erasable read only memory, and the 2112A chips are random-access memories (RAM). Lastly, the system clock is the MC6875 chip. It has an on-board crystal to maintain accurate timing.

Whenever the MPU wants to communicate with any section of the microcomputer (i.e., memory, input, or output), it selects that particular chip by means of the 74LS155 decoder. Notice that MPU address lines A_{15} and A_{12} are connected to input pins A and B on the decoder. The levels on these input pins determine which output pin $2 Y_0$–$2 Y_3$ is made low. That is, if the A and B inputs are both low (binary zero) output $2 Y_0$ is low. If $B = 0$ and $A = 1$, output $2 Y_1$ is low. Notice that the decoder outputs are connected to the chip enables on the memory, input, and output chips. So when the MPU outputs an address with $A_{15} = 1$ and $A_{12} = 0$, output $2 Y_2$ of the decoder goes low, enabling the tristate buffer, and thereby causing the external inputs, to be fed to the data bus. This action is highlighted on Fig. 11-25.

Suppose you want to test the microcomputer of Fig. 11-25, which is all built on a single printed circuit board. Where do you begin? It is common practice to mount the large-scale chips (shaded), like the MPU, EPROM, and RAM, in sockets, while the small- and medium-scale chips are flow soldered into the PC board. This allows you to test the smaller components independently of the MPU and memory. Let's assume that you have the PC board with the small- and medium-scale chips, as well as all resistors and capacitors, soldered in, and the LSI chips removed from their sockets.

One of the first chips you should check is the clock. Of course, you should have the manufacturer's spec sheets available so that you will know what the typical waveforms should look like. But basically, you look at the output pins, labeled Ø1 and Ø2, and see if they are good rectangular pulses, with normal rise and fall time, and, of course, at the specified clock frequency (in this case 1 MHz). You would also test the reset button to see if the reset output pin goes low when the switch is pressed.

If the clock seems to be working properly, check the decoder circuitry to see that the various chips are enabled properly. For example, since the MPU is not in its socket, you could insert thin test leads into the MPU pin sockets of A_{15} and A_{12}. Then you would ground both free ends of these leads, making $A = 0$ and $B = 0$ on the decoder inputs. Pins 2 and 14 of the decoder must also be made low to enable the decoder. Next you would test the $\overline{CE}$ pins on the RAM sockets with a logic probe to make sure they are low.

Likewise, test the input buffer by setting up some bit pattern on the external inputs, perhaps by means of toggle switches. When MPU address lines are set at $A_{15} = 1$ and $A_{12} = 0$, the 74LS244 should be enabled, allowing you to observe the external inputs on the data bus (buffer outputs $2 Y_4$–$1 Y_1$). These can easily be observed with a logic monitor clipped onto the buffer chip.

This procedure is continued until all small- and medium-scale circuitry is tested. Any inverters, gates, flip-flops, counters, etc., can easily be tested using a logic probe and digital pulser, as discussed in chapter 10.

Once the small- and medium-scale circuits are checked out, the LSI chips (or at least a few of them) are plugged in, and the system is put into operation. Suppose the system does not operate normally with the LSI chips plugged in, now what? You can use some very sophisticated pieces of test equipment, such as logic analyzers and in-circuit emulators, which will help you find the problem. However, these tools are very expensive and require that you understand a great deal about microprocessor systems.

There is another piece of equipment, called a *signature analyzer*, which can be used effectively by testers and troubleshooters who do not have a thorough background in computers. In fact, it is very convenient for field service personnel, because it is small, about the size of a bench-top DVM, and therefore quite portable. Fig. 11-26 shows a Hewlett-Packard Model 5005A signature multimeter. This instrument is a multifunction unit, which can be used to measure voltage, resistance, frequency, time interval, and logic levels, as well as being a signature analyzer. Only the signature analyzer is discussed here.

The signature analyzer is used much the same as an oscilloscope is used in troubleshooting linear equipment. You simply look at the outputs of a circuit, then back up to the inputs. When you find a circuit with bad outputs and good inputs, you have localized the fault. But rather than displaying waveforms at various points, the signature analyzer changes the "waveform" or bit pattern at a particular point, into a four-digit *number*.

For example, Fig. 11-27 shows typical waveforms you would expect to see with an oscilloscope at the various input and output pins of a normally operating 7490 decade counter. The signature

11-23 **Inexpensive frequency counter using an LSI chip.**

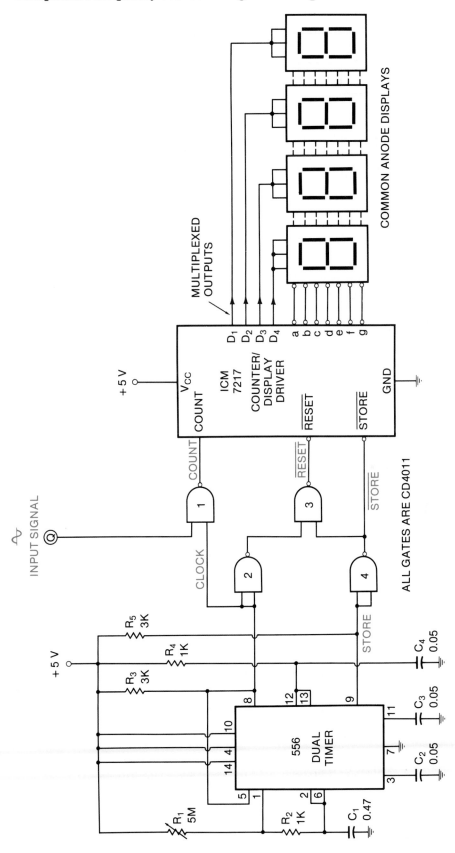

11-24 Timing diagram for circuit of Fig. 11-23.

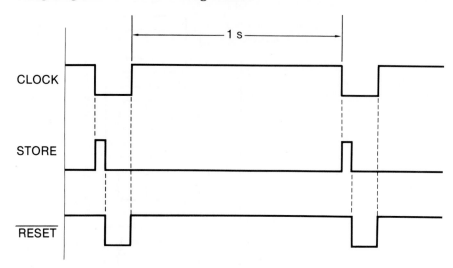

analyzer takes the high or low levels at a particular point, say output *A*, and shifts them into an internal shift register on each clock transition. It begins shifting the pattern in when it receives a START signal, and it stops shifting when it receives a STOP signal. In this manner, it stores the "serial bit stream," appearing at point *A*, in an internal register.

Next, rather than displaying the bit pattern on an oscilloscope, the signature analyzer changes the bit pattern into a unique *number*. The technique for the number generation is somewhat complex, but we really do not need to know how it is done. The important thing is that each different waveform, or bit pattern, will have a unique number associated with it. The number, or *signature* as it is called, is displayed on the front of the instrument, and it may consist of numbers 0–9, as well as letters, A, C, F, H, P, or U. Fig. 11-28 shows the set up and typical signatures that you could read if you tested the various points of the 7490.

Getting back to the microprocessor system of Fig. 11-25, the signature analyzer can be used to test various points in the system. Of course, before the signature analyzer can be used, a good, working system must be signaturized by a technician who records the signatures corresponding to each pin. The signatures are then usually recorded on the logic diagrams, so that subsequent testers can compare their readings to those of a good, working system. Usually test loops are designed into the computer's ROM, so that various sections of the system can be tested separately. Also, a diagnostic flowchart describing exactly the points to check and the order to check them is extremely helpful. Once a section is found with normal input signatures and abnormal output signatures, begin swapping chips and checking nearby discrete components.

PROBLEMS

For the next ten problems, refer to Fig. 11-2.

11-1. When S_2 is pressed, $\overline{\text{START}}$ goes (a)high. (b)low.

11-2. When S_2 is pressed, ENABLE goes (a)high. (b)low.

11-3. When CLOCK and ENABLE are both high, $\overline{\text{COUNT}}$ is (a)high. (b)low.

11-4. When S_1 is pressed, RESET goes (a)high. (b)low.

11-5. To change the maximum measurable time to 99.9 s, change the clock frequency to (a)10 Hz. (b)1000 Hz.

11-25 A small Motorola Model MC6800 microcomputer system.

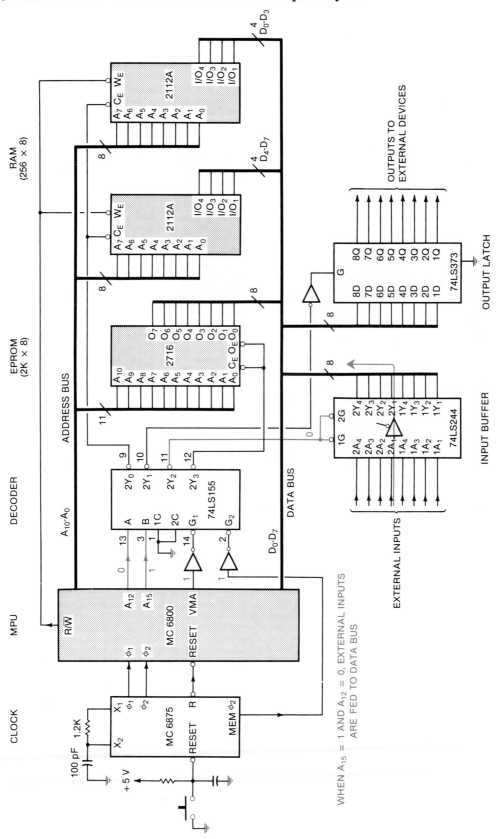

11-26   Hewlett-Packard Model 5005A signature analyzer. Courtesy of Hewlett-Packard Co.

11-6.   To measure time to the nearest millisecond, change the clock frequency to (a)10 Hz. (b)1000 Hz.

11-7.   Describe what you would observe on the second display digit if the connection between the *A* output and the *B* input of the middle 7490 became open.

11-8.   Referring to problem 11-7, describe what you would see on the most significant digit display.

11-27   **7490 decade counter and waveforms.**

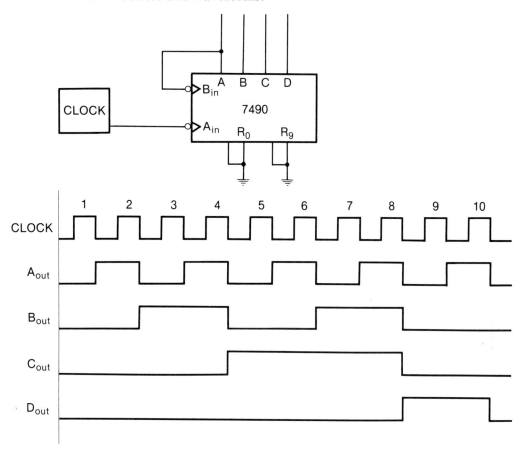

11-28   **Testing a 7490 with a signature analyzer.**

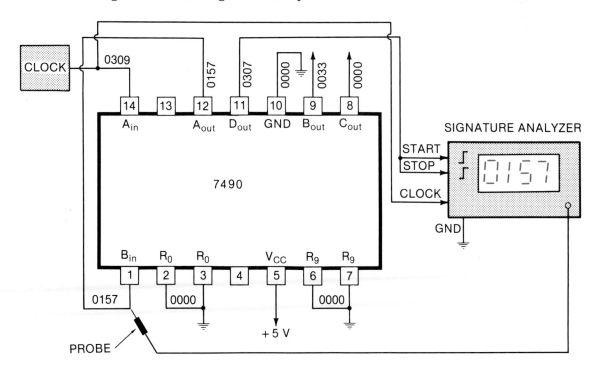

11-9.   Suppose the most significant digit always displays a zero, but the other digits work normally. Which should be the next step? (a)Replace the left most 7490. (b)Replace the most significant display. (c)Replace the most significant 7447. (d)Apply pulses to the A input of the left most 7490.

11-10.   If you did as described in problem 11-9 and still got the same results, you should next check (a)S_1. (b)R_1. (c)the R_0 inputs to the left most 7490. (d)the signal at the C output of the middle 7490.

11-11.   Refer to Fig. 11-5, and fill in Table 11-1.

11-12.   Refer to Fig. 11-5, and fill in Table 11-2.

11-13.   Refer to Fig. 11-10, and fill in Table 11-3.

11-14.   Refer to Fig. 11-10, and fill in Table 11-4.

For the next six problems, refer to Fig. 11-19.

11-15.   If you apply a 0.1-Vp-p sine wave at a frequency of 600 Hz to the input jack, what should you see at pin 2 of the LM-339?

11-16.   To test the 74221, you could apply a short-duration negative pulse to pin 1. If the circuit is working normally, what should you see at pin 13?

11-17.   Referring to problem 11-16, what should you see at pin 5?

11-18.   How would the signal at pin 13 change if C_2 were increased to 0.5 μF?

11-19.   How would the signal at pin 13 change if C_3 became open?

11-20.   How would the signal at pin 5 change if C_3 became open?

Table 11-1

	Z_0	Z_1	Z_2	Z_3	Y_0	Y_1	Y_2	Y_3	b_0	b_1	b_2	b_3	X_0	X_1	X_2	X_3	W_0	W_1	W_2	W_3	STORE Z	STORE Y	LOAD X	LOAD W
1													0	0	1	1	0	1	0	0	1	0	1	0
2													1	1	0	1	1	0	1	0	1	0	1	0
3													0	1	0	0	0	1	1	0	0	1	1	0
4													1	1	0	1	0	1	0	1	0	1	0	1
5													0	0	0	0	1	0	1	0	1	0	0	1
6													1	0	1	1	1	0	0	1	1	0	1	0
7													1	1	0	0	0	1	1	1	0	1	0	1
8													1	1	0	1	1	1	1	1	1	0	0	1

Table 11-2

Z_0	Z_1	Z_2	Z_3	Y_0	Y_1	Y_2	Y_3	b_0	b_1	b_2	b_3	X_0	X_1	X_2	X_3	W_0	W_1	W_2	W_3	STORE Z	STORE Y	LOAD X	LOAD W	
												0	1	0	1	0	0	0	0	1	0	1	0	1
												1	1	0	1	1	0	0	0	1	0	0	1	2
												0	0	0	1	0	1	0	0	0	1	1	0	3
												1	1	1	1	1	1	0	0	0	1	0	1	4
												0	0	0	0	0	0	1	0	1	0	1	0	5
												1	0	0	1	1	0	1	0	1	0	1	0	6
												1	0	1	0	0	1	1	0	0	1	0	1	7
												1	0	0	0	1	1	1	0	1	0	0	1	8

Table 11-3

b_0	b_1	b_2	b_3	X_0	X_1	X_2	X_3	W_0	W_1	W_2	W_3		LOAD $\overline{X}$	LOAD $\overline{W}$		A_0	A_1	A_2	STROBE	
				0	1	0	1	1	0	0	0					0	1	0	1	1
				1	0	0	0	0	1	0	0					0	1	0	0	2
				1	0	1	1	1	0	1	0					1	0	0	0	3
				0	0	0	0	0	1	1	0					0	1	0	0	4
				0	1	1	1	1	0	1	1					0	1	1	0	5
				1	1	1	1	1	1	1	0					1	1	1	0	6
				0	1	0	0	1	0	1	1					0	1	1	1	7
				1	0	1	0	0	0	1	1					0	1	1	0	8

Table 11-4

	b_0	b_1	b_2	b_3	X_0	X_1	X_2	X_3	W_0	W_1	W_2	W_3		$\overline{\text{LOAD } X}$	$\overline{\text{LOAD } W}$		A_0	A_1	A_2	STROBE	
					1	1	1	1	1	0	1	0					0	1	0	0	1
					0	0	0	0	0	1	1	0					0	1	0	1	2
					1	0	0	0	0	1	0	1					0	1	1	0	3
					0	1	0	0	0	1	1	1					0	1	1	1	4
					1	1	0	0	1	1	1	0					1	1	0	0	5
					0	0	1	0	1	1	0	1					1	0	1	0	6
					1	0	1	0	1	0	1	0					0	1	1	0	7
					0	1	1	0	1	0	1	1					0	1	0	0	8

For the next four problems, refer to Fig. 11-23.

11-21. If the CLOCK signal is normal, but there is no STORE signal, which of the following could be the cause? (a) Open R_1. (b)Open C_1. (c)Defective 556. (d)Open R_4.

11-22. If the output of gate 4 was always high, but gates 2 and 3 were operating normally, what would you observe on the displays?

11-23. One way to test the 7217 would be to leave the CD4011 out of its socket and apply signals to the COUNT, $\overline{\text{RESET}}$, and $\overline{\text{STORE}}$ inputs. In what order should these signals be applied? (a)COUNT, $\overline{\text{RESET}}$, $\overline{\text{STORE}}$. (b)$\overline{\text{RESET}}$, $\overline{\text{STORE}}$, COUNT. (c)$\overline{\text{STORE}}$, COUNT, $\overline{\text{RESET}}$. (d)$\overline{\text{RESET}}$, COUNT, $\overline{\text{STORE}}$.

11-24. If the seven-segment displays have random segments lit in meaningless patterns, which of the following would be the most likely cause? (a)The displays are in the wrong sockets. (b)The digit lines D_1–D_4 are interchanged. (c)The segment lines a–g are scrambled. (d)The displays are common cathode instead of common anode types.

EXPERIMENT 11-1   DESIGNING A TEST PROCEDURE

This will be your most challenging and most original experiment. To do this experiment well, you will have to call on all the knowledge and experience you have gained in this course.

Here's the problem. The testing of modules and/or systems that have just been assembled in production is one of the most common tasks of technologists. Whether the modules have been hand assembled or whether parts have been inserted into PC boards by a computer-aided manufacturing (CAM) process, the modules must be tested by a competent technologist. Your job, as supervisor of a test department, is to design a thorough test procedure for a small system.

Here are your requirements.
1.   Choose either the time interval counter of Fig. 11-2 or the frequency counter of Fig. 11-23.
2.   Design a test procedure for the chosen system. Write a detailed, step-by-step procedure for test technicians to follow, based on the test equipment you have available (logic monitor, logic probe, digital pulser, digital multimeter [DMM], etc.).
3.   Draw a detailed flowchart for the test procedure. Your procedure should tell the technician the following:

   a.   What to test
   b.   Where to apply inputs—specify chip and pin numbers
   c.   What kind of inputs to use
   d.   What to expect at the outputs
   e.   What to do next, if normal outputs are found
   f.   What to do next, if normal outputs are not found

You may consider leaving certain chips out of their sockets until some tests are made. You also have the option of leaving certain leads open for initial testing, then inserting jumper wires across the opens to continue testing.

Your final report for this project should include:
1. A logic diagram of the system to be tested, identifying test points and all pin numbers.
2. A flowchart showing the order of testing.
3. A written, detailed test procedure.

> **NOTE**   If done well, a neat, typewritten copy of this report would make an excellent document for you to carry with you to a job interview and to show to a prospective employer. It could just give you the extra edge you need to land the job. Or if you are already employed, show it to your supervisor at your next review.

EXPERIMENT 11-2   USING A TEST PROCEDURE

Using the test procedure you wrote for Experiment 11-1, build and test your system to see whether the procedure works.

> **OPTION**   Use another student's test procedure, and build and test the system.

Your report for this project should be a written, step-by-step explanation of the tests you performed according to directions. Describe the results of each test and where you were led based on the test. Explain whether, in your opinion, the test procedure was adequate for complete testing. Did you find any fault? Do you have any recommendation for improving the test procedure?

Sensors and Transducers 12

A *transducer* is a device that changes energy from one form to another. In this chapter, we will study some special types of transducers used as *sensors*, that is, transducers that are used to sense some change in a physical quantity, such as heat or light, and then produce an electrical signal proportional to that change. These sensors are used in countless industrial and consumer applications. Some examples include:

- Controlling the temperature in a furnace or a home oven with a temperature sensor.
- Controlling the exposure time in an automatic camera with a light sensor.
- Weighing tractor trailer rigs with load cells.
- Measuring the flow of gasoline into a tank with a flowmeter.

These are just a few of the countless applications where electronic sensors are used. In this chapter, we will not attempt to catalog all of the various types of sensors used, nor all of the possible applications of sensors. Our purpose here is to get familiar with some commonly used sensors and their important characteristics, to show a few typical circuits and applications, and to get familiar with the kinds of problems you may encounter when troubleshooting systems that use sensors.

TEMPERATURE SENSORS

Temperature is measured for a wide variety of reasons and applications. We need to know the temperature inside food freezers and inside ovens. We need to accurately measure the temperature of the human body for diagnosing health problems. We constantly want to know the temperature of the outside air. And for a variety of control and safety reasons, we need to measure the temperature of various points in automobile engines, aircraft systems, and in industrial processes such as in chemical plants and steel mills. Temperature is, without a doubt, one of the most often measured physical quantities.

Since the range of temperatures and surrounding conditions are so diverse, there are several methods used for measuring temperature. Four common types of temperature sensors are the *thermocouple*, the *resistance temperature device* (RTD), the *thermistor,* and the *integrated circuit* (IC) *sensor.* Table 12-1 summarizes the main advantages and disadvantages of each type. Let's look at each type to learn its characteristics and uses.

THERMOCOUPLES

If two dissimilar metals are joined together at both ends and one of the junctions is heated, a continuous current will flow in the wires. This is shown in Fig. 12-1A. The amount of current flow depends on the difference in temperature between the hot junction and the cold junction. If only

Table 12-1
Common Temperature Transducers

	Advantages	Disadvantages	Temp. Range (°C)
Thermocouple	Self-powered, rugged, inexpensive, wide temperature range.	Low output voltage, reference needed.	−270 to +2000
RTD	Most stable, most accurate.	Expensive, current source needed.	−200 to +800
Thermistor	High output, fast, inexpensive.	Very nonlinear.	−80 to +150
IC	Most linear, highest output.	Power supply needed.	Below +200

12-1   Basic operation of a thermocouple.

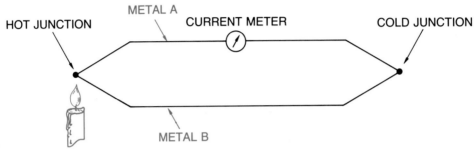

(A) Thermoelectric current flows when
junction of two dissimilar metals is heated.

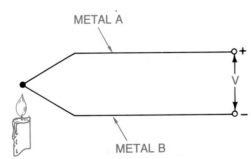

(B) Voltage across open ends is proportional
to temperature of heated junction.

one end of the wires is joined while the other is left open, a small voltage (called *Seebeck voltage* after its discoverer) will be developed across the open ends. This is shown in part B of the figure. The magnitude of the voltage depends on the temperature difference, as well as on the specific metals used. For a small change in temperature, the relationship between the temperature and voltage is linear, and is given by equation 12-1.

$$V = s \times \Delta T \qquad\qquad 12\text{-}1$$

where V = open-circuit voltage
  s = Seebeck temperature coefficient, usually expressed in microvolts per degree Celsius
  ΔT = difference in temperature in °C between the hot and cold ends

Table 12-2
Seebeck Coefficients of Various Thermocouples

Type	Alloys	Coefficient (μV/°C)	Operating temp range (°C)
E	Chromel-constantan	62	−270 to +1000
J	Iron-constantan	51	−210 to +750
K	Chromel-alumel	40	−270 to +1370
S	Platinum-rhodium	7	0 to +1760
T	Copper-constantan	40	−270 to +400

All dissimilar metals exhibit this characteristic, but some work better than others. Table 12-2 shows five commonly used types of thermocouples and their characteristics. The type letter (E, K, etc.) identifies the alloys used to make the wires.

EXAMPLE 12-1   A chromel-alumel (type K) thermocouple junction is heated to 30°C. What is the thermocouple voltage, assuming that the cold ends are at 0°C?

SOLUTION

$$V = s \times \Delta T = 40 \text{ μV/°C} \times 30°C = 1200 \text{ μV} = 1.2 \text{ mV}$$

As you can see, the thermocouple voltage is quite small, so either a very sensitive voltmeter must be used to measure it, or a good quality, high-gain amplifier, such as an op amp, must be used to build up the level to a more usable value.

For large changes in temperature, the equation is not very accurate due to nonlinearity of the thermocouple. However, the National Bureau of Standards (NBS) has published tables accurately listing the voltage generated by standard thermocouples over their entire useful range in 1° increments. All you have to do to find the temperature is to read the voltage generated by the thermocouple, and then look up the temperature in the tables. When using a computerized data acquisition system, there are mathematical methods to account for the nonlinearity. So when using a thermocouple sensor interfaced to a computer, it is not necessary to use the tables. The computer simply reads in the voltage through an analog-to-digital converter, then converts the reading into temperature.

NEED FOR A REFERENCE JUNCTION

Getting back to Fig. 12-1B, although the open-circuit voltage generated by the thermocouple is proportional to the temperature of the heated junction, we cannot simply connect a voltmeter to the open ends if we expect an accurate measurement. The reason is that the voltage developed by the thermocouple depends on the difference in temperature between the hot junction and the cold junction. The tables published by the NBS assume that the cold junction is at 0°C. If we simply connect a voltmeter to the open ends of the wire, where is the cold junction? Well, as shown in Fig. 12-2, two junctions (J_2 and J_3) are actually formed when the copper leads of a voltmeter are connected to the thermocouple. No voltage is generated at J_3, since both wires are copper. Therefore J_2 becomes our cold junction. If J_2 is at room temperature and not at 0°C, the voltage measured will be less than what it would be if J_2 were at 0°C. Therefore the temperature we find in the lookup tables will be incorrect.

One way to correct the problem is to hold J_2 at 0°C. In order to maintain a constant temperature of 0°C, a container of ice water can be used to hold the *reference junction*. As long as ice and water coexist in the same container, the temperature of the water will be 0°C. The setup, shown in Fig. 12-3, is sometimes used in research labs, but it is rather impractical to keep an ice-water bath handy all the time for everyday use.

A practical alternative to the ice water is to use a small device called an *electronic ice-point reference*. Inside the box, shown in Fig. 12-4, is an electronic circuit which measures the temperature

12-2 Voltmeter leads connected to thermocouple form two new junctions.

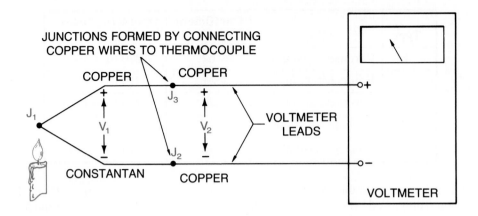

12-3 Using a reference junction at 0°C.

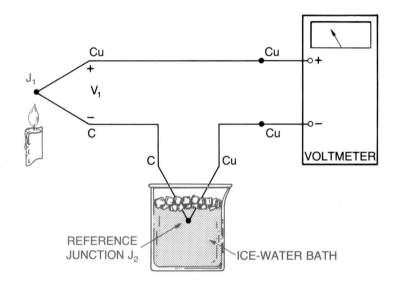

12-4 Using an electronic ice-point reference.

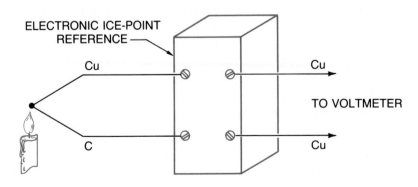

of the box (ambient temperature) and automatically balances an internal bridge circuit which generates a voltage to cancel out the effect of junction J_2 not being at 0°C. Commercially available ice-point references have built-in batteries, so they are quite convenient to use. You simply connect the box in series between the thermocouple and the voltmeter, as shown in the figure. However, the electronic ice-point reference generates a correction voltage which is specific to each type of thermocouple. Therefore a box made for a type K thermocouple will not work for a type J thermocouple, and so forth.

Another way to eliminate the ice bath is to use a technique shown in Fig. 12-5. Connections to the voltmeter are made on a device called an *isothermal block*. The block is made of an electrical insulating material which is a good heat conductor, so both terminals are at the same temperature. Inside the block is embedded a temperature-sensitive resistor (R_T). By measuring the resistance of R_T, we can determine the temperature of the isothermal block, and then calculate a correction factor for the measured voltage. This method is often used in computerized data acquisition systems. The resistance of R_T is measured by the computer; then a correction factor is calculated by means of software. Incidentally, the thermocouple does not have to have any copper leads. No matter what type of thermocouple is used, a correction factor can be calculated that corrects for the fact that the cold junction(s) is (are) not at 0°C.

You might wonder why we bother to use a thermocouple at all, since we have to determine the temperature of the isothermal block in order to calculate a correction factor. Well, the reason is that thermocouples have a much wider operating temperature range than the temperature-sensitive resistor. The temperature of the block is normally room temperature, yet the thermocouple may actually be located in an environment of +1000° or more. We can also connect several thermocouples, monitoring temperatures in several places, to a common block. The computer then makes the appropriate corrections, even if different types of thermocouples are used. No hardware changes are needed.

Whenever the thermocouples are located some distance from the measuring device, the extension wires used should be of the same material as the thermocouples themselves, to prevent additional junction voltages from being generated. This wire is available from thermocouple manufacturers and is called *extension-grade* wire. It may have a lower temperature insulation than the thermocouple, and it is usually advisable to use extension wires of larger diameter than the thermocouple so that the resistance of the wires is negligible and so that pulling the wires through a conduit will not damage them.

The thermocouples themselves can be purchased already made up in various wire sizes and with various insulations, including armor-coated. They are available encased in ceramic tubes, stainless steel tubes, and other fixtures, for protection. Thermocouple wire can also be purchased as pairs in a single-insulation coating on large spools, so you can make whatever length you need. The ends can be soldered together as long as the operating temperature is going to be less than the melting point of the solder. Normally, the ends are welded together.

TYPICAL PROBLEMS WITH THERMOCOUPLE SYSTEMS

Obviously an open thermocouple or a short somewhere between the hot junction and the measuring device can mean catastrophic failure. But here are a few other problems which can lead to errors:

12-5 Using an isothermal block to compensate for J_2 not being at 0°C.

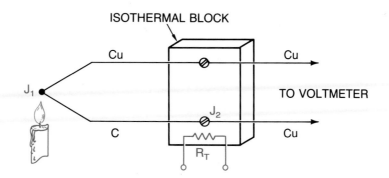

No Reference Junction Compensation

We already discussed why the compensation is needed. Failure in the compensator will cause errors in the measured voltage and, thus, in the recorded temperature. A simple test can be made to eliminate the thermocouple voltage from the measuring device and thereby test the compensator. Place a short circuit (piece of copper wire) directly across the thermocouple input terminals to the compensator (ice-point reference or isothermal block). If the compensator is working properly, the temperature indicated on the readout should be room temperature.

Wrong Extension Leads

If the thermocouple is mounted some distance from the measuring device, extension leads are needed. The leads should be of the same material as, and preferably of larger diameter than, the thermocouple being used. Incorrect leads (wrong materials) will introduce additional voltages to the system and cause errors.

Noise Pickup

If the thermocouple is mounted a long distance from the measuring device, the leads may pick up noise, such as a-c line noise. This may cause errors in the readout. When the thermocouple is remote, use shielded extension leads to minimize noise pickup.

Temperature Gradient

If the thermocouple is used to measure the temperature of a very hot fluid, such as molten metal, the junction should be encased in a metal sleeve to protect it from abrupt temperature change along its length, which can lead to early failure. Actually, thermocouples are commercially available in a wide variety of packages, including those in simple exposed wires, enclosed in metal sleeves, enclosed in ceramic insulators, and even enclosed in hypodermic needles used by veterinarians. Fig. 12-6 shows several commercially available thermocouples and packages.

12-6   **Thermocouple assemblies. Courtesy of Omega Engineering, an Omega Technologies Company.**

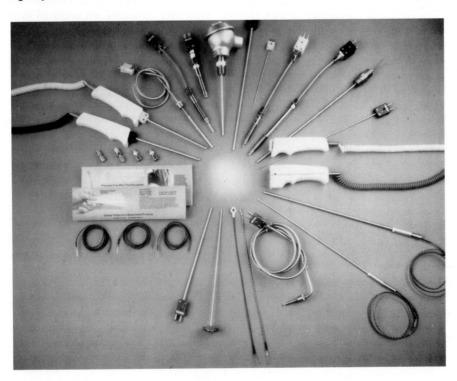

Change in Thermocouple Characteristics

When thermocouples are exposed to very high temperatures or to hot gases, the metals in the thermocouple can change, causing changes in the output voltage for a given temperature. This usually happens slowly over a period of time. One way to predict possible trouble with a thermocouple is to periodically measure its resistance. By keeping an accurate log of the resistance of the thermocouple, you can see any significant changes in its characteristics. Particularly, you should watch for sudden changes in resistance, indicating that some probable failure has occurred, such as wire damage or insulation breakdown.

RTDs

In your early studies of electricity, you undoubtedly learned that the resistance of conductors varies with temperature. This phenomenon is widely used to measure temperature. The conductor most commonly used is platinum, which is quite stable over a wide temperature range. To make the sensor, a thin platinum wire is wound in a coil on a ceramic form. Sometimes platinum metal film is used to make sensors, which are very small and economical, but it is not as stable as pure platinum wire.

RDTs can be obtained in resistance values from 10 Ω to several kilohms, but by far the most commonly used value is 100 Ω. The temperature coefficient for platinum (Pt) is 0.00385/°C, and is called α (alpha). This means that a 100-Ω sensor will increase resistance by 0.385 Ω for each degree increase in temperature. Equation 12-2 shows how to find the resistance for a small temperature change.

$$R = R_0(1 + \alpha \, \Delta T)$$ 12-2

where R = resistance at new temperature
R_0 = resistance at reference temperature (0°C)
α = temperature coefficient of 0.00385/°C
ΔT = temperature change from 0°C

EXAMPLE 12-2 If the resistance of a sensor is 100 Ω at 0°C, what will be its new resistance at 20°C?

SOLUTION

$$R = R_0(1 + \alpha \, \Delta T) = 100(1 + 0.00385 \times 20) = 107.7 \, \Omega$$

Although RTDs are more linear than thermocouples, they are still not completely linear over wide temperature ranges. The NBS has published temperature-versus-resistance tables for platinum, so quite accurate temperature measurements can be made.

It is difficult to measure small changes in resistance accurately with an ohmmeter. So two of the most common methods of using an RTD are shown in Fig. 12-7. Part A of the figure shows the bridge circuit method. By balancing the bridge after the RTD has been connected, the resistance

12-7 Methods of using an RTD.

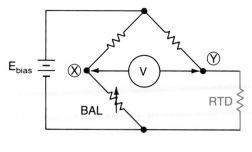

(A) RTD bridge.

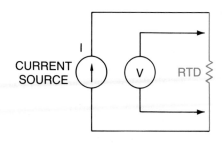

(B) RTD driven by constant-current source.

of the connecting leads does not affect the reading for the change in resistance of the RTD. When the temperature of the RTD changes, its resistance changes, causing an unbalance in the bridge. A voltage is developed across points X and Y by this unbalance. Even for small changes in temperature, the small voltage developed can be amplified, and thereby the temperature change can be determined quite accurately.

Another way of using the RTD is to drive a constant current through it, as shown in Fig. 12-7B. In this way, the voltage change across the RTD is proportional to the change in temperature. Note that the voltmeter leads *do not* carry the RTD current. This is important for accuracy. In either case, notice that some current flows through the RTD; in fact, the larger the amplitude of current flow, the greater the output voltage for a given temperature change. So it might seem that you would like a fairly large current to flow through it. However, a problem results from this current flow. As you know, due to current flow through the RTD (as in any resistor), some heat is developed in the RTD. This heat further increases its temperature, which further increases its resistance, thus causing an error. So a correction factor must be used to account for the self-heating. The correction factor, called T_c, is 0.5 °C/mW in free air.

EXAMPLE 12-3 Suppose you are using an RTD whose resistance is 100 Ω 0°C. A bias current of 10 mA flows through it. If at some higher temperature the voltage measured across the RTD is 1.5 V, what is the correction factor T_c?

SOLUTION

The power dissipation in the RTD is

$$P = V \times I = 1.5 \times 0.01 = 0.015 \text{ W}$$

So the correction factor is

$$T_c = 15 \text{ mW} \times 0.5°C/mW = 7.5°C$$

You should subtract 7.5°C from your measured temperature. The less current that flows through the RTD, the less error there will be. Of course, when used with computerized equipment, the correction factor can easily be accomplished by the software. RTDs are packaged in many of the same kinds of assemblies that thermocouples are housed in, like some of the ones shown in Fig. 12-6.

THERMISTORS

Thermistors are simply resistors with a high *negative* temperature coefficient. That is, unlike the platinum RTD, as the temperature of the thermistor goes up, its resistance goes down.

Thermistors are usually made of semiconductor material, and are in the form of small rods or beads, with two leads, and are used like resistors in other circuits. That is, they can be used as part of a voltage divider, or in a bridge arrangement, and so on.

Having a *high* temperature coefficient means that there is a large change in resistance for a relatively small change in temperature. They are quite inexpensive and easy to use, so they are extensively used in temperature-compensating circuits, such as to regulate the bias of an amplifier stage under changing temperature conditions.

Thermistors are available in a wide variety of nominal resistances, from a few hundred ohms to many thousands of ohms. Manufacturers publish curves showing the resistance variation with temperature. Fig. 12-8 shows a typical characteristic curve of a thermistor with a nominal resistance of 5 K Ω at 0°C.

It's easy to test a thermistor; simply measure its resistance with an ohmmeter. When you hold it in your fingers and warm it, you'll notice the resistance decreases. However, if you are trying to measure its nominal resistance, don't hold it in your hand. Conversely, if you spray it with coolant, its resistance will go up.

12-8 Resistance versus temperature for a typical thermistor.

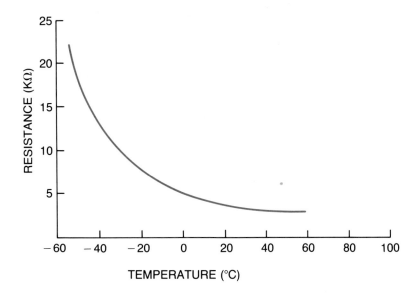

INTEGRATED CIRCUIT (MONOLITHIC) SENSORS

IC temperature sensors are specially manufactured chips which give a linear output voltage or current that is proportional to the change in temperature. Typical values are 1 μA/°K and 10 mV/°K. (A change of 1 degree kelvin is equivalent to a change of 1°C.)

Fig. 12-9 shows two typical circuit arrangements for the IC types. Note that in both cases, the output voltage is 10 mV/°K.

Although IC sensors are only useful at temperatures below +200°C, they are often used in calibration and reference circuits, such as in the cold-junction compensation circuits for thermocouples.

LIGHT SENSORS

Light sensors are used for two primary purposes. One is to measure the intensity of light, and the other is to sense the presence or absence of light for counting or detection purposes. We will now look briefly at two types of sensors that can be used for either application.

12-9 Using IC temperature sensors.

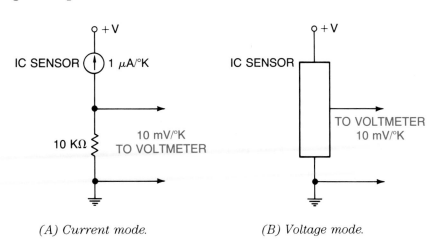

(A) Current mode.   *(B) Voltage mode.*

PHOTOVOLTAIC CELLS

Photovoltaic cells are thin flat sheets covered with a special semiconductor material that generates a voltage when light shines on it. A typical cell of 2.5 × 5 cm in area will produce about 0.4 V at about 180 mA when exposed to direct sunlight. Series-parallel combinations of these cells are used to make up solar batteries, which are used to power satellites.

Although photovoltaic cells are occasionally used to detect the presence or absence of light, their primary purpose is for power generation. So we won't cover them any more at this point.

PHOTOCONDUCTIVE CELLS

A *photoconductive cell* is basically a light-dependent resistor. Made of light-sensitive material, like cadmium sulfide (CdS), the photocell has a high resistance when dark and drops to a very low resistance when light shines on it. Typical values range from less than 50 Ω in light to over 1 MΩ in darkness.

Fig. 12-10A shows a typical package for a CdS cell, which measures about 0.5 inch in diameter. When used to measure light intensity, the cell can be connected into a circuit like the one shown in Fig. 12-10B. With no light shining on the photocell, its resistance is very high. So the voltage at point X is very low, depending on the value of R_A. Then when light shines on the cell, its resistance goes down, increasing the voltage at point X accordingly. The op amp circuit is a typical amplifier whose gain is adjustable by means of potentiometer R_F. By the proper choice of R_A and the amplifier gain, the output voltage can be used to represent the intensity of light.

Another application of the photoconductive cell is shown in Fig. 12-11. Part A of the figure shows a simple counting circuit which is activated by the breaking of a beam of light. Light from the light source (not shown) shines into the photocell, making its resistance low. The photocell is part of a voltage divider along with resistor R_1. Since the cell resistance is low, the voltage at the base of the transistor is low, causing the transistor to be turned off. Then when the beam of light is broken, say by a person walking through the beam, or a part passing by on a conveyer, the resistance of the cell goes high. This causes the voltage at the base of the transistor to go up, turning on the transistor and activating the electromechanical counter.

A more sensitive variation of the scheme is shown in Fig. 12-11B. Assume that light shining into the photocell keeps its resistance low and keeps the voltage at point X lower than the voltage at point Y. Since the voltage at the noninverting input to the comparator is lower than the voltage at its inverting input, the output of the comparator is low. This keeps transistor Q_1 cut off. Then, when the light level shining into the cell decreases sufficiently, the voltage at point X rises above the voltage at point Y. This causes the output of the comparator to switch high, thus turning on the transistor and sounding the sonic alarm.

You will notice that the sensitivity can be adjusted by means of the 10-KΩ pot. That is, by setting the pot to where the voltage at point Y just keeps the transistor turned off, a slight decrease in light level will cause the alarm to sound. This type of circuit can be used to sense the presence of smoke in the air. Another use could be to have the light beam shine through a clear pipe through

12-10   Photoconductive cell.

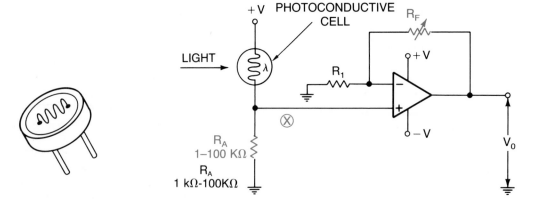

(A) Photoconductive cell package.   (B) V_0 is proportional to light intensity.

12-11 Photoconductive cells used in alarm circuits.

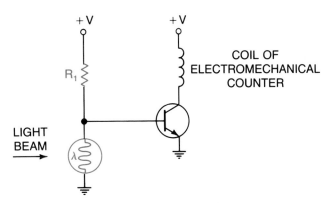

(A) Simple circuit for bright beam.

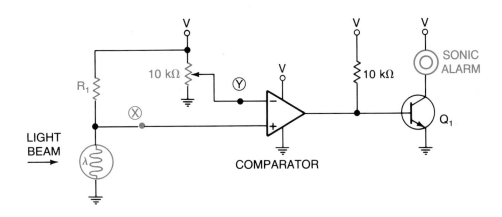

(B) Sensitive circuit to sense decrease
in light intensity.

which water, or some other fluid, flows. An increase in turbidity, or cloudiness of the fluid, would set off the alarm.

Photoconductive cells are simple to test. Simply measure the resistance of the cell with an ohmmeter. As you cover the face of it with your hand (blocking light), its resistance should go up.

If the cell is to be mounted a long distance from the amplifier circuitry, you should use shielded wires to minimize noise pickup. Also, connecting a capacitor of 0.1 μF or so across the cell will reduce noise pickup in the circuit of Fig. 12-11B. Notice that in this type of circuit, the light beam is not going to be broken at any high frequency. You are simply looking for a slow change in light level.

PHOTODIODES

You are already familiar with light-emitting diodes (LEDs). An LED *emits* light when it is forward biased. Photodiodes work in the opposite mode. That is, *when light shines on a photodiode, it becomes forward biased.* It is used to sense the presence or absence of light beams, or the rate at which light beams are broken. Generally, a fairly bright light (several footcandles) is needed to forward bias a photodiode.

Photodiodes are usually about the same physical size as LEDs, and they have windows through which incoming light is sensed. Photodiodes have a very fast response to a change in light, so they are often used in punched card or punched paper tape readers. The setup is shown in Fig. 12-12. Part A of the figure shows how bytes, or binary code patterns, are punched in vertical columns across the tape.

12-12 Using photodiodes to read punched tape.

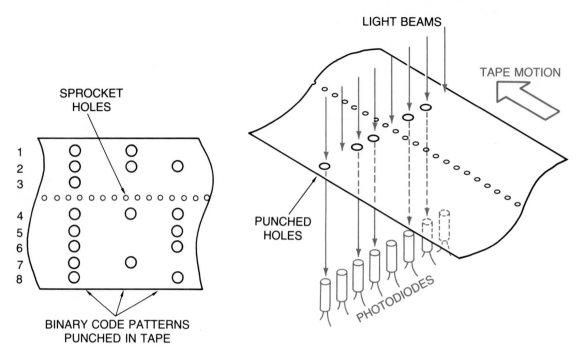

(A) Punched paper tape.

(B) Reading out one byte.

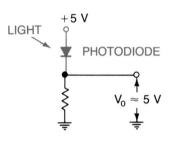

(C) Output voltage is
HIGH for any
forward-biased diode.

The punched tape is made to pass over an array of eight photodiodes, as shown in part B of the figure. Light from one side of the tape passes through holes in the tape to the photodiodes on the other side. Depending on the pattern of holes in the tape, only *some* of the photodiodes will be forward biased. These forward-biased diodes act like closed switches, while the dark (reverse-biased) diodes act like open switches. The circuit for each diode could be like that shown in part C of the figure, so that any diode with light shining on it causes a HIGH output voltage for that particular bit of the pattern. A digital circuit examines the HIGH or LOW voltages for each diode and thus determines the binary bit pattern punched into the tape.

Testing a photodiode is simple. Just connect an ohmmeter across it and if the diode is good you should read a high resistance (reverse-biased diode) when the diode is dark. But when you shine light on the diode, the resistance should drop to a low value (that of a forward-biased diode). Of course, you must observe polarity, just as with any diode.

PHOTOTRANSISTORS

When measuring light beams that are broken at high rates of speed, or when the light levels are low, a *phototransistor,* rather than a photoconductive cell or a photodiode is used. The phototransistor acts much like any silicon bipolar junction transistor, except that instead of base current causing collector current to flow, light shining into its window causes collector current to flow.

Typical ratings for phototransistors are much like low-power transistors, that is, maximum collector-to-emitter voltages of 30 to 50 V and maximum collector currents of 25 mA or so.

Some phototransistors are sensitive mainly to infrared (IR) light, which makes them useful in applications such as burglar detectors and remote controls for TV sets or VCRs. Since they have relatively fast switching speeds, usually less than 10 μs, they are useful for high-speed switching applications.

An application of an IR phototransistor is shown in Fig. 12-13. It monitors the tiny droplets falling through an intravenous (IV) administration set which feeds a saline solution to a patient in a hospital bed. In this set up, a plastic bag filled with a saline solution is hung on a stand near the patient. A plastic tube carries the fluid from the bag to a hypodermic needle stuck into the patient's vein. In series with the plastic tube there is a small clear cylinder through which the droplets fall. The nurse administering the IV sets the drip rate by means of a small control valve below the cylinder.

The circuit shown uses an IR LED which shines through the cylinder into the window of an IR-sensitive phototransistor. As each droplet falls, a spike is generated at the collector of the phototransistor. These spikes are passed to another ordinary silicon transistor, which further amplifies the spikes and sends them to a digital counter circuit which monitors the drip rate. The digital circuit can then sound an alarm if the drip rate falls below some predetermined value.

Although the IR phototransistor is sensitive mostly to IR light, it also is sensitive to visible light of higher intensity. So to test the phototransistor, all you need to do is to connect a resistor, say 1 KΩ or so, from its collector to a source of d-c voltage of 3 to 10 V. The emitter, of course, must be grounded. Then measure the voltage at the collector as you block and unblock light from shining into its window. If the voltage at the collector goes up when the window is blocked, the transistor is probably OK.

12-13 Using a photo transistor in an IV monitor.

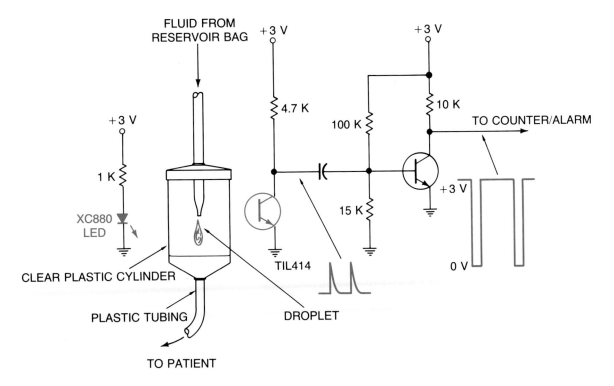

IMPORTANT CHARACTERISTICS FOR REPLACEMENT LIGHT SENSORS

When replacing a defective light sensor, you normally try to get an exact replacement. If you can't, however, then you must consider certain characteristics of the replacement device. The most obvious of these are voltage and current ratings, power dissipation, and size. These will normally be supplied by the manufacturer and can be found in a data book or catalog. There are a few other things you need to consider, so let's discuss them.

The normal human eye is sensitive to a range of light frequencies with wavelengths from about 400 to 700 nanometers (nm). These wavelengths correspond to colors of light including violet (about 400 nm), blue, green, yellow, orange, and red (about 700 nm). The normal eye is most sensitive to (sees best) light in the yellow-green region (about 550 nm). Light with wavelengths of less than about 400 nm is invisible to the human eye and is called *ultraviolet* (UV) light. Similarly, light with wavelengths longer than 700 nm is also invisible to the human eye, and is called *infrared* (IR) light.

As you might expect, light sensors are also color-sensitive. So the color sensitivity, also called *spectral response,* of a device is one of the characteristics a design engineer must consider for each application. Likewise, it is a characteristic that you must consider in a replacement part. For example, one manufacturer lists the peak spectral response of a certain type of CdS photoconductive cell at 550 nm. This closely resembles the human eye response. So this cell would be a good choice for a device that is designed to "see" what the human eye sees.

Sometimes the spectral response is specified in *Angstroms* (Å) rather than nanometers. There are 10 Å in 1 nm. The response of the device mentioned in the previous paragraph would be listed as 5500 A, meaning 5500 Å.

For photoconductive cells, the manufacturer usually gives the resistance of the device at a specified light intensity. For example, the rating might say $R = 9$ kΩ at 2 fc. The *2 fc* means 2 *footcandles.* A footcandle is a unit of measure of light intensity. Basically, a footcandle is the amount of light falling on the area of 1 ft^2 at a radius of 1 ft from a standard candle. The thing to remember is that the replacement should have approximately the same resistance at the same light level. This point can sometimes be stretched a bit if the signal developed by the cell is fed through an amplifier whose gain can be adjusted.

Phototransistors similarly have certain colors of light that they are more sensitive to. In addition to spectral response, rather than specifying a resistance, the manufacturer usually specifies the amount of *collector current* that will flow at a given light level. Both of these factors should be taken into account when you select a replacement device.

PROBLEMS WITH LIGHT-SENSING SYSTEMS

We have already discussed the characteristics of two commonly used sensors, how to test them to see if they work, and what to look for in replacement parts. There are, however, a few other points worth keeping in mind when troubleshooting light sensing systems.

Burned Out, Weak, or Obstructed Light Source

When working with a system having a beam of light that shines into a photocell, be sure to check the light source. If the source or receiver has lenses to concentrate the beam, be sure that they are clean and dust free. Look for obstructions in the light path. Lastly, if the system uses an IR light source, you may not be able to see the light with the naked eye. You will need to check it with a device sensitive to the band of light it emits. Sometimes the light source and sensor have IR filters over them. You may be able to see the light by removing the filter. Be sure to keep the filters clean of dust, dirt, and oil.

Shielding

Many light sensors use hoods or other shields to keep out unwanted light. Check to see if the shield may have been removed or bumped, so that it does not point in the right direction. Also, be on the lookout for extraneous light sources, like a new lamp that someone just installed that happens to shine into the sensor.

MECHANICAL SENSORS

Mechanical sensors are used to measure force, motion, and position. We will now take a look at a few commonly used types.

STRAIN GAGES

Strain gages are sensors used to measure a change in length or some other deformation of a solid piece of material. The material may be a piece of sheet metal covering a section of aircraft body, or it may be a structural member of a building. By knowing how much (by what percentage) a piece of metal is deformed when a certain load is applied, engineers can determine the safety limits of the structure. Strain gages are also used to build *load cells,* which are devices used to measure forces and weights from just a few pounds, up to and including the heavy weight of a tractor-trailer rig.

The term *strain,* when applied to mechanics, refers to the change in length per unit length of a solid. Although we normally think that solid metal bars don't stretch much, all metals are somewhat *elastic.* That is, a piece of metal *does* stretch when a force is applied, but it returns to its original dimensions when the force is removed, as long as the force has not exceeded the *elastic limit* of the metal. Fig. 12-14A shows a bar of metal with no forces applied to it. Call its original length L_0. When a tension force is applied, the bar stretches to a new length $L_0 + \Delta L$. Strain, then, is defined as

$$\text{Strain} = \frac{\Delta L}{L_0} \qquad\qquad 12\text{-}3$$

where ΔL = change in length
L_0 = original length

Strain on a solid can be caused by a compression force, as well as by tension, and is measured in the same way, but of course the new length will be less than the original length if the force is compressional.

It is obvious that a bar with a large cross-sectional area would not stretch as much as a bar with a smaller cross-sectional area for the same applied force. Therefore we work with the applied *force per unit area,* called *stress.* That is,

$$\text{Stress} = \frac{F}{A}$$

where F = total applied force
A = cross-sectional area

Now it turns out that the *ratio* of stress to strain is a constant value for a given material, such as steel or aluminum, regardless of the dimensions of the material. This ratio of stress to strain, called *Young's modulus,* has been tabulated for many different materials and can be found in engineering reference books. So by making tests on a given structure of known material, engineers can predict safe loading limits on other structures made with the same material, even if the dimensions are different. Also, by continuously monitoring the strain in a given piece of material,

12-14   Applying a tensile force to a bar of metal stretches it slightly.

(A) Unstretched bar.

(B) Stretched bar.

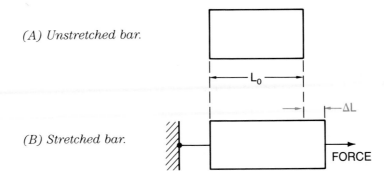

say in the tail section of an airplane, they can observe when the piece is approaching its elastic limit. If the metal is stretched beyond its elastic limit, it may deform or even fracture.

We cannot get into calculations involving Young's modulus in this text, but we can discuss how to measure the strain. Tests show that most metals won't stretch more than about 0.5% of their original length before permanent deformation occurs. This corresponds to a strain of 0.005 inch/inch. Since the change in length is so small, we obviously can't measure strain with a meter stick. In other words, if we want to measure the strain in a metal bar, we need some device that will give us an accurate measurement even when the change in length is very small. The strain gage is such a device.

You may recall from your early studies of electricity that the resistance of a conductor (say a piece of wire) can be determined by the equation $R = \rho L/A$. In this equation, R is the resistance in ohms, ρ (rho) is the resistivity of the material, L is the length of the conductor, and A is the cross-sectional area. Now suppose we have a piece of wire whose resistance is 100 Ω. Next we cement the wire to a bar of metal which is anchored to the ceiling, as shown in Fig. 12-15. (The wire is electrically insulated from the metal bar.) When a force is applied to the metal, say by hanging a weight to it, the metal stretches and so does the wire. If the metal and the wire stretch, say, 0.4% longer than the original length, the resistance of the wire will increase by about 0.4% due to the increase in length. So by measuring the change in resistance, we can determine the change in length of the metal. That is basically how a strain gage is used.

Rather than use a single strand of wire for the strain gage, commercially available gages are made of metal or semiconductor foil woven back and forth to increase the length. The gage is then bonded to a plastic-like base material for easy handling and mounting. Fig. 12-16 shows several different gages. Some assemblies have two, three, or four gages in one package. Notice that the patterns of some of the multiple gages point in different directions so that the strain in those particular directions can be monitored.

Commercially available strain gages are sold in resistance values from 30 to 3000 Ω. But the two most commonly used values are 120 and 350 Ω.

Due to changes in resistivity as well as in length when a gage is stressed, commercially available strain gages actually exhibit a greater percentage change in resistance than the change in length. This property is called the *gage factor* (GF). For example, if a 1% change in length causes a 2% change in resistance, the gage is said to have a gage factor of 2.

Equation 12-4 shows how to find the resistance of a strain gage if its original resistance, change in length, and gage factor are known.

$$R = R_0 \left(1 + \frac{\Delta L}{L_0} \times GF\right) \qquad \qquad 12\text{-}4$$

12-15 Simplified strain gage.

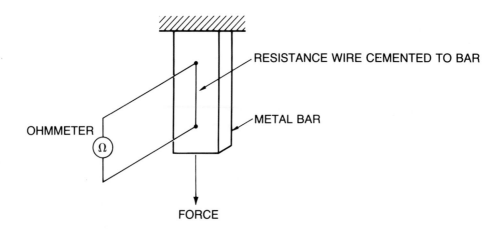

12-16 **Assortment of strain gages. Courtesy of Omega Engineering, an Omega Technologies Company.**

where R = resistance of gage under stress
R_0 = original resistance of gage
ΔL = change in length
L_0 = original length
GF = gage factor

EXAMPLE 12-4 Suppose a 350-Ω strain gage with a gage factor of 2 is mounted to a metal bar originally 1 m long. The bar is then stretched 3 mm. What is the new resistance of the gage?

SOLUTION

$$R = R_0\left(1 + \frac{\Delta L}{L_0} \times GF\right) = 350\left(\frac{1 + 0.003}{1} \times 2\right) = 352.1\ \Omega$$

Since we can easily measure the change in resistance, we usually want to know the change in length.

EXAMPLE 12-5 A 350-Ω strain gage with a gage factor of 4 is mounted to a metal bar 0.5 m long. The bar is then stretched, and the resistance of the gage is measured to be 351.4 Ω. What is the change in length of the bar?

SOLUTION Rearranging Eq. 12-4 to solve for the change in length, we have

$$\Delta L = \frac{L_0(R - R_0)}{R_0 \times GF} = \frac{0.5(351.4 - 350)}{350 \times 4} = 0.0005\ \text{m or } 0.5\ \text{mm}$$

Strain gages are normally used in either of two circuit configurations. They are either mounted in a bridge circuit, or they are driven by a constant-current source. Since they have resistance values similar to those of RTDs, they often use much of the same circuitry. In addition, often the same computer interface module that is used for RTDs can also be used for strain gages.

The circuits of Fig. 12-17 can be used to monitor strain gages. But Fig. 12-17 shows some variations in bridge circuits that you might encounter. Fig. 12-17A is the familiar bridge circuit using one gage. Although the bridge balancing components are not shown, normally the bridge would be balanced so that the voltage measured on the voltmeter is zero when there is no stress on the gage. This, of course, occurs when the following relationship is true: $R_1/R_2 = R_g/R_3$, where R_g is the resistance of the gage.

Any change in R_g unbalances the bridge and causes a voltage to be measured on the voltmeter. So the voltage read on the voltmeter is proportional to how much the gage (and the metal to which it is bonded) has stretched.

As indicated in the figure, when the gage is mounted some distance from the rest of the bridge components, it is good practice to run a separate lead from the voltmeter to the junction of R_g and R_3 and connect it as close to R_g as possible. This arrangement, sometimes called a three-wire bridge, is used to prevent current flow through the connecting leads from affecting the voltage read by the

12-17 Various bridge circuit arrangements.

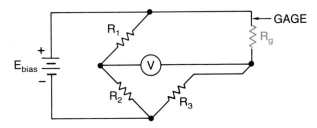

(A) Quarter bridge (three-wire bridge).

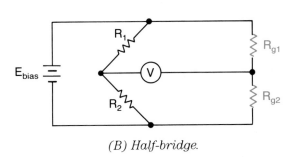

(B) Half-bridge.

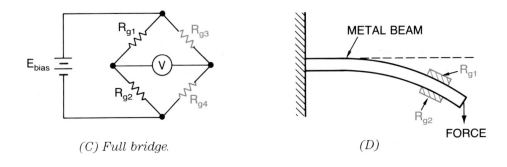

(C) Full bridge.   *(D)*

meter. The lead connected to the top of R_g must have a very low resistance. It may possibly be the system ground connection.

Two other methods of using strain gages in a bridge arrangement are shown in Fig. 12-17B and C. In part B of the figure, *two* gages are used. In this arrangement, usually one of the gages is connected in the tension mode while the other is connected in the compression mode, perhaps in a setup like the one shown in part D of the figure. Here we have a beam which is being bent by exerting a force at the end. Gage R_{g1} is stretched when the beam is bent, causing its resistance to increase, while R_{g2} is compressed, causing its resistance to decrease. The net effect is that the voltage measured by the voltmeter is *twice* as great as it would be for one sensor. Similarly, the circuit of Fig. 12-17C gives an output signal four times as great as the circuit of part A, because R_{g1} and R_{g4} are wired so as to *increase* in resistance while R_{g2} and R_{g3} are wired so as to *decrease* in resistance when the beam is bent.

Another advantage of using multiple sensors is that it minimizes the effects of variations in gage resistance caused by temperature variations. For example, suppose that a single strain gage were cemeted to a beam to measure strain in a circuit like that of Fig. 12-17A. The bridge would then be balanced with no strain on the beam. If the temperature of the beam changed, the resistance of the gage would change. You would not know whether the voltage measured across the bridge was caused by stress or by temperature variation. However, if two gages are used as in part B, both gages will be at the same temperature at all times. So even if the temperature should change, there will be no corresponding change in the ratio of R_{g1} to R_{g2} and, therefore, no bridge signal caused by temperature variation. Similarly, a four-gage circuit is more temperature sensitive than a one-gage circuit.

Although all of the circuits show a voltmeter across the bridge, the signal amplitude will, of course, be very small, usually in the order of millivolts. So some form of high-quality, high-impedance amplifiers are usually connected across the bridge, such as the op amp bridge amplifiers you studied earlier.

PROBLEMS WITH STRAIN GAGE CIRCUITS

You can easily test a strain gage by simply measuring it with an ohmmeter. It should read the nominal value (say 120 Ω) with no stress applied to it. However, don't try to stretch the gage with your fingers to see its resistance change. As mentioned before, the usual change in length is very small, so the change in resistance is also small and probably not noticeable on an ohmmeter. You can damage the gage by stretching it too much.

Besides the obvious problems of opens or shorts occurring in the gages themselves or in the connecting leads, there are a few other things to watch out for.

Temperature Change

As mentioned before, if your installation has the gage mounted where the temperature of the gage may vary, particularly in an outdoor environment, be sure that some method of temperature compensation is used.

Bonding

Manufacturers sell a special bonding cement to be used with their gages. This cement should be used to ensure that the gage undergoes the same strain that the test piece undergoes. Otherwise you'll get erroneous readings. Also watch out for loose bonds in an older installation. These will give false measurements.

Noise

Noise pickup might be a problem, especially when the connecting leads are long or pass near electrically noisy machinery. If so, use shielded cables and connect filter capacitors across the bridge circuit at the point of measurement.

Unregulated Bias Supply

The bias supply voltage shown in the circuits of Fig. 12-17 provides current flow through the bridge; the higher the supply voltage, the greater the output voltage for a given change in gage resistance. It is obvious that the supply voltage must be the correct voltage specified by the equipment manufacturer and must be regulated and well filtered. Any variation in the supply voltage will give errors in the readings.

Bridge Loading

The amplifier connected across the bridge must have a very high input impedance. Loading across the bridge will change the output voltage and give errors.

CURRENT LOOP TRANSDUCERS

The term *current loop* refers not to a *type of transducer* but to the *method of using* a transducer. For example, we have discussed how strain gages are used in bridge circuits. The voltage across the bridge is proportional to the stress or load. The *output voltage*, then, is what we use to determine stress. However, the same strain gage bridge can be used in a current loop mode, in which the *output current* is what we use.

Fig. 12-18 shows the principle involved here. In part A of the figure we see a simple inverting op amp circuit. You will recall that V_{id} is approximately zero volts. Therefore $I_1 = V_{in}/R_1$. Since no current flows into the op amp inverting terminal, the feedback current I_F is equal to I_1. Keep in mind that the two currents are equal regardless of the value of resistor R_F (up to the point where the op amp goes into saturation).

In part B of the figure, we see how this principle is used to make a current-mode device. The input signal from the transducer (or transducer bridge) is fed to the inverting input of the op amp, causing current I_1 to flow. The amplitude of this current depends on the value of voltage V_{in}. For example, if V_{in} is 1 V, I_1 will be 1 mA, and so on. As a result, current I_F flows toward resistor R_x

12-18   Principle of current-mode operation.

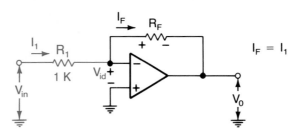

(A) Simple inverting amplifier.

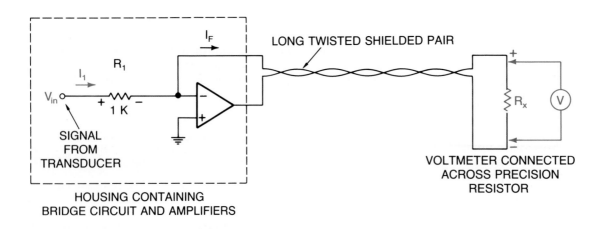

(B) Using current-mode transducer.

through the twisted-shielded pair of wires. When current I_F flows through the precision resistor R_x, the voltage developed across R_x will be proportional to the value of the current and, hence, to the transducer voltage V_{in}.

The advantage of a current loop system is this: The bridge circuit and op amp are mounted in a rugged metal housing very close to the physical test point. Then relatively low impedance wires in the form of a twisted pair are routed, perhaps hundreds of feet, to where the actual voltmeter is located. The transducer can be mounted in a hot, dusty, or oily environment, and yet the voltmeter (or other computer-based equipment) can be mounted far away in a clean, cool location. Likewise, many transducers can be scattered around a chemical plant or steel mill and all of the leads routed to a single clean room where the precision electronic equipment is located. This allows the monitoring of several processes at a single control station.

Since the current I_F depends only on I_1 and not on the actual value of resistance in the feedback lead, it doesn't make any difference how long the wires are or what the resistance value is (within limits, of course).

Many heavy-duty industrial transducers are made to operate in the current loop mode. For example, you can buy pressure gages to measure air or fluid pressure, and flowmeters to measure the rate of flow of fluids, which use a current loop transducer. The transducer for the pressure gage, incidentally, is often the very same strain gage bridge we studied earlier. Pressure is a measure of force/area. So, since the strain gage bridge can measure force (or weight), all we have to do to make a pressure gage is to know the area to which the measured force is applied, and calibrate the output voltage to read pounds per square inch (psi) rather than simply pounds.

The current loop transducer is used so often, in fact, that standard values of output current are used; hence, standard readout equipment can be used with a variety of gages. The most common range of currents used is 4 to 20 mA. In the case of a pressure transducer, 4 mA represents a zero reading (0 psi) and 20 mA represents full-scale reading. The full-scale readings can be anywhere from 5 to 5000 psi or more, depending on the application. But the full-scale value is always represented by 20 mA.

You might wonder why 4 mA rather than 0 mA is used to represent a zero pressure reading. The advantage of using a nonzero value is for troubleshooting. Suppose you used 0 mA to represent 0 psi. If the wires connecting the op amp output to the test resistor R_x in Fig. 12-18 should open, the current through R_x would fall to zero. You would not know if the zero reading resulted from zero pressure or from a broken wire. But if 4 mA represents a zero reading, a broken wire can easily be identified.

> **EXAMPLE 12-6** A 4–20 mA current loop is used to measure air pressure from 0 to 50 psi. What value of air pressure does a 10-mA current represent?
>
> **SOLUTION** The current span of $20 - 4 = 16$ mA represents a pressure span of $50 - 0 = 50$ psi. Therefore, each milliamp above 4 mA represents $50/16 = 3.125$ psi. Our reading of 10 mA is 6 mA above the zero pressure reading. So the pressure must be
>
> $$P = 6 \times 3.125 = 18.75 \text{ psi}$$

Figure 12-19 shows a simplified diagram of how a 4–20 mA loop can be used to indicate the output of a pressure transducer bridge. The signal from the transducer bridge is fed to the usual differential amplifier A_1. This signal is fed to the inverting input of amplifier A_2. Notice that the noninverting input of amplifier A_2 is held at a constant -4 V by zener diode D_1.

Let's assume that the gain of the differential amp is adjusted so that 0 psi on the pressure transducer gives an output voltage from A_1 of 0 V, and 100 psi gives an output voltage of $+16$ V. When the pressure is 0 psi, the -4-V level at the input of A_2 causes a 4-V drop across resistor R_1. Thus the current through R_1, and hence through the loop, is 4 mA (the zero reference level). Then, as the pressure increases, the output voltage from amp A_1 increases. When the output from A_1 reaches $+16$ V, the total voltage across R_1 will be 20 V. The current through R_1, and of course through the loop, will be 20 mA, indicating a full-scale reading.

A typical industrial diagram for a 4–20 mA transducer hookup is shown in Fig. 12-20. The external bias power supply actually supplies d-c power to the bridge and op amps through the

12-19 Generating 4–20 mA from bridge signal.

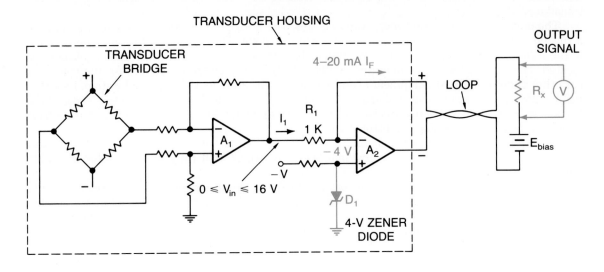

12-20 Hookup for 4–20 mA loop.

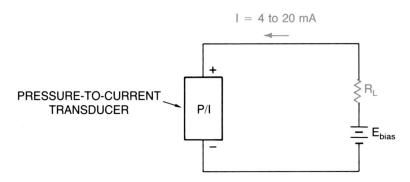

same leads that carry the 4–20 mA signal current. The manufacturer specifies the range of voltages for the power supply, as well as the range of total load resistance R_L (including lead resistance). A typical device might be rated at 24–30 V and 800–1200 Ω. In other words, the supply voltage is not critical, nor is the value of load or lead resistance critical. In this example, only two wires need to be connected to the device. Sometimes an additional ground lead is needed.

Although we discussed only a pressure transducer, other transducers are available in the 4–20 mA loop packages. Another common one is a *flowmeter,* used to measure the rate of flow of fluids through pipes. A common type uses a paddle wheel turned by the moving fluid. The rate at which the wheel revolves is proportional to the rate of fluid flow. An internal circuit converts the pulses generated by the paddle wheel into a current. As you might expect, 4 mA represents no flow, and 20 mA represents maximum flow rate for a given device, say a flow of 10 ft/min.

Although Figs. 12-18 and 12-19 show a voltmeter being used to indicate the signal amplitude, standardized readout devices are available to convert the 4–20 mA into appropriate physical units. That is, the readouts display pressure in psi or flow rate in ft/min, and so on. The 4 mA representing a zero reading is automatically nulled out at the readout, so 0 psi is read as zero, and so on. The same readout device can be used to measure and display several different quantities at different full-scale values. The actual circuitry still only has to measure current from 4 to 20 mA. Usually, there is some form of alarm if the current drops to zero.

PROBLEMS WITH CURRENT LOOP DEVICES

Current loop transducers are quite rugged, being housed in heavy metal enclosures. And, as already mentioned, they have built-in detection for broken leads. One thing that might cause problems is loss of, or out-of-range, bias supply voltage. If the voltage gets too high, it could damage the transducer or circuitry. If it drops too low, the calibration will be off. Accuracy of the device will also be off if the total resistance in series with the device gets higher than the manufacturer's recommended value. This could result from corrosion of connecting leads or from careless routing of the leads.

PROBLEMS

12-1. Of the four types of temperature sensors, which would be the best choice for each of the following applications? (a)Measuring temperatures above $+1000°C$. (b)Very accurate, stable measurements in the region around $+300°C$. (c)Inexpensive sensor for temperature compensation of an amplifier, where accuracy is not critical.

12-2. What is the voltage generated by a type J thermocouple at a temperature of 150°C, assuming that the reference junction is at 0°C?

$V =$

12-3. You find a thermocouple, but you are not sure whether it is a type J or a type K. So you place a reference junction at 0°C, and you put the other junction in boiling water. Then you measure the terminal voltage of the thermocouple to be 4 mV. What type is it?

Type $=$

12-4. Suppose you are using an ice-bath reference as shown in Fig. 12-3. If after some period of time all of the ice melted and the water warmed up several degrees, would the thermocouple voltage be higher or lower than normal for a given J_1 temperature?

12-5. You have a thermocouple connected through an electronic ice-point reference to a voltmeter. The reading on the voltmeter seems to be much different from the expected reading, so you suspect a problem. You don't know whether the problem is in the thermocouple, ice-point reference, or voltmeter. So you take a screwdriver and short the two input terminals of the ice-point reference together. The reading on the voltmeter reads a value corresponding to room temperature. Where is the problem?

12-6. A certain control system needs a voltage of $+5$ V to represent a temperature of $+200°C$ and 0 V to represent 0°C. If the sensor is a type J thermocouple, how much amplifier gain is needed?

$A_v =$

12-7. Using a platinum RTD with a resistance of 100 Ω at 0°C, what will be its resistance at 55°C?

$R =$

12-8. Suppose that the current source in Fig. 12-7 drives 1 mA through the platinum RTD, whose nominal resistance is 100 Ω. If the voltage read on the meter is 135 mV, what is the temperature of the RTD? (Ignore self-heating.)

$T =$

12-9. Recalculate the temperature of the RTD in problem 12-8, but this time take into account the self-heating caused by the bias current. Use a correction factor of 0.5°C/mW.

$T =$

12-10. RTDs have a (negative, positive) temperature coefficient, while thermistors have a (negative, positive) temperature coefficient.

12-11. You have a thermistor whose characteristics are shown in Fig. 12-8. If you started measuring its resistance at 0°C, at what temperature would its resistance double?

$T =$

12-12. Is it a good idea to measure the resistance of a thermistor while holding it in your hand? Why?

12-13. In the circuit of Fig. 12-10B, suppose that R_A = 1 KΩ, R_1 = 1 KΩ, R_F = 5 kΩ, $+V$ = 12 V, and the resistance of the photocell is 100 KΩ. What is the value of the output voltage V_0?

$V_0 =$

12-14. Refer to problem 12-13. Now assume that more light shines on the photocell so that its resistance drops to 10 KΩ. What is the value of the output voltage V_0?

$V_0 =$

12-15. In the circuit of Fig. 12-11B, R_1 = 10 KΩ, V = +10 V, and the slider of the pot is set so that the voltage at point Y = +4 V. (a)What voltage at point X will cause the alarm to sound? (b)What is the minimum value of resistance of the photocell that will cause the alarm to sound?

$V =$

$R =$

12-16. Suppose that the photocell in the circuit of Fig. 12-11, which operates from an infrared beam, becomes damaged and has to be replaced. The specs on the original unit where 1 KΩ at 2 fc, 1 MΩ dark. You can't get an exact replacement, but you have a choice of three possible cells. Here are the specs: (a)Cell 1: 0.6 KΩ at 2 fc, 1.5 MΩ dark, and peak response at 5150 Å. (b)Cell 2: 100 KΩ at 2 fc, 20 MΩ dark, and peak spectral response at 700 nm. (c)Cell 3: 0.8 KΩ at 2 fc, 2 MΩ dark, and peak response at 7150 Å. Which of the three cells would be the best replacement?

12-17.   What type of cell makes a better unit to read a high-speed pulse-coded beam, such as those used to read the Universal Product Codes on boxes in a supermarket? (a)a photoconductive cell. (b)a phototransistor.

12-18.   The circuit of Fig. 12-11B is used to check the clarity of fluid being pumped through a plastic pipe. A beam of light shines through the fluid and into the photocell. Which of the following conditions might cause a false alarm? (a)R_1 becomes open. (b)R_1 becomes shorted. (c)Photocell lens becomes dirty. (d)E_{bias} decreases in value by a few tenths of a volt.

12-19.   A rod is made of a certain type of metal which is listed as having a maximum strain of 0.004 inch/inch. How far can the rod be safely stretched if the original length is 20 cm? Give your answer in millimeters.

$L =$

12-20.   Assume that you are using a strain gage whose nominal resistance is 120 Ω. The gage factor is listed as 4. The gage is bonded to a metal bar that is stretched 0.2% of its original length. What is the new resistance of the gage?

$R =$

12-21.   A strain gage with a nominal resistance of 120 Ω is used in a quarter bridge circuit like that shown in Fig. 12-17A. R_1 and R_2 are both 1-KΩ resistors, and R_3 = 120 Ω. The bias supply voltage is 10.0 V. With no strain on the gage, what is the voltage read on the voltmeter?

$V =$

12-22.   Referring to problem 12-21, suppose that the gage is stretched so that its resistance is increased to 121 Ω. What is the voltage read on the voltmeter?

$V =$

12-23.   Suppose the full bridge circuit (using four gages) is used to measure the bending of a beam. As the beam bends, R_{g1} and R_{g4} increase in value from 120 to 121 Ω, while R_{g2} and R_{g3} decrease in value from 120 to 119 Ω. With a bias voltage of 10.0 V, what is the voltage measured across the bridge?

$V =$

12-24.   In the circuit described in problem 12-22, if the temperature of the gage increases, will the voltmeter reading change? (Yes, no)

What about the circuit described in problem 12-23, will the voltmeter reading change if the temperature changes? (Yes, no)

12-25.   In the circuit of problem 12-22, suppose that the bias supply voltage decreased in value to 9.0 V. What would happen to the voltmeter reading across the bridge? It would (a)go up. (b)go down. (c)remain the same.

12-26. In the circuit of Fig. 12-18B, if V_{in} = 6.7 V, what is the value of current through resistor R_x?

I =

12-27. Assume that the circuit of the previous problem is used to measure air pressure. The readout is calibrated so that 4 mA represents 0 psi and 20 mA represents 100 psi. What pressure will be indicated when V_{in} = 6.7 V?

P =

12-28. Referring to the previous two problems, suppose that the voltmeter reads 0 V. Which of the following is the most likely cause? (a)Air pressure has fallen to below zero psi. (b)Air pressure is above 100 psi. (c)There is a break in the connecting leads.

EXPERIMENT 12-1   TEMPERATURE MEASUREMENT WITH THERMOCOUPLES

Thermocouples are used to measure temperatures from near absolute zero to above +2000°C. In this experiment you will construct and calibrate a circuit which will measure temperatures well below the freezing point to well above the boiling point of water. You can use any common type of thermocouple, that is, iron-constantan, chromel-alumel, and others. These thermocouples are available in a variety of assemblies, including simple welded wires, insulated types, and those connected to washers for measuring temperatures under metal bolts of engine blocks, and so on. It doesn't matter which type you use, since you will calibrate it in this experiment. Thermocouples can often be found in surplus houses, but new ones can be purchased from several different supply houses, including Omega Engineering of Stamford, Connecticut.

EQUIPMENT
- +15- or +12-VDC dual-power supplies
- voltmeter (preferably one with a low-voltage scale, such as 0–1.5 V)
- (1) LF353 dual BiFET op amp
- (2) thermocouples (iron-constantan or any)
- (2) 100-KΩ, ½-W resistors
- (1) 10-KΩ, ½-W resistor
- (2) 1-KΩ, ½-W resistors
- (2) 50-KΩ pots
- ice bath (container with ice and water)
- heat source (container with boiling water)

PROCEDURE

1. Build the circuit of Fig. E12-1. Note that the thermocouples shown are iron-constantan types. The constantan wire of J_1 is connected to the constantan lead of J_2. Any available types of thermocouples can be used, but they must both be of the same type, and you must connect two leads of the same material together. These leads may be twisted together or lightly tacked together with solder.

Keep your lead lengths short to minimize noise pickup. If you are working in an electrically noisy environment, you may have to connect a capacitor (5–10 µF) from the left end of resistor R_1 to ground, to filter out noise pickup. The best way to check for noise pickup is to observe the amplifier output with a scope after applying power.

E12-1   Circuit for Experiment 12-1.

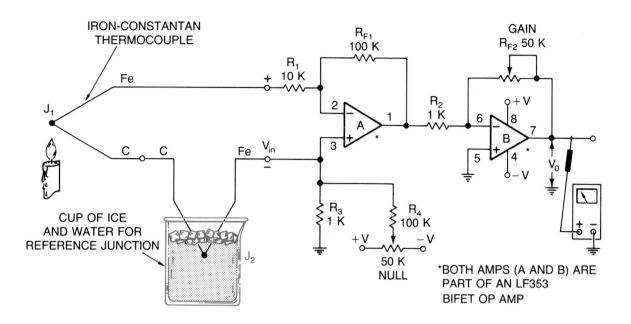

2. Put both thermocouples in the ice-water bath. Now apply power to the circuit. At this point, the voltage generated by the thermocouples should be zero, so the output voltage V_0 should be zero.

Connect the voltmeter to the output of amplifier B, as shown in the figure. Adjust the gain control to near maximum; then adjust the null control back and forth a few times. You should be able to adjust the output from some slightly negative voltage to a slightly positive voltage. (You will recall that the null control compensates for any output offset of the op amps when the input signal voltage is zero). Adjust the null control until the output voltage is exactly zero. Record V_0.

$V_0 =$

> **NOTE**   If you cannot null the output to zero volts, check the circuit wiring. If your null control works OK, go on to the next step.

3. Take junction J_1 out of the ice bath. Immerse J_1 into a boiling pot of water while keeping the reference junction J_2 in the ice reference bath. The boiling water is at 100°C.
4. After a few minutes, adjust the gain control of amplifier B so that V_0 reads exactly 1.00 VDC. You have now calibrated the instrument to read temperature in degrees Celsius. A temperature of 0°C reads 0 V, a temperature of 100°C reads 1.00 V, so a temperature of 50°C will read 0.50 V, and so on.
5. Remove thermocouple J_1 from the boiling water and allow it to cool to room temperature (several minutes). Read the voltage on the meter and record your room temperature.

$T =$

6. If possible, measure and record the temperatures of several other objects or devices. Examples could include the inside temperature of a refrigerator, the surface temperature of a power transistor in a power supply, and the temperature of the tip of a soldering iron. Remember to wait a few minutes after attaching the thermocouple to any device before you record the temperature so that the temperature of the thermocouple stabilizes.

1. What type of thermocouples did you use? (Specify metals.)

2. According to Table 12-1, what is the temperature coefficient of your thermocouples?

3. Using the temperature coefficient, what voltage should your thermocouple generate at +100°C? (This is the voltage V_{in} that appears at the input to amplifier A.)

4. What is the gain of amplifier A?

5. Over what range of values can the gain of amplifier B be adjusted?

6. Since your output voltage was +1.00 V at a temperature of +100°C, what was the gain of the overall amplifier (including A and B)?

7. In order to get the overall gain calculated in question 6, what was the gain setting of amplifier B?

8. If you wanted the output voltage to read +10.00 V at +100°C, what should be the gain of amplifier B?

9. Amplifier A has a fairly high input impedance (10 KΩ). What would be the disadvantage of using an amplifier with a very low input impedance (say 25 Ω or less)?

10. Can you think of any problems that might occur if the leads from the thermocouples were very long (say 1000 ft)?

EXPERIMENT 12-2   THERMISTOR BRIDGE

The circuit of Fig. E12-2 shows a typical application of a difference amplifier monitoring changes in the resistance of a transducer. In this particular application, the circuit represents a temperature control for an oven. Amplifier A_4 drives the heater control. When the LED is lit, it represents power being applied to a heater coil inside the oven. The transducer in this case is a thermistor mounted inside the oven to sense the oven temperature.

> **NOTE**   When you are finished with this experiment, do not disassemble the circuit. It will be used again for Experiment 12-3.

EQUIPMENT
-   +15- or +12-VDC dual-power supplies
-   voltmeter
-   (4) 741 or LF351 op amps OR (2) LF353 dual op amps
-   (1) LED (any color)
-   (1) thermistor, preferably with approximately 10 KΩ nominal resistance at room temperature
-   (1) 10-KΩ pot
-   (2) 100-KΩ + 1% resistors
-   (2) 10-KΩ + 1% resistors
-   (1) 100-KΩ + 5% resistor
-   (2) 10-KΩ + 5% resistors
-   (1) 5.1-KΩ + 5% resistor
-   (1) 1.5-KΩ + 5% resistor

E12-2 Circuit for Experiments 12-2 and 12-3.

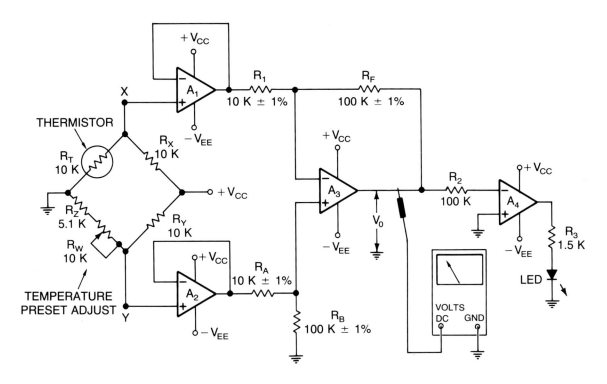

Notes: All op amps 741 or LF 351.

All resistors $\frac{1}{2}$W $\pm$ 5% unless otherwise noted.

V_{CC} = 12 to 15 VDC

1. Build the circuit of Fig. E12-2. Apply power and adjust pot R_W back and forth over its entire range a few times while watching the difference amp (A_3) output with a voltmeter. If your circuit is working properly, the amp's output should swing from some positive voltage to some negative voltage as the pot is varied. Explain what happens to the LED as the pot is varied.

Why?

2. Shut off the power supplies and disconnect the end of R_W from point Y. Then, using an ohmmeter, adjust R_W so that $R_W + R_Z = 8$ KΩ. Reconnect R_W to point Y again and apply power. Measure V_0.

$V_0 =$

What is the condition of the LED (on or off)?

3. Now, using a soldering iron, slowly heat the thermistor for a couple of minutes while watching the output of A_3. (Do not melt the thermistor.) What happens to V_0 and the LED?

4. Remove the heat source and watch V_0 and the LED again. What happens to V_0 as the thermistor cools? What happens to the LED?

SUGGESTIONS FOR TROUBLESHOOTING

Have your lab partner or instructor open (disconnect) or short any resistor in the circuit. Then try to locate the faulty component by measurements with a voltmeter.

> **CAUTION** Do not short either power supply directly to any pin on the chips. This may damage the IC.

QUIZ

1. When balance pot R_W was adjusted back and forth, the output of A_3 swung from positive to negative. When the output of A_3 was at 0 V, the bridge was balanced. That means that (a)point Y was at 0 V with respect to ground. (b)point X was at 0 V with respect to ground. (c)point Y was at the same voltage as point X, so there was no difference voltage applied to A_1 and A_2.

2. Amplifiers A_1 and A_2 are used (a)as high-gain inverting amplifiers. (b)as noninverting amplifiers with high voltage gain. (c)as voltage followers with very high input impedance to prevent loading on the bridge.

3. The voltage gain of difference amplifier A_3 is (a)1. (b)10. (c)100.

4. One reason for using a difference amplifier is to (a)get higher voltage gain than can be obtained from a single-ended amp. (b)eliminate the effect of the common-mode voltage across the bridge.

5. In step 2 of the experiment, when the resistance of $R_W + R_Z$ was adjusted to 8 kΩ, that was (more, less) than the nominal resistance of the thermistor. This caused point X to be (positive, negative) with respect to point Y. The voltage *between* points X and Y is called the (common-mode, difference) signal.

6. Refer to question 5. Amplifier A_3's output was driven (positive, negative) with respect to ground.

7. Amplifier A_4 is being used as (a)a zero crossing detector. (b)a noninverting amp. (c)an impedance matching device.

8. Pot R_W is used to balance the bridge and to preset the desired oven temperature. When the oven temperature is below the preset temperature, the resistance of the thermistor is (greater than, less than) the resistance of $R_W + R_Z$.

9. Refer to question 8. The LED is (on, off), indicating that power (is, is not) being applied to the heater.

10. As the temperature inside the oven rises slightly above the preset value, the resistance of the thermistor becomes (greater, less) than the resistance of $R_W + R_Z$. This causes A_4's output to switch (positive, negative), thus turning the heat (on, off).

EXPERIMENT 12-3   PHOTOCONDUCTIVE DETECTOR

This experiment will not be as formal as previous ones. You will use your imagination to come up with an interesting application of the photoconductive cell.

Using the same circuit that you used for Experiment 12-2, replace the thermistor with a photoconductive cell. The bridge amplifier will detect changes in light intensity shining on the cell and give an output voltage proportional to the change.

You can probably use almost any available photoconductive cell, such as a VT-801 from VACTEC, Inc. The VT-801 is listed as having a dark resistance of 250 KΩ, and a resistance of about 3 KΩ with 2 footcandles of light on it. The thing to keep in mind when setting up your circuit is that with *some* value of light shining on the cell, its resistance will be somewhere between the two extremes.

So in order to balance the bridge (if you want to), the resistance of $R_W + R_Z$ should equal the resistance of the sensor. This means that you may have to select a value for the balance pot R_W which more closely approximates the cell resistance.

Using this type of circuit, you can measure light intensity by interpreting the voltage at the output of A_3 as a measure of intensity.

Another common application for this type of circuit is as an object (or person) detector. By shining a light from some source, say a flashlight located across the room, into the cell, the output of A_4 will be normally low and the LED will be off. Then if something or someone passes through the beam, the output of A_4 switches high and the LED lights, possibly indicating an alarm or simply generating a *count* pulse.

Here is one additional hint when you work with the photoconductive cell. Outside light or room lights may interfere with the operation by flooding the cell or by introducing 60-Hz or 120-Hz ripple into the output signal. You may have to try to block out extraneous light by placing a light shield or tube around the cell. The tube should point toward the intended light source so that the cell effectively only "sees" the light you want it to see.

So go ahead and build a circuit. Experiment with it and invent your own application.

Power Control Devices

We have seen transistors used in some power applications, such as audio power amplifiers. In this chapter, we will look at some devices which control power through other types of loads, such as motors, lighting, and heating systems. In these loads, the power required may reach values of hundreds or even thousands of watts.

Bipolar junction transistors are sometimes used to control power in heavy-duty loads, but these transistors need d-c power supplies. Similarly, many of the newer MOSFETs can handle very high currents and high voltages, and so they are used to control d-c loads, such as servos or other d-c motors. But most high-power loads operate from the a-c line voltage. Therefore a different type of power control device is needed, one that can operate from an a-c source.

SCRs

A *silicon controlled rectifier* (SCR) is a solid-state device that can control the current through a-c loads, and is occasionally used to control d-c loads as well. The SCR belongs to a group of solid-state devices known as *thyristors,* which means that the device has characteristics similar to those of a *thyratron* tube. Thyratons have been used for decades to control the current through a-c loads. We won't study the thyratron here, but we will concentrate on the SCR.

Basically, the SCR is used as a *switch,* that is, an on-off device, rather than as a variable resistance, like a MOSFET. The SCR acts like a reverse-biased diode (a very high resistance) until it is fired (turned on). Then it switches to a very low resistance, acting like a closed switch. The two most important ratings of an SCR are its maximum forward current and its maximum voltage rating. The maximum voltage rating tells you how much voltage can be applied of either polarity without damaging the SCR or without causing it to turn on accidentally. Some small SCRs are rated at 0.5 A at up to 30 V maximum. Heavy industrial types can go as high as a maximum forward current of up to 3000 A, and a maximum reverse voltage rating of up to 3000 V. Fig. 13-1 shows some typical SCR packages.

We will get a better understanding of the SCR by studying how it is constructed. You will recall that a bipolar junction transistor is basically an NPN or PNP sandwich. Well, as shown in Fig. 13-2A, the SCR is a four-layer device, consisting of alternate layers of PNPN material. Although it has four layers, it only has three external terminals, called the *anode* (A), the *cathode* (K), and the *gate* (G). The equivalent circuit, shown in part B of the figure, shows that it acts as an NPN section connected to a PNP section. In fact, it acts as if there were two transistors connected as shown in part C of the figure. Study the figure until you see the equivalence.

Fig. 13-3 depicts the operation of the SCR. Refer to part A of the figure. Assume that initially $V_{in} = 0$ V, so that I_g (gate current) is zero when supply voltage E is applied. This initially makes the base current of Q_1 zero, so Q_1 is initially off. With Q_1 off, no base current flows into Q_2, so Q_2 is also off. With both transistors off, the device acts like a reverse-biased diode, that is, practically an open circuit from A to K. Therefore, no current flows through the load.

13-1   SCR packages.

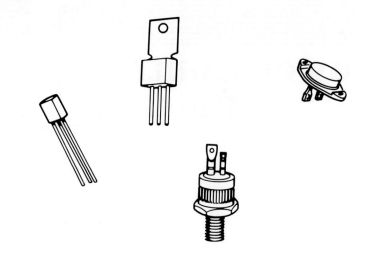

13-2   How an SCR works.

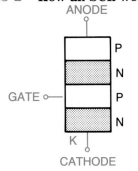

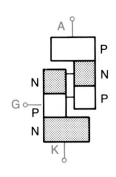

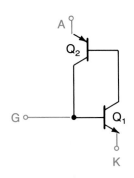

(A) SCR is a PNPN sandwich.   *(B) Equivalent circuit.*   *(C) Transistor equivalent.*

Now, suppose that we apply a voltage V_{in}, which causes some I_g to flow into the base of Q_1. This base current begins to turn on Q_1, causing collector current I_{c1} to flow. As you can see, I_{c1} is also the base current of Q_2, so Q_2 begins to turn on. As Q_2 turns on, *its* collector current I_{c2} flows into the base of Q_1, turning it on harder. More collector current through Q_1 increases the base current of Q_2 turning it on harder still. The action snowballs, each transistor being turned on harder and harder until both transistors are driven rapidly into saturation. In this state, the SCR acts almost like a closed switch.

Fig. 13-3B shows the schematic symbol for the SCR. Note that it looks like a diode with an extra terminal connected to the cathode. The reason for the diode (rectifier) symbol is that the SCR will only conduct in one direction after being triggered, hence the name silicon controlled rectifier. As shown in the figure, the forward voltage drop across the SCR when conducting is approximately 1 V.

Once the SCR fires (turns on), the gate loses control of it. Even if V_{in} is reduced to zero volts, the SCR remains on, due to the collector current of Q_2 supplying the base current of Q_1. The only function of the gate is to trigger (turn on) the SCR. The only way to stop current from flowing through the SCR, once triggered, is to momentarily break the anode circuit by means of a switch in series with the load, or by forcing the anode-cathode voltage to zero. Forcing the anode-cathode voltage to zero can be accomplished by either reducing supply voltage E to zero or by momentarily shorting a lead from anode to cathode.

Fig. 13-4 shows one *d-c* application for the SCR. The circuit shows a simple intrusion alarm. Assume that the light beam is shining into the photoconductive cell (making its resistance low)

13-3   Current flow in an SCR.

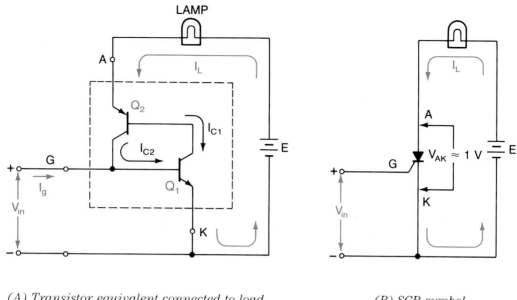

(A) Transistor equivalent connected to load.

(B) SCR symbol.

13-4   Intrusion alarm using SCR.

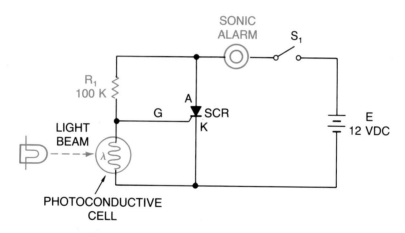

when supply voltage E is turned on. Since the resistance of the cell is much lower than the resistance of R_1, the voltage from gate to cathode is very small and the SCR remains off. When an intruder walks through the light beam, the resistance of the cell goes up to many thousands of ohms, causing the voltage at the gate to rise to a high enough value to trigger the SCR.

Once the SCR fires, it carries a high current through the sonic alarm, causing it to sound. Even if the intruder gets out of the beam of light, the SCR will remain on because the gate has lost control of the SCR. The alarm keeps sounding until switch S_1 is opened. Switch S_1 would, of course, be located in some remote or hidden place where the intruder couldn't get at it. Also, the light source used could be an infrared (IR) source which shines into an IR-sensitive photocell, making the beam invisible to the intruder.

Although in the alarm application the latching of the SCR is exactly what we want, the loss of control by the gate is one of the reasons that the device is not often used in d-c circuits. Special *commutating*, or turn-off, components must be used in d-c applications. SCRs are used in a-c circuits far more often than in d-c circuits.

A-C OPERATION OF SCRs

When used to control power to a-c loads, SCRs are used in either of two modes. The first mode is sometimes called *zero voltage switching*, which means that the SCR is turned on when the a-c voltage passes through (or just slightly above) zero volts. The second mode is called *phase control*, which means that the phase angle, or timing, of the trigger is delayed until the a-c line voltage has passed through some portion of its cycle. We will first look at zero voltage switching.

In the circuit of Fig. 13-5, the SCR is connected in series with its load, a lamp, to the a-c power line. Note that switch S_1 in series with R_1 is open. Therefore, no voltage is applied between gate and cathode regardless of the value of the line voltage. As a result, the SCR remains off, so V_{AK} is equal to the line voltage and $V_L = 0$. This is shown in parts B and C of the figure.

Part D of Fig. 13-5 shows S_1 closed. Let's look at the waveforms of parts E and F to see what happens now. Suppose the line voltage began to increase before S_1 was closed, as shown in the first half-cycle of waveform E. V_{AK} begins to rise, as shown from t_0 to t_1. Then at time t_1 switch S_1 is closed, perhaps by an operator flipping the switch. Due to the voltage divider made up of R_1 and R_2, a voltage V_g is applied between gate and cathode. If V_g is high enough (usually 2 or 3 V), the SCR fires (turns on). This is seen by the voltage V_{AK} dropping rapidly to near zero at time t_1.

Notice in the waveform of Fig. 13-5F that the voltage across the load was zero until the SCR fired; then it switched rapidly to a value equal to the line voltage. The shaded portion of the waveform indicates when current is flowing through the load.

13-5   SCR in an a-c circuit.

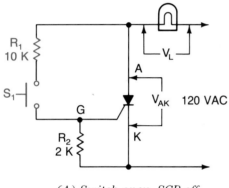

(A) Switch open, SCR off.

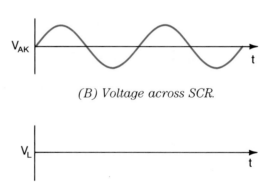

(B) Voltage across SCR.

(C) Voltage across lamp.

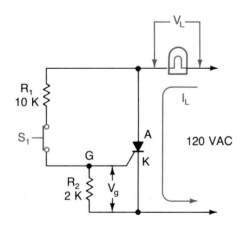

(D) Switch closed, SCR on.

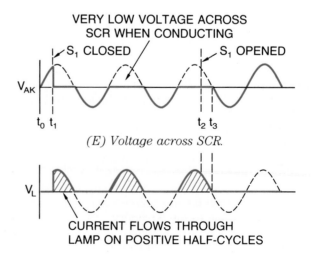

VERY LOW VOLTAGE ACROSS
SCR WHEN CONDUCTING

(E) Voltage across SCR.

CURRENT FLOWS THROUGH
LAMP ON POSITIVE HALF-CYCLES

(F) Voltage across lamp.

When the line voltage reduces to zero, the SCR *turns off* and remains off for the entire negative half-cycle. However, as soon as the line voltage passes through zero in the positive direction (actually slightly above zero), the SCR fires again and remains on for the entire positive half-cycle. The SCR will continue to turn on for each positive half-cycle as long as S_1 remains closed.

Now suppose that S_1 is opened at time t_2. Note that the SCR does not immediately turn off, since the gate has lost control once the SCR fired. Rather, the SCR remains on until the line voltage drops to zero at time t_3. At that time, the SCR turns off and remains off until S_1 is closed again.

An important point to note is that the heavy line current does not flow through control switch S_1. Switch S_1 is just used to turn on the SCR, and then the load current flows through the load and the SCR in series. Remember that once the SCR fires, it acts almost like a closed switch, so not much of the a-c line voltage is wasted across the SCR; hence, practically all of it appears across the load.

As you have probably observed, since current only flows through the SCR on positive half-cycles, the power delivered to the load is only half of what it would be if current were allowed to flow for the full cycle. One way to solve this problem is to use a full-wave bridge rectifier between the load and the SCR, as shown in Fig. 13-6. In this circuit, the load is an a-c motor, such as those used in electric drills. However, the load could just as well be a heater, lighting, or some other type of device. When S_1 closes, the SCR switches on and the motor runs at full speed. When S_1 opens, the SCR switches off and the motor stops. Switch S_1 does not carry the heavy motor current. In fact, S_1 may be mounted a long way from the motor, and the connecting leads need not carry any heavy current.

TRIACs

Since full-wave control of a-c power is needed so frequently, SCR manufacturers have developed another device, called the *TRIAC*, which can conduct in both directions. The TRIAC, whose symbol is shown in Fig. 13-7, acts like two SCRs in parallel but facing in opposite directions. Like the SCR, the TRIAC remains off until triggered by a voltage applied between its gate and terminal T_1, which is the equivalent of the cathode. However, it doesn't matter whether T_2 is positive or negative with

13-6 Full-wave SCR motor control.

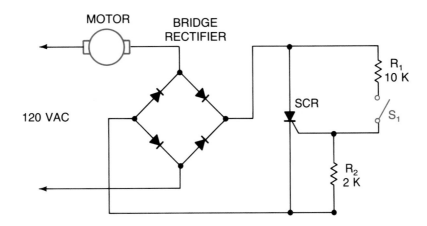

13-7 TRIAC symbol.

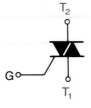

respect to T_1. Generally, when T_2 is positive with respect to T_1, a positive voltage at the gate with respect to T_1 is used to trigger, just as with an SCR. However, when T_2 is negative with respect to T_1, a negative voltage at the gate with respect to T_1 will trigger the TRIAC.

You may wonder why we bother using SCRs at all, since the TRIAC is obviously more versatile. Well, the reason is that TRIACs can't handle as much current as heavy-duty SCRs. Small TRIACs begin with ratings of about 0.8 A of current with a maximum voltage of 50 V, but the largest TRIACs can handle up to 25 A, and have voltage ratings of up to 600 V. As you can see, these are quite a bit lower than the ratings of heavy-duty SCRs. Nevertheless, there are many applications, such as variable-speed electric drills, mixers, sewing machines, and so on, that can get by fine with currents less than 25 A, so TRIACs are used quite extensively.

Fig. 13-8 shows a common method of using a TRIAC to carry an a-c load current. In this case, the TRIAC is controlled by a low-voltage d-c circuit consisting of a transistor and a *reed relay*. The

13-8 Controlling a TRIAC with a reed relay.

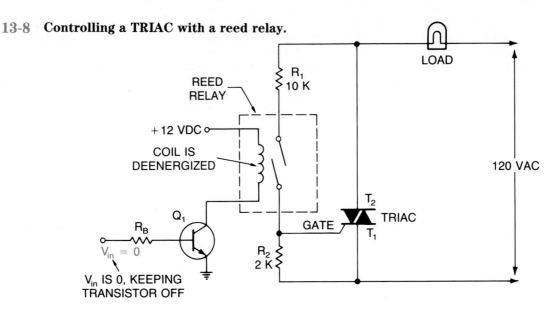

(A) Transistor off, reed switch open, TRIAC off.

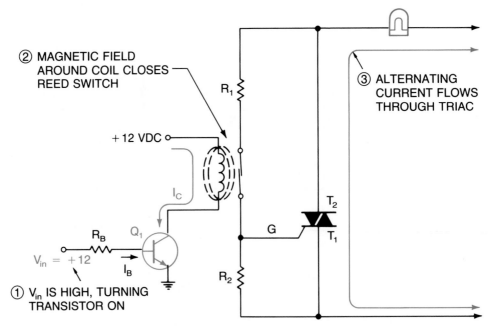

(B) Transistor on, reed switch closed, TRIAC on.

transistor itself could be driven by some logic or computer circuit. Here's how it works: In part A of the figure, the input voltage V_{in} at the base of the transistor is 0 V. Therefore the base current of the transistor is zero, causing the transistor to be off. Since the collector current of the transistor is zero, the relay is deenergized, causing the reed switch to be open. Of course, with the switch open the TRIAC is off and so is the load.

In part B of the figure, a voltage is applied to the base of the transistor. Base current flows, turning on Q_1. With collector current I_{c1} flowing through the coil of the reed relay, a magnetic field builds up, which closes the reed switch. Once the switch closes, the TRIAC fires and the load current flows through the TRIAC.

Notice that the controlling device (the transistor) is operating from a low-voltage *d-c* supply, but it indirectly controls *alternating current* through the load from the high-voltage a-c line. There is no electrical connection between the a-c circuit and the d-c circuit; that is, they are isolated electrically but coupled magnetically.

If you don't know how a reed relay operates, examine Fig. 13-9. The switch portion of the relay consists of two magnetic reeds (thin flat strips of metal) separated a small distance from each other by spring tension. In one type, the reeds are enclosed in a thin glass tube, about an inch long and a little more than ⅛ inch thick. Then a coil is wrapped around the tube. When a small current (a few milliamps) is made to flow through the coil, the magnetic field set up by the coil causes the two flat reeds to attract each other, thereby closing the switch. When the current stops flowing through the coil, the reed switch opens. Tiny reed relays are often used in computer-controlled circuitry. The relays can be purchased in small DIP packages the size of integrated circuit logic chips.

Since this technique of controlling an a-c power circuit with a low-voltage d-c circuit is used so frequently, special packages called *solid-state relays* are available, which contain both the reed relay and the TRIAC in one module. Fig. 13-10A shows a typical solid-state relay (SSR) package. The equivalent circuit is shown in part B. Note that only four terminals are accessible to the user. Two of them are connected to the d-c control circuit, the other two to the a-c load and line voltage, as shown in part C of the figure. As indicated in part A of the figure, a d-c input voltage of anywhere between 3 and 32 V is needed to activate the reed relay. That is, input voltage is not critical. So the same unit can be used in logic systems which operate on a +5-V power supply, or they can be used in higher-voltage circuits with no changes needed.

An alternative to the reed relay is an *optocoupler,* also called *optoisolator,* circuit, as shown in Fig. 13-11. In this circuit, the MOC3010 couples the low-voltage d-c circuit to the power TRIAC. Here's how it works: Suppose that a logic 1 (HIGH) is applied to the data input of the *D*-type flip-flop and the CLK input goes HIGH. The $\bar{Q}$ output of the flip-flop then goes to the same level as the *D* input, causing the $\bar{Q}$ output, labeled $\overline{\text{RUN}}$, to go LOW. This forward biases the LED inside the MOC3010, causing it to light. Also inside the MOC3010 is a *light-activated SCR* (LASCR). When light shines on the LASCR, it turns on, just as if a gate trigger voltage had been applied. With the LASCR conducting, the gate of the power TRIAC is pulled toward T_2 (away from T_1), causing the

13-9 Reed relay.

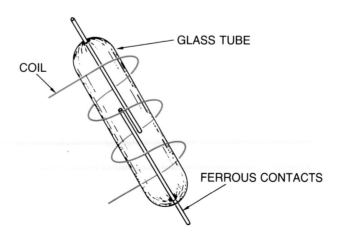

13-10 Solid-state relay.

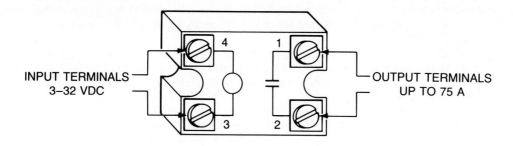

(A) SSR package.

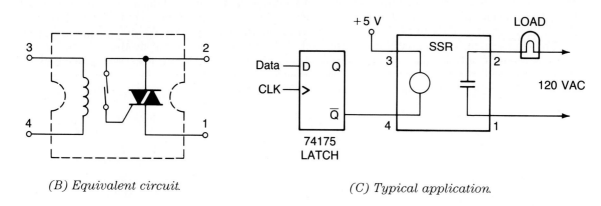

(B) Equivalent circuit.

(C) Typical application.

13-11 Controlling a power load from a low-voltage driver through an optocoupler.

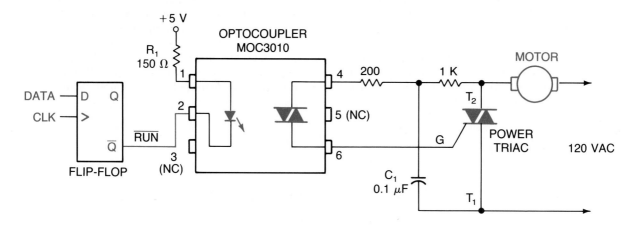

power TRIAC to turn on. The motor then runs. The power TRIAC will fire each half-cycle, causing the motor to run continuously until the $\overline{\text{RUN}}$ signal goes inactive HIGH, which turns off the LED.

If the load were very small, requiring less than 0.1 A, the LASCR itself could handle the load current. But it is assumed here that the load requires many amperes, which the MOC3010 can't handle.

Some SSR packages which look the same as the one in Fig. 13-10A use optocoupler devices rather than reed relays to couple the low-voltage d-c circuit to the a-c power device. However, they are used essentially the same as the reed relay types.

PHASE CONTROL

Phase control of an SCR or TRIAC allows us to smoothly control the amount of power delivered to a load, rather than switching it on or off. Phase control is commonly used in lamp dimmers and in variable-speed motor controls.

Although there are variations in the components used, the circuit in Fig. 13-12 is a typical example of a TRIAC phase control circuit. Besides the TRIAC, it uses another device called a *DIAC* to trigger the TRIAC.

A DIAC is a two-terminal device that acts like two diodes in parallel, facing in opposite directions, somewhat like a TRIAC. The difference is that the DIAC does not have a gate. The DIAC acts like a reverse-biased diode when voltage is applied in either direction. However, when the voltage reaches a critical breakdown value, the diode resistance drops sharply to a small value. Some DIACs have a breakdown voltage V_B of about 30 V, while others will switch at 8 or 9 V. In a typical application, the DIAC is used to produce trigger spikes to the gate of a TRIAC, causing the TRIAC to turn on.

Let's see how the circuit of Fig. 13-12 works. Assume that S_1 closes, but the TRIAC is initially off. The lamp does not light. However, a small current does flow through the lamp, R_1, and R_2, and begins charging capacitor C_1. Remember, the line voltage is 120 VAC, so the waveform across C_1 would look something like Fig. 13-13A. Note that V_C rises more slowly than the line voltage because it has to charge through R_1 and R_2. The actual charging rate depends on the setting of the variable resistor R_2. If R_2 is adjusted to a relatively low value, V_C will rise rather quickly, as shown.

As long as V_C is less than the DIAC breakdown voltage V_B, the diode acts like an open circuit, so the voltage at the gate of the TRIAC is zero and the TRIAC is off. However, when the voltage across the capacitor reaches V_B, the DIAC switches to a low resistance. This immediately couples the capacitor voltage to the gate of the TRIAC, and the capacitor dumps its charge through the DIAC, producing a trigger spike at the gate of the TRIAC. This trigger spike is shown in Fig. 13-13B. As soon as the TRIAC is triggered, it turns on and conducts for the remainder of the half-cycle. As a result, the line voltage appears across the load, and the load current flows for the remainder of the half-cycle, as shown in Fig. 13-13C.

Since triggering occurs early in the cycle, load current flows for almost the entire half-cycle and the lamp burns brightly. Now let's see what happens when we adjust R_2 to a larger value. As shown in Fig. 13-13D, capacitor voltage V_C rises much more slowly than it did when R_2 was small. It doesn't reach breakdown voltage V_B until much later in the cycle. Therefore the trigger spike occurs much later in the cycle, as shown in part E of the figure. As a result, the firing of the TRIAC is delayed, and current only flows through the load for a short portion of each half-cycle, as shown by the shaded portion of the waveform in part F of the figure. In this case, the lamp burns dimly.

By varying control pot R_2, you can advance or retard the trigger spike, effectively controlling the phase angle or the portion of the cycle that current flows through the load. This circuit is

13-12   DIAC-TRIAC lamp dimmer.

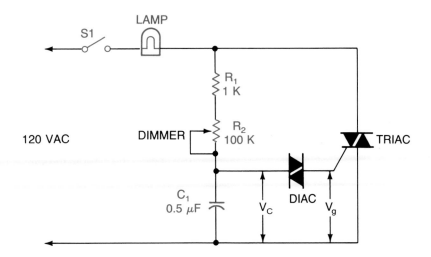

13-13 Phase control waveforms.

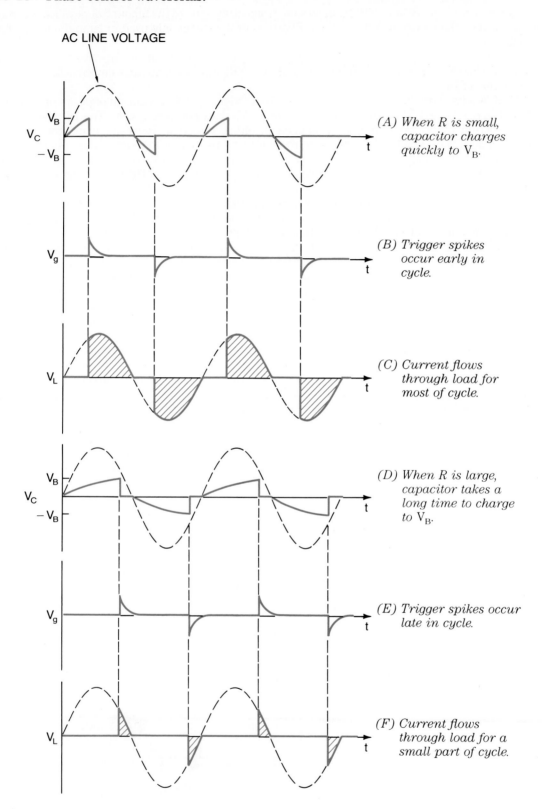

AC LINE VOLTAGE

(A) When R is small, capacitor charges quickly to V_B.

(B) Trigger spikes occur early in cycle.

(C) Current flows through load for most of cycle.

(D) When R is large, capacitor takes a long time to charge to V_B.

(E) Trigger spikes occur late in cycle.

(F) Current flows through load for a small part of cycle.

commonly used in lamp dimmers in homes, as well as in the speed control on portable electric drills. In the drill, pot R_2 is varied by means of the speed control trigger on the drill.

Phase control can also be used with SCRs, but a full-wave bridge rectifier is needed to allow full power to be fed to the load.

PROBLEMS WITH SCR AND TRIAC CIRCUITS

SLOW TURN-ON

When turning on an SCR or TRIAC, it is important to turn it on *rapidly*. That is, the trigger voltage must quickly jump to the appropriate value (say 2 or 3 V). This is the reason that the DIAC is used in the phase control circuit. If the DIAC were not used, and the capacitor were directly tied between gate and cathode, the gate voltage would rise *gradually* to the trigger point. There would be no application of a trigger *spike* to the gate. As a result, the TRIAC would turn on *slowly*. When this happens, the TRIAC may overheat and could possibly be ruined. Therefore, the triggering circuit must always hit the gate with the full amplitude trigger voltage all at once!

INDUCTIVE LOADS

When working into inductive loads, like motors or solenoids, there is a tendency for the SCR or TRIAC to *remain on* after it should have been shut off. That is, the SCR continues to conduct even after the anode-cathode voltage goes to zero. In order to understand why this is true, let's review what happens in an inductive circuit when you try to stop current from flowing through it.

Refer to Fig. 13-14A. Assume that no current was initially flowing through the coil when S_1 is closed. Current then starts to build up in the coil. As a result of the increasing current, there is a counter emf (cemf) developed across the coil which opposes the applied voltage E.

Next, assume that current has been flowing through the coil for some time when switch S_1 opens. As shown in part B of the figure, the polarity of voltage across the coil *reverses* to try to keep current flowing in the same direction. This voltage is often called a *kickback* voltage. The kickback voltage can rise to a high value rapidly, since the resistance of the open switch is very high (an open circuit). Fig. 13-14B shows that the voltage V_s across the switch is the sum of E plus the kickback voltage.

Well, if we replace S_1 with an SCR, there will be a rapidly rising, high-amplitude voltage appearing across it at the instant it tries to turn off. If the anode-cathode voltage increases at a high enough rate, the SCR may turn on again. Normally, however, the SCR will turn on again only if the voltage across it increases very rapidly. That is, if the *rate of change* of voltage (dv/dt) is high.

Fortunately, there is a way of preventing the rate of change of voltage across the SCR from being high enough to turn it on. This is accomplished by placing a small RC network across the SCR, as shown in Fig. 13-15. Since voltage across a capacitor can't change rapidly, the capacitor

13-14 Voltage developed across inductor due to changing current.

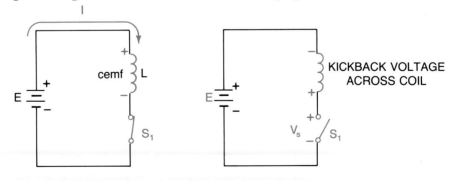

(A) *Cemf across coil prevents rapid buildup of current through coil when S_1 is first closed.*

(B) *When S_1 opens, voltage V_S appears across switch.*

13-15 Using an RC network to reduce dv/dt across SCR.

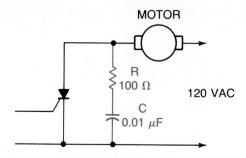

keeps the voltage across the SCR from changing too fast when the SCR tries to shut off the current through the motor. The resistor is in the circuit to prevent the possibility of the capacitor forming a resonant circuit with the inductance of the motor.

PROBLEMS

13-1.   Refer to Fig. 13-16. Assuming that E and V_{in} are within the normal range of values, in which circuit(s) will the lamp light?

13-2.   Assume that an SCR is triggered on by applying a trigger voltage of positive 3 V to the gate. (a)Will the SCR turn off when V_g goes back to zero? (b)Will the SCR turn off when V_{AK} goes to zero?

Refer to the intrusion alarm of Fig. 13-4 for the next four problems.

13-3.   When first applying power to the circuit, should the light source be turned on before or after S_1 is closed, or does it matter which comes first?

13-4.   Suppose that with the light shining into the photoconductive cell, the resistance of the cell is 1 KΩ. (a)What is the approximate voltage between gate and cathode with the light on the cell? (b)If the gate trigger voltage for the SCR is +3 V, will the SCR turn on with light shining on the cell?

13-5.   Suppose that the dark resistance of the cell is 100 KΩ. Will the SCR fire when the beam is broken?

13-6.   If the resistance of the sonic alarm is 50 Ω, how much current flows through the SCR after it fires?

13-16 Circuit for Problem 13-1.

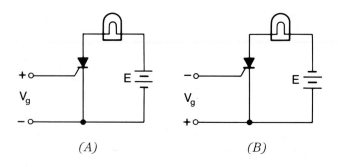

(A)   *(B)*

For the next four problems, refer to Fig. 13-5.

13-7. Before S_1 is closed, the SCR must remain off. What is the minimum value of anode-cathode voltage rating that the SCR must have?

13-8. Part E of the figure shows that S_1 was closed at about 60° of the positive half-cycle. The SCR fired immediately when the switch was closed. Explain what would happen if S_1 had closed during the negative portion of the cycle.

13-9. Part E of the figure also shows that S_1 was reopened at about 120° of a positive half-cycle, yet the SCR did not immediately turn off. Why?

13-10. Although this method of switching is sometimes called zero voltage switching, the line voltage does have to go slightly above zero volts before the SCR fires. Assuming that the gate trigger voltage is $+2$ V, what will be the instantaneous line voltage when the SCR fires?

13-11. Refer to problem 13-1. If the SCR in Fig. 13-16 were replaced with a TRIAC, in which circuit(s) will the lamp light?

For the next four problems, refer to Fig. 13-8.

13-12. With the transistor off, what is the gate-to-cathode voltage of the TRIAC, regardless of the value of the line voltage?

13-13. When the transistor is on, will the TRIAC conduct on the positive half-cycle, the negative half-cycle, or both half-cycles?

13-14. Suppose the lamp is lit continuously, even when $V_{in} = 0$ V. Which of the following could be the cause? (a)Q_1 is shorted. (b)Relay coil is open. (c)Relay contacts are welded closed. (d)R_1 is open.

13-15. Suppose the lamp never lights, regardless of whether V_{in} is high or low. Which of the choices of problem 13-14 could be the cause?

13-16. Refer to Fig. 13-10C. Suppose that the D input to the latch is a logic zero, and the CLK input is driven HIGH. Will this cause the load to be energized?

13-17. Suppose you want to test the flip-flop and optocoupler in the circuit of Fig. 13-11. You disconnect the 200-Ω resistor and TRIAC gate from the MOC3010 and connect an ohmmeter across pins 4 and 6. Next you connect a logic HIGH to the D input of the flip-flop and clock it. Should the ohmmeter read a high or a low resistance? Explain.

13-18. Referring to the setup of problem 13-17, you next connect a logic LOW to the D input of the flip-flop and clock it. Should the ohmmeter now read a high or low resistance? Does it make a difference which lead of the ohmmeter (positive or negative) you connect to pin 4 of the MOC3010?

13-19. In the phase control circuit of Fig. 13-12, will the lamp be brighter at the maximum or minimum resistance setting of R_2?

13-20. Suppose that the lamp in the circuit of Fig. 13-12 burns at near full brightness all the time, regardless of the setting of R_2. Which of the following could be the cause? (a)shorted capacitor. (b)open capacitor. (c)shorted TRIAC. (d)open DIAC. (e)open R_2.

EXPERIMENT 13-1   TESTING SCRs WITH AN OHMMETER

In Chapter 2 you learned how to test diodes with an ohmmeter. Remember that the meter has an internal battery which can forward bias the diode if the positive lead of the meter is connected to the anode and the negative lead is connected to the cathode. The diode then reads a low resistance. When the meter leads are reversed, the diode reads a high resistance. In this experiment, you'll learn how to test an SCR or TRIAC using an ohmmeter in a similar fashion.

EQUIPMENT

- VOM (ohmmeter)
- (1) or more low-power SCRs or TRIACs

BACKGROUND INFORMATION

You have learned that an SCR is a PNPN "sandwich," as shown in Fig. E13-1B. If you connect an ohmmeter from anode to cathode as shown, you will read a high resistance. Even if you reverse

E13-1   Measuring resistance of a PNPN "sandwich."

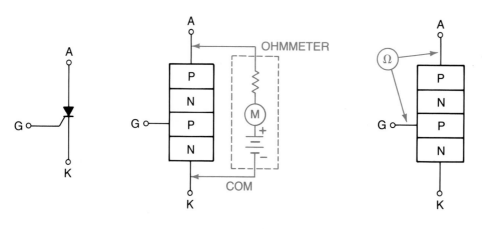

(A) Symbol.   (B) Measuring anode-cathode resistance.   (C) Measuring anode-gate resistance.

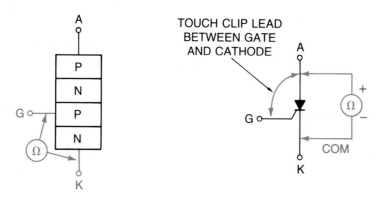

(D) Measuring cathode-gate resistance.   (E) Triggering the SCR.

the leads of the meter, you will still read a high resistance. Likewise, if you connect an ohmmeter from gate to anode, as shown in part C, you will read a high resistance in both directions. However, if you connect an ohmmeter between gate and cathode, as in part D of the figure, you will read a high resistance in one direction and a low resistance in the other direction. This gives you a simple way of identifying the gate lead and cathode lead. The anode lead is usually mounted to the stud or heat sink of the SCR and is easy to recognize.

PROCEDURE

1. Obtain a low-power SCR and make a sketch of it in the space below. With your ohmmeter on a medium resistance scale, say $R \times 100$ or so, identify the anode, cathode, and gate leads and label them on your drawing.

2. You will now perform a simple test to see whether the SCR is in working condition. With your ohmmeter set to the $R \times 1$ scale, connect the meter from anode to cathode, as shown in Fig. E13-1E. Be sure to have the positive lead of the meter connected to the anode. The meter should read a high resistance (open circuit) with the gate lead unconnected.

Do you read a very high resistance?

3. Now connect a clip lead from anode to gate. The resistance of the SCR should drop to a low value because the SCR fires when the gate is made positive with respect to the cathode. The battery in the meter makes the gate positive.

Do you read a low resistance?

4. Now remove the clip lead from anode to gate, but keep the meter attached from anode to cathode. Does the SCR remain on?

Here's an important point. Touching the clip lead from the gate to the positive terminal triggered the SCR. Once the SCR was triggered, the signal at the gate was no longer needed. The current from the meter keeps the SCR conducting. However, there is a minimum amount of current (called *holding current*) which must flow through the SCR to hold it in conduction. The amount of holding current is usually small, on the order of milliamps, but to supply this current the meter must be set on the low resistance scale.

This test only works for low- to moderate-power SCRs (up to 20 A or so), because the gate drive and holding currents for a high-power SCR are more than the meter can supply. This test works for TRIACs as well as for SCRs.

EXPERIMENT 13-2   DIAC-TRIAC PHASE CONTROL

You will now work with a simple DIAC-TRIAC phase control circuit which can be used to vary the brightness of a lamp or vary the speed of a small motor. For safety purposes, it is recommended that you use an isolation transformer between your circuit and the a-c line or that you connect your measuring instruments to the various test points before plugging your circuit into the a-c line.

EQUIPMENT

- oscilloscope, preferably dual-trace
- TRIAC 2N6151 or equivalent
- DIAC 1N5761 or equivalent
- capacitor, 0.5 μF, 200 V, paper or plastic
- capacitor, 0.1 μF, 200 V, paper or plastic
- potentiometer, 100 kΩ, 1 W
- resistor, 1 KΩ, 5%
- resistor, 100 Ω, 5%
- 100-W, 120-V lamp
- (optional) small ¼-inch electric drill or similar a-c motor

1. Build the circuit of Fig. E13-2, but do not apply power yet.

> **WARNING:** *You will be working with lethal voltages in the experiment. Be very careful about what you touch. If you do not use an isolation transformer, do not touch the chassis of your scope.*

E13-2 **Circuit for Experiment 13-2.**

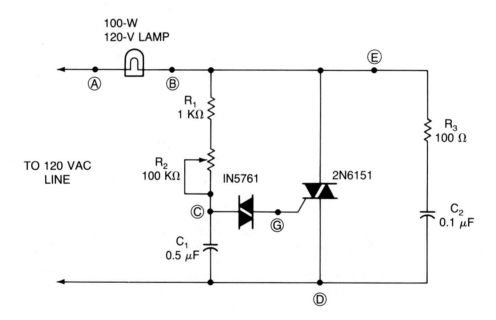

2. Connect the common of your scope to point D, which is terminal T_1 of the TRIAC. Connect one probe to point A and the other to point C.

> **NOTE** Sync your scope on waveform A. If you have a single-trace scope, use an external trigger input from point A and connect your probe to point C.

3. Apply a-c power and vary the resistance of R_2 back and forth a few times. You should observe a waveform at point C similar to that of Fig. 13-13A. Does the brightness of the lamp change as R_2 is varied?

4. Shut off the a-c power, remove your probe from point C, and connect it to point G, the gate of the TRIAC. Now reapply a-c power. Vary R_2 back and forth a few times and observe the waveform at point G. Observe the trigger spikes that fire the TRIAC. The spikes should advance or retard as you vary R_2. Note how the brightness of the lamp changes with the position of the spikes.

5. If you have an electric drill or small motor, shut off the a-c power and replace the lamp (between points A and B) with the motor. Reapply power and note how the motor speed varies with the setting of R_2.

6. Shut off the power again and disconnect the top of R_3 from point E. Then apply power and see if varying R_2 still allows you to vary the motor speed.

> **NOTE** Sometimes, depending on the motor inductance and current amplitude, removing R_3-C_2 has no effect, but try it anyway to see if yours does.

QUIZ

1. In step 3, increasing the resistance of R_2 caused C_1 to charge (faster, slower).

2. When C_2 took a longer time to charge, the lamp burned (brighter, dimmer).

3. In step 4, increasing the resistance of R_2 caused the trigger spikes to occur (earlier, later) in the cycle.

4. When the spikes occurred earlier in the cycle, the lamp burned (brighter, dimmer).

5. If C_1 became disconnected from point C, the trigger spikes (would never occur, would always occur early regardless of the setting of R_2).

6. If R_2 were replaced with a 10-KΩ pot, could you still adjust the lamp to maximum brightness? (yes, no)

7. Referring to question 6, could you still adjust the lamp to the same minimum brightness? (yes, no)

8. If you were able to replace the lamp with a motor, increasing the resistance of R_2 (increased, decreased) the motor speed.

9. Did removing R_3 from point E have any effect? (yes, no)

10. Even if you did not observe any effect of disconnecting R_3, in general the effect of disconnecting it would be to make the motor (run slower, race too fast even if R_2 is adjusted to a high resistance).

Appendix A

ANSWERS TO ODD-NUMBERED PROBLEMS

CHAPTER 2

2-1. $V_L = 6$ V, $V_D = 0$ V
2-3. $V_L = 5.3$ V, $V_D = 0.7$ V
2-5. $V_{ave} = 28$ V, $V_{rip} = 4.18$ V
2-7. $V_{ave} = 27.3$ V, $V_{rip} = 4.18$ V
2-9. $V_P = 14.1$ V, PRV $= 28.2$ V
2-11. $V = 7.06$ V, PRV $= 7.06$ V
2-13. a
2-15. c
2-17. d
2-19. b
2-21. d
2-23. c

CHAPTER 3

3-1. $I_C = 900$ mA
3-3.   a, d
3-5. $I_B = 20$ μA, $I_C = 1.2$ mA, $V_{CE} = 9$ V
3-7. $R_B = 300$ KΩ
3-9. $I_C = 2$ mA, $V_C = 10$ V
3-11. $R_E = 6$ KΩ
3-13. $A_v = 400$
3-15. $A_v = 171$, $v_0 = 514$ mV
3-17. b, The transistor is cut off.
3-19. a
3-21. b
3-23. d, All other faults would cause lower than normal V_c.
3-25. a
3-27. b, The low d-c resistance of the dynamic microphone pulls the base voltage down.

CHAPTER 4

4-1.   b, c
4-3. $V_{GS} = -3$ V
4-5. $g_m = 5$ mA/2 V $= 2500$ μmho

4-7. $R_S = 5\ V/7.5\ mA = 666\ \Omega$

4-9. $V_{DS} = 7.5\ V$

4-11. $A_v = 5$

4-13. To make $I_D = 4\ mA$, $V_{GS} = +3\ V$, since $V_{GS} = R_2/(R_1 + R_2) \times V_{DD}$.
$R_1 = R_2\ (V_{DD} - V_{GS})/V_{GS} = 2\ M\Omega\ (20\ V - 3\ V)/3\ V = 11.3\ M\Omega$.

4-15. $r_{in} \cong R_1 \parallel R_2 = 11.3\ M\Omega \parallel 2\ M\Omega = 1.7\ M\Omega$

4-17. $A_v = g_m R_D = 4 \times 10^{-3} \times 2 \times 10^3 = 8$

4-19. a, c, f

4-21. b

4-23. c

4-25. c

4-27. b

4-29. b

4-31. b

CHAPTER 5

5-1. $A_{v1} = 2K/25 = 80$, $A_{v2} = 3K/20 = 150$, $A_{v(tot)} = 80 \times 150 = 12{,}000$, $v_0 = v_{in} \times A_{v(tot)}$
$= 1.5\ mV \times 12{,}000 = 18\ V$

5-3. Measure V_{C2}

5-5. c

5-7. a, See equation 5-2.

5-9. $C_c = 1/(2\pi R_t f_1) = 1/(6.28 \times 3 \times 10^3 \times 50) \cong 1\ \mu F$

5-11. b

5-13. b, d

5-15. R_{124} is R_F, R_{123} is R_E

5-17. $R_F = R_{109}$ and $R_E = R_{106}$, so $g_m = 12\ K\Omega/220\ \Omega + 1 \cong 55$

5-19. $I_C = 1.5\ mA$

5-21. b

5-23. b, c

5-25. b

5-27. $I_E = \frac{1}{2}\ V_{EE}/R_E = 2\ mA$

5-29. With $r_e = 25\ mV/1.5\ mA = 16.7$ and $A_v = 240$, $V_D = 120\ mV$

5-31. $r_{in} = \beta R_E = 100 \times 600 = 60\ K\Omega$

5-33. $r_L = R_{129} \parallel (R_{130} + R_{131}) = 390 \parallel 340 \cong 182\ \Omega$, so $r_{in} = \beta\ r_L = 100 \times 182 = 18.2\ K\Omega$

5-35. a

5-37. b, X_{106} will turn on harder.

5-39. c

5-41. c

CHAPTER 6

6-1. $V_A = 5\ V$, $I_{C1} = 1\ mA$

6-3. $V_C = 15\ V$, $V_D = 15\ V$

6-5. $V_X = 24\ V$

6-7. $V_{rms} = 17\ Vrms$, $P_D = 18\ W$

6-9. a, Q_3 will conduct less, driving the base of Q_1 up.

6-11. a

6-13. a

6-15. b

6-17. $V_X = 15\ V$

6-19. d

6-21. b

6-23. a, 2N3704 would turn on hard.

6-25. b

6-27. $V = 0.62\ V$, same as at base of TR_4

CHAPTER 7

7-1. $n = 9$

7-3. $v_s = 13.3\ V$

7-5. $i_s = 1.33$ A
7-7. $i_p = 148$ mA
7-9. $r_{ref} = 1024$
7-11. $r_{ref} = 800$
7-13. $v_C/v_{in} = 32, v_C = 160$ mV, $v_L = 40$ mV
7-15. c
7-17. b
7-19. $f_r = 225$ KHz, $X_L = 707$ Ω, $Q = 47$, $Z_t = 33.3$ KΩ
7-21. b, e
7-23. a
7-25. a, d
7-27. $I_C = 18$ mA
7-29. a

CHAPTER 8

8-1.   a, c, e, g
8-3.   a
8-5. $A_v = -3, V_0 = -7.5$ V, $R_{in} = 20$ KΩ
8-7.   Since $1 \leq A_v \leq 25, -0.2 \leq V_0 \leq -5.$
8-9. A_v varies from 1 to 4, so V_0 varies from 3 to 12 V.
8-11. $f = 1$ MHz
8-13. X_c should be $\leq 0.1(10$ KΩ $+ 40$ K$\Omega) = 5$ KΩ. Then $C \geq 1/2\pi f X_c = 1/6.28 \times 50 \times 5000$
       $= 0.64$ μF.
8-15. $75 = 20$ log CMRR, $75/20 =$ log CMRR, $3.75 =$ log CMRR, CMRR $=$ antilog $3.75 = 5623$
8-17. $V_0 = 0.2 \times A_v = 0.2 \times -10 = -2$ V
8-19. $A_{cm} = 20/25,000 = 8 \times 10^{-4}$, so $V_{cmo} = 8 \times 10^{-4} \times 4 = 3.2$ mV
8-21. V_0 is a rectangular wave similar to that of problem 8-20. But this time V_0 is positive whenever v_{in} is more positive than -3 v_0.
8-23. $\Delta V/\Delta T = 5$ V/s, so $R_1 = V_{in}/(C \times \Delta V/\Delta T) = 800$ KΩ.

CHAPTER 9

9-1. $V_0 = 18$ V, $I_S = 128$ mA, $I_L = 90$ mA, $I_Z = 38$ mA
9-3. $V_0 = 18$ V, $I_S = 191$ mA, $I_L = 90$ mA, $I_Z = 101$ mA
9-5. $P_Z = 0.68$ W
9-7. $V_0 = 12$ V, $I_L = 150$ mA, $P_d = 0.9$ W
9-9. $I_L = 300$ mA, $I_Z = 7.8$ mA
9-11. $I_S = 9.8$ mA, $I_B = 2.63$ mA, $I_Z = 7.17$ mA, $P_Z = 194$ MW
9-13. c
9-15. a, the base of X_2 would be pulled more positive by R_1, but the circuit would not regulate. That is, the voltage at TP1 would vary with changes in load current.
9-17. b
9-19. a. I, b. I, c. S, d. I, e. I, f. D, g. D
9-21. b, f
9-23. $R_{SC} = 0.47$ Ω
9-25. b
9-27. b
9-29. a. D, b. S, c. I, d. I, e. I
9-31. a. I, b. I
9-33. a, c

CHAPTER 10

10-1. a. H, b. L, c. U, d. U, e. H, f. L
10-3. b
10-5.

C	B	A	G	H
0	0	0	1	1
0	0	1	1	1
0	1	0	1	1
0	1	1	0	1

1	0	0	1	0
1	0	1	1	0
1	1	0	1	0
1	1	1	0	1

10-7.   *C*

10-9.   *B*

10-11. *A*

10-13. High. Unconnected inputs float high.

10-15. I_A = 8 mA (5 unit loads of 1.6 mA each)

10-17. d, When high, a TTL gate looks like a back-biased diode and does not source much current.

10-19. See Fig. A10-32

A10-32   Answer to problem 10-19.

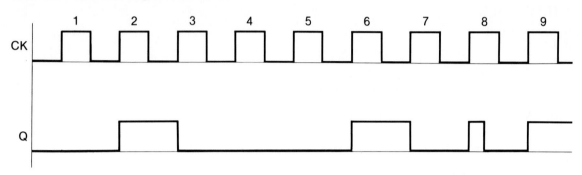

10-21. a

10-23. 0

10-25. a, d

10-27. c

CHAPTER 11

11-1.   b

11-3.   b

11-5.   a

11-7.   The middle display would toggle between 0 and 1.

11-9.   d

11-11. See Table A11-1

11-13. See Table A11-3

11-15. a square wave of 600 Hz and 5 Vp-p amplitude

11-17. a short duration positive pulse

11-19. no change

11-21. c, d

11-23. d

Table A11-1

	Z_0	Z_1	Z_2	Z_3	Y_0	Y_1	Y_2	Y_3	b_0	b_1	b_2	b_3	X_0	X_1	X_2	X_3	W_0	W_1	W_2	W_3	STORE $\underline{Z}$	STORE $\underline{Y}$	LOAD X	LOAD W
1	0	0	0	0	0	0	1	1	1	1	0	0	0	0	1	1	0	1	0	0	1	0	1	0
2	0	0	0	0	0	1	1	0	1	0	0	1	0	1	0	1	1	0	1	0	1	0	1	0
3	0	1	0	0	0	0	0	0	1	0	1	1	0	1	0	0	0	1	1	0	0	1	1	0
4	0	1	0	1	0	0	0	0	1	0	1	0	1	1	0	1	0	1	0	1	0	1	0	1
5	0	0	0	0	1	0	1	0	0	1	0	1	0	0	0	0	1	0	1	0	1	0	0	1
6	0	0	0	0	1	0	1	1	0	1	0	0	1	0	1	1	1	0	0	1	1	0	1	0
7	0	1	1	1	0	0	0	0	1	0	0	0	1	1	0	0	0	1	1	1	0	1	0	1
8	0	0	0	0	1	1	1	1	0	0	0	0	1	1	0	1	1	1	1	1	1	0	0	1

	b_0	b_1	b_2	b_3	X_0	X_1	X_2	X_3	W_0	W_1	W_2	W_3		LOAD $\overline{X}$	LOAD $\overline{W}$		A_0	A_1	A_2	STROBE
1	1	1	1	1	0	1	0	1	1	0	0	0		1	1		0	1	0	1
2	1	0	0	1	0	0	0	0	1	0	0	1		1	0		0	1	0	0
3	1	1	1	1	1	0	1	1	1	0	1	0		1	1		1	0	0	0
4	0	1	1	0	0	0	0	0	0	1	1	0		1	0		0	1	0	0
5	0	1	1	1	0	1	1	1	1	0	1	1		0	1		0	1	1	0
6	1	1	1	1	1	1	1	1	1	1	1	0		1	1		1	1	1	0
7	1	1	1	1	0	1	0	0	1	0	1	1		1	1		0	1	1	1
8	1	0	1	0	1	0	1	0	0	0	1	1		0	1		0	1	1	0

CHAPTER 12

12-1.   a. Type K thermocouple, b. RTD, c. Thermistor.
12-3.   Type K
12-5.   Thermocouple
12-7.   121.2 Ω
12-9.   90.83°C
12-11. -30°C
12-13. $V_0 = 0.7128$ V
13-15. V_0 slightly $> +4$ V, $I = 6/10K = 0.6$ mA, so $R = 4/0.006 = 6.67$ KΩ
13-17. b
13-19. 0.8 mm
13-21. 0 V
13-23. 83.3 mV
13-25. b
13-27. Span $= 16$ mA, 100 psi/16 mA $= 6.25$ psi/mA, so $6.25 \times 2.7 = 16.875$ psi.

CHAPTER 13

13-1.   A
13-3.   Turn on light source first.
13-5.   yes
13-7. $E_{max} > 1.41 \times 120 > 170$ V. Use a 200-V SCR.
13-9.   The gate lost control. SCR turns off when V_{AK} goes to zero.
13-11. TRIAC will light in both circuits.
13-13. both
13-15. b, d
13-17. You read a low resistance because the LASCR is conducting.
13-19. minimum

Appendix B

LIST OF PARTS AND EQUIPMENT FOR LAB EXPERIMENTS

PARTS AND EQUIPMENT

Test Equipment
- oscilloscope—preferably triggered time base, $f_{max} \geq 1$ MHz
- (VOM–D'Arsonalval or digital type
- (adjustable d-c power supply 0–20 V at 500 mA
- fixed d-c power supply 20 V at 1 A
- 9-V transistor radio battery
- signal generator, sine-square wave
- 0- to 100-μA meter movement

Solid-state Components
- (4) NPN general purpose transistors, 2N2222 or equivalent
- (1) PNP transistor, 5447 or equivalent
- (1) NPN power transistor, TIP 31
- (1) PNP power transistor, TIP 32
- (1) N-channel JFET, general purpose
- (4) silicon diodes, 1N4001
- (2) LED, 10- to 15-mA type
- (1) 5.1-V, 1-W zener

Integrated Circuits
- (4) 741 op amps or
- (4) LF351 BIFET op amps
- (1) 7805 regulator
- (1) 7400 quad 2-input NAND
- (1) 7476 dual J–K flip-flop
- (1) 555 timer
- (1) LF353 dual op amp

Capacitors
Electrolytic, all ≥25 WVDC
- (1) 500 μF
- (1) 200 μF
- (2) 100 μF
- (2) 50 μF
- (2) 20 μF
- (1) 10 μF
- (2) 5 μF
- (1) 1 μF
- (2) 0.1-μF disc
- (1) 0.1 μF at 200 VAC
- (1) 0.5 μF at 200 VAC

Potentiometers
½ W, 1 each
- 1 MΩ
- 100 KΩ
- 25 KΩ
- 10 KΩ
- 5 KΩ
- 2.5 KΩ

Special Components
- (2) toggle switches DPDT
- (4) toggle switches SPDT
- (1) momentary pushbutton SPST normally closed
- (1) thermistor, 10 KΩ at 20°C
- (1) power transformer V_P = 120 V, VS = 12.6 V at 1 A
- (1) power transformer V_P = 120 V, V_S = 12.6 V at 1 A
- (1) IF transformer Graymark Intl. 62718 or equivalent
- (1) lamp 120 VAC at 25 W with socket
- (1) line cord
- (1) 8-Ω speaker at 6 W
- (2) thermocouples (iron-constantan or any)
- (1) photoconductive cell VACTEC VT-801 or equivalent
- (1) silicon controlled rectifier (any low-power type)
- (1) TRIAC 2N6151 or equivalent
- (1) DIAC 1N5761 or equivalent

Resistors
ohms ½ W ± 5%

(2) 1 M	(1) 2.2 K	(1) 20
(1) 220 K	(6) 2 K	(1) 10
(2) 100 K	(1) 1.5 K	(2) 1
(2) 75 K	(3) 1 K	
(1) 47 K	(1) 510	(2) 100 K Ω, ½ W ± 1%
(1) 33 K	(1) 330	(2) 10 K Ω, ½ W ± 1%
(2) 15 K	(2) 270	
(2) 12 K	(2) 220	(1) 62 Ω 1 W ± 5%
(2) 10 K	(1) 150	(1) 27 Ω 1 W ± 5%
(1) 7.5 K	(1) 100	(1) 8 Ω 1 W ± 5%
(1) 6.8 K	(1) 51	(2) 1 Ω 1 W ± 5%
(3) 5 K	(1) 47	
(2) 4.7 K	(1) 22	(1) 8 Ω 10 W ± 10%

Appendix C

BASING AND LEAD IDENTIFICATION OF TYPICAL SOLID-STATE COMPONENTS

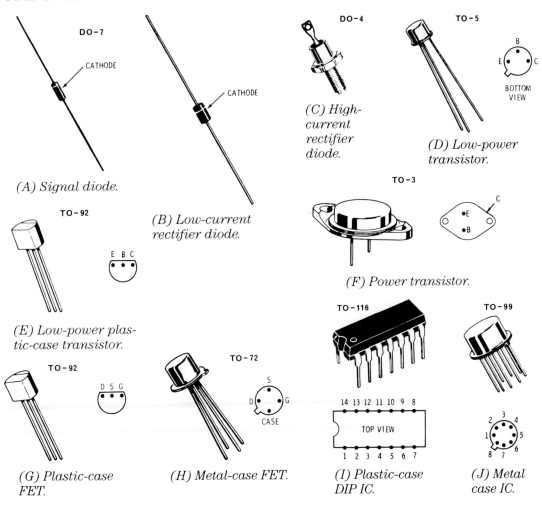

(A) Signal diode.

(B) Low-current rectifier diode.

(C) High-current rectifier diode.

(D) Low-power transistor.

(E) Low-power plastic-case transistor.

(F) Power transistor.

(G) Plastic-case FET.

(H) Metal-case FET.

(I) Plastic-case DIP IC.

(J) Metal case IC.

Index